The MARINE BIOLOGY COLORING BOOK

Second Edition

Coloring Concepts books:

The Zoology Coloring Book, by L.M. Elson
The Human Evolution Coloring Book, Second Edition, by A.L. Zihlman
The Botany Coloring Book, by P. Young
The Human Brain Coloring Book, by M.C. Diamond, A.B. Scheibel, and L.M. Elson
The Biology Coloring Book, by R.D. Griffin
The Microbiology Coloring Book, by I.E. Alcamo and L.M. Elson
The Biology Illustrated Coloring and Resource Book, by C.L. Elson
The Physics Coloring Book, by Richard N. Stuart and L.M. Elson

Coloring Concepts Inc.
1732 Jefferson Street, Suite 7
Napa, California 94559
1-800-257-1516
www.coloringconcepts.com

The MARINE BIOLOGY COLORING BOOK

Second Edition

by Thomas M. Niesen

Illustrations by
Wynn Kapit
Carla J. Simmons
Lauren Hanson

Tom Niesen received his Ph.D. in biology from the University of Oregon in 1973, specializing in the ecology of marine invertebrates. He has taught at San Francisco State University since 1973, where he is currently Professor of Marine Biology. In addition to a career-long fascination with the ecology of the rocky intertidal zone, Dr. Niesen's research emphasis has been on the ecology of introduced marine invertebrates in San Francisco Bay. He has also conducted research in the Caribbean and Bering seas. His hobbies are surfing, diving, swimming, photography, and gardening. Tom Niesen lives in Half Moon Bay, California, with his wife, Anne.

This book was produced by Coloring Concepts Inc.
1732 Jefferson St., Suite 7, Napa, CA 94559

Developmental editor: Joan W. Elson
Prepress digital layout: Christopher L. Elson
Copy editor: Jennifer Warrington
Digitization: Mark Jones
Film production: Kevin Frye and Rob Frye, Frye's Printing
Digital cover production: Mehdi Stephens

The Coloring Concepts name and logo is a registered trademark of Coloring Concepts Inc.

THE MARINE BIOLOGY COLORING BOOK, SECOND EDITION. Copyright © 1982, 2000 by Coloring Concepts Inc. All rights reserved. Printed in the United States of America. No part of this book may be used or reproduced in any manner whatsoever without written permission except in the case brief quotations embodied in critical articles and reviews. For information address HarperCollins Publishers Inc., 10 East 53rd Street, New York, NY 10022

HarperCollins books may be purchased for educational, business, or salespromotional use. For information please write: Special Markets Department, HarperCollins Publishers Inc., 10 East 53rd Street, New York, NY 10022

ISBN: 978-0-06-273718-2

13 14 15 20 19 18

DEDICATION

This book is dedicated to the memory of Matt Luerken, a good friend gone too soon; also to my family, Anne, Amy, Andy, and Maggie, and to all my students over the past thirty years who have never failed to keep it interesting.

ACKNOWLEDGMENT

First, I would like to thank my able production crew for all their hard work. Carla Simmons was a joy to work with as she molded my crude ideas into beautifully rendered art. Joan Elson was tireless and ever-cheerful in her chief editorial duties, and was ably assisted by Jennifer Warrington. Christopher Elson was a virtual jack-of-all-trades and indispensable in the production process. My daughter, Amy, who dutifully colored all the plates eighteen years ago in the first edition as a seven-year-old, added her new-honed skills as a communication professional in a final review.

Thank you all.

Many colleagues have aided in putting this book together. Ralph Larson never tired of my questions about fish, Ken Goldman shared his great white shark research findings with me and trusted me with his slides, and Heidi Dewar was very helpful in explaining the subtle workings of the new generation of fish tags. I am grateful for their generosity.

Finally, I would like to thank my wife, Anne, for all the time and emotional and financial support she has lavished on me over the years. I could not have done any of this without her.

TABLE OF CONTENTS

PREFACE

Since the first edition eighteen years ago, I have continued teaching and learning about marine biology. This book is a distillation of what I consider basic, necessary information about life in the ocean for the informed layperson. The new edition adds several habitat descriptions, new coverage of marine birds and sound production in the sea, an expanded section on symbiotic relationships among marine organisms, and updated and increased coverage of marine invertebrates, fish, and reptiles. I have also added two plates on the new technology available for ocean study, including remotely controlled underwater vehicles and fish tags that contain tiny computers capable of communicating vital data to shore via satellite. I hope you enjoy learning from this book.

COLORING INSTRUCTIONS

1. This is a book of illustrations (plates) and related text pages. You color each structure the same color as its name. The structure and its name are linked by identical letters (subscripts: a, b, c, etc.). Later you will be able to relate identically colored names (titles) and structures at a glance.

2. You will need coloring instruments. Colored pencils or colored fine to medium felt-tip pens are best. Twelve pens or pencils will work, but more is better.

3. The organization of the contents is based on the author's overall perspective of the subject and may follow the order of presentation of a formal course of instruction. To achieve maximum benefit, you should color the plates in order, at least within a group or section.

Once you begin coloring a plate in order of presentation of the titles and reading the matched text, the illustrations will have greater meaning, and relationships of different parts will become clear.

4. As you come to each plate, look over the entire illustration and note the arrangement and order of titles. You may count the number of subscripts to find the number of colors you will need. Then scan the coloring instructions (printed in boldface type) for further guidance. Be sure to color in the order given by the instructions. Most of the time this means starting at the top of the plate with the first title, a, and coloring in alphabetical order. Contemplate a number of color arrangements before starting. In some cases, you may want to color related forms with different shades of the same color; in other cases, contrast is desirable. Where a natural appearance is desirable, the coloring instructions may guide you or you may choose colors based on your own knowledge and observations. One of the most important considerations is to link the structure and its title (printed in large outline or blank letters) with the same color. If the structure to be colored has parts taking several colors, you might color its title as a mosaic of the same colors. It is recommended that you color the title first and *then* its related structure.

5. In some cases, a plate of illustrations will require more colors that you have in your possession. Forced to use a color twice or thrice on the same plate, you must take care to prevent confusion in identification and review by employing them on separate areas well away from one another. On occasion, you may be asked to use colors on a plate that were used for the same structure on a previous related plate. In this case, color their titles first regardless of where they appear on the plate. Then go back to the top of the title list and begin coloring in the usual sequence. In this way, you will not use a color already specified for another structure.

6. Symbols used throughout the book are explained below. Once you understand and master the mechanics, you will find room for considerable creativity in coloring each plate. Now turn to any plate and note:

a. Areas to be colored are separated from adjacent areas by heavy outlines. Lighter lines represent background, suggest texture, or define form and should be colored over. If the colors you use are light enough, these texture lines may show through, in which case you may wish to draw darker or heavier over these lines to add a three-dimensional effect. Some boundaries between coloring zones may be represented by a dot or two or dotted lines. These represent a division of names or titles and indicate that an actual structural boundary may not exist or, at best, is not clearly visible.

b. As a general rule, large areas should be colored with light colors and dark colors should be used for small areas. Take care with very dark colors: they obscure detail and texture lines or stippling. In some cases, a structure will be identified by two subscripts (for example, a+d). This indicates you are looking at one structure overlying another. In this case, two light colors are recommended for coloring the two overlapping structures.

c. In the event structures are duplicated on a plate, as in left and right parts, branches, or serial (segmented) parts, only one may be labeled with a subscript. Without boundary restrictions or instructions to the contrary, these like structures should all be given the same color.

d. If you will look at Plate 6 you will see the following symbols. The ✿ means to color the titles gray. A title with (a) next to it means that only the titles following it are colored in the art. This usually indicates a category (such as coralline algae) and only the two representations (encrusting and articulated) are to be colored. When you see titles with a^1, a^2, etc., this indicates the parts so labeled are sufficiently related to receive *shades* of the same color (for example, the two types of coralline algae).

7. In the text, certain words are set in italics. According to convention, the generic name and species of an animal or plant are set in this way (for example, *Chaetocerus denticulatus*). In addition, the title of a structure to be colored is set in italics where the word *first* appears.

8. For further guidance on color use, see Getting the Most Out of Color.

The

MARINE BIOLOGY COLORING BOOK

Second Edition

1 OCEAN CURRENTS AND GLOBAL WEATHER

The ocean is the largest habitat for life on earth, and is vital to the land habitats as well. The ocean regulates the earth's weather patterns.

The main force influencing global climate is the solar energy of the sun. Because the earth is a sphere, its curved surface is not heated equally by incoming sunlight. Near the equator, the earth's surface receives more sunlight because it is closer to the sun and the sunlight strikes at a direct angle. At the poles sunlight strikes at an oblique angle, so much of the sunlight is reflected away. With this difference in incoming solar radiation you would expect the water at the equator to boil and the poles to freeze. But this doesn't happen because solar energy falling on the equator is redistributed around the globe by the ocean and the atmosphere above.

Color the sun and its radiating energy. Color the earth. Color the three weather cells with contrasting colors.

Solar radiation striking the ocean near the equator warms the surface water and causes evaporation. Approximately half of the incoming *solar energy* is utilized in converting water from the dense liquid to the lighter gaseous state. As this water vapor rises into the atmosphere and away from the equator it cools, condenses, and falls as rain or dew. With the light-weight, gaseous water vapor removed, the dry air is heavy and more dense and begins to sink. The area of lighter air rising near the equator creates a low pressure zone, whereas the areas of heavy, sinking air to the north and south of the equator create high pressure zones.

Together a high pressure zone and a low pressure zone form a weather cell. Three weather cells are found in each hemisphere: the *equatorial cell* just described, a *mid-latitude cell*, and a *polar cell*. These global atmospheric weather cells act as heat pumps, driven by solar radiation and the evaporation of sea water. When water vapor in the atmosphere condenses, the stored heat required for its evaporation is released and it warms the air above the higher latitudes. Weather cells redistribute two thirds of the equatorial solar radiation to the rest of the globe. The remaining third is redistributed by ocean currents.

Color the illustrations of planetary winds. Note that the winds are flowing in different directions and should be given contrasting colors.

Air pressure over the earth attempts to equalize, thus air flows from areas of high pressure to areas of low pressure. We refer to the flowing air as wind. The winds in the equatorial weather cells blow towards the equator. Winds are deflected relative to the earth's spinning surface in a phenomenon known as the Coriolis effect. The winds flowing towards the equator from both hemispheres receive a westerly component. In contrast, the winds flowing in the mid-latitude weather cells are given an easterly push. The polar cell winds receive a westerly push. These winds that flow in the global weather cells are called planetary winds.

Winds are named for the direction from which they originate. The planetary winds blowing towards the equator in the equatorial weather cells are the *eastern trade winds*. The winds of the mid-latitude weather cells are known as the *westerlies*.

Color the flow of surface water in the Pacific Ocean.

The eastern trade winds blowing across the surface of the ocean near the equator set the warm water in motion as a surface current. The water flows from east to west and when the currents run into the continents, they split and flow north or south, eventually completing an ocean wide loop or gyre and returning to the equator. In the Pacific Ocean, the northern hemisphere ocean surface current gyre is a mirror image of the southern hemisphere gyre.

Now color North America. Color the California Current a cool color and the Gulf Stream a warm color.

Global scale surface currents influence local weather. In the ocean basins of the northern hemisphere, surface water circulates in a large clockwise eddy (gyre). In the Pacific, water warms at the equator, flows northward along the coastline of Asia, and eastward in the chilly northern latitudes. When the water collides with North America, it begins its southward flow along the California coastline as the *California Current*. All the heat captured at the equator has been released. Californians experience the cold water of the California Current flowing south towards the equator to begin the cycle all over again. This cold water cools the atmosphere above it and tempers the coastal climate.

On the east coast of the United States, the warm water of the northerly flowing *Gulf Stream* comes near shore in the summer, bringing warm water up from the equator and the Gulf of Mexico. The warm water near shore contributes to warm summer temperatures.

OCEAN CURRENTS AND GLOBAL WEATHER

SUN a
 SOLAR ENERGY a1
EARTH b
EQUATORIAL WEATHER CELL c
MID-LATITUDE CELL d
POLAR CELL e

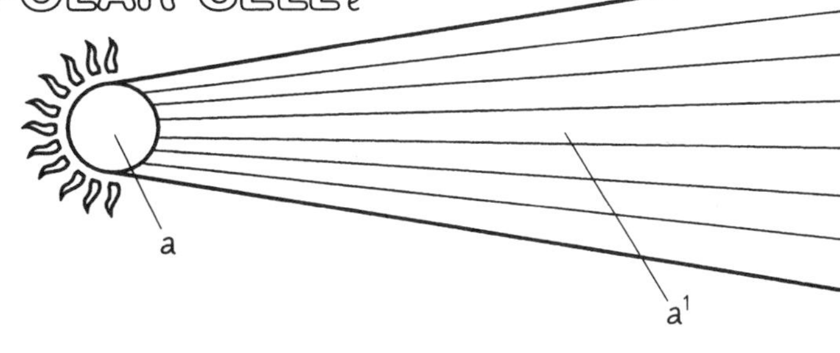

PLANETARY WINDS *
 EASTERN TRADE WINDS f
 WESTERLIES g
 POLAR WINDS h
DIRECTION OF EARTH'S ROTATION i

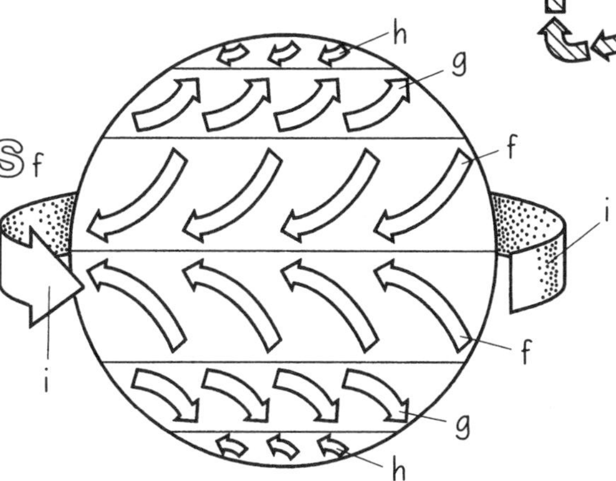

PACIFIC OCEAN BASIN *
 N. HEMISPHERE CURRENT GYRE j
 S. HEMISPHERE CURRENT GYRE k

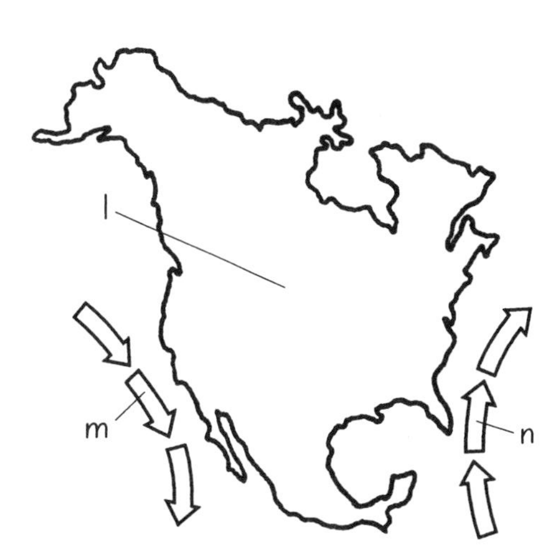

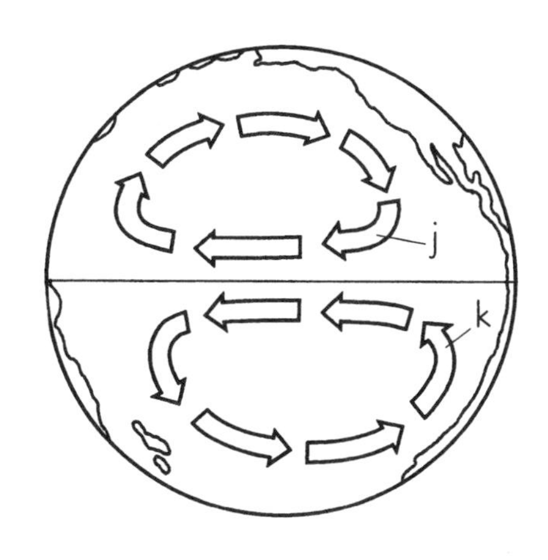

NORTH AMERICA l
CALIFORNIA CURRENT m
GULF STREAM n

2
UPWELLING AND EL NIÑO

Oceanic surface current flow was discussed in the previous plate. These global patterns represent the average annual directions of wind and water movement. Local seasonal patterns can also be found. In the early spring on the east side of the ocean basins, stable high pressure zones develop in the mid-latitudes. These stationary highs may persist through the summer. The winds from these high pressure zones blow along the coast and initiate a process known as upwelling.

Color the illustration of upwelling. Color the surface water a warm color and the upwelled water a cool color.

During spring in the northern hemisphere of the Pacific Ocean basin, a stationary *high pressure zone* develops off the Pacific Northwest. Winds from this high pressure zone blow along the coastline out of the northwest. Upwelling begins when the *northwest wind* blows along the shoreline and pushes against the upper *surface layer* of water, causing it to flow southward. As the surface water layer flows, it is affected by the Coriolis effect (Plate 1) and is deflected away from the shore. The offshore-flowing surface water is replaced by colder water which *wells up* from below.

In the unlit depths, bacteria and other marine decomposers have broken down plant and animal tissue and have released a high level of basic nutrients. Upwelling brings this cold water to the upper, sunlit portion of the ocean. These nutrient fertilizers foster a tremendous level of photosynthesis in the single-celled marine plants called phytoplankton. A high level of plant production persists through the spring and into the summer and greatly enriches the food chain. Upwelling zones are the richest marine environments in the world.

Color the abbreviated upwelling food chain.

Along California and Oregon in the northern hemisphere, and Peru and Equador in the southern hemisphere, great schools of small *plankton-feeding fishes* such as anchovies and sardines feast on the abundance of *plankton* produced by upwelling. These fishes are eaten by *larger fishes* and marine *birds* and *mammals*. *Man* also has reaped this harvest, although not wisely. Unsound fishing practices have caused the fisheries to crash, and only small populations of fishes occur where mighty schools once prospered.

Color the final two illustrations which indicate patterns of wind direction and surface water flow in the central south Pacific under normal and El Niño conditions.

Upwelling is a predictable, annual process; however, it is occasionally disrupted by changes in the global weather patterns. In the south central Pacific Ocean, the *easterly trade winds* that blow just below the equator occasionally drop in intensity or shift direction. The *warm surface water* that is ordinarily pushed westward by the wind moves eastward instead where it bumps into South America and continues southward along the coast. This unusual warm water current was named "*El Niño*" (The Child) after the Christ child by the fishermen of Peru because they usually first encountered the warm water at the beginning of the year, just after Christmas.

El Niño was not welcomed by the fishermen because this always meant poor fishing. Along with the warm water current, the wind pattern that produces the spring and summer upwelling along the west coast of South America is disrupted. Upwelling occurs only sporadically or not at all during El Niño and the food chain that is dependent on the rich supply of nutrients in the upwelled water collapses.

Oceanographers and climatologists now realize that El Niño events are not limited to the south Pacific Ocean. Initiated by a change in atmospheric pressure and wind pattern in the south Pacific, called the Southern Oscillation, El Niño changes affect the tropical jet streams and go on to modify weather around the globe. These global events are sometimes referred to as ENSO events (El Niño-Southern Oscillation). Upwelling fails worldwide and surface sea water temperatures rise. Areas that are normally dry can experience monsoonlike rains. Tropical typhoons, hurricanes, and strong winter storms can occur in areas that normally do not experience severe weather.

No two El Niño events are the same. A very strong El Niño/ENSO event occurs on average once every seven to ten years, and lesser events can occur several years in a row. In 1982–83, the severe El Niño/ENSO was estimated to have caused over eight billion dollars in damage.

Scientists use buoys and satellites to study wind speed and direction, as well as surface sea water temperature in the eastern and central southern Pacific. Developing El Niños may be detected so the world can prepare for them, as with the 1997–1998 ENSO. Many scientists believe an increase in the number of El Niño events is part of an overall pattern of weather change brought on by global warming.

UPWELLING AND EL NIÑO

UPWELLING*
- HIGH PRESSURE ZONE$_a$
- NW WINDS$_{a^1}$
- SURFACE WATER$_b$
- UPWELLING WATER$_c$

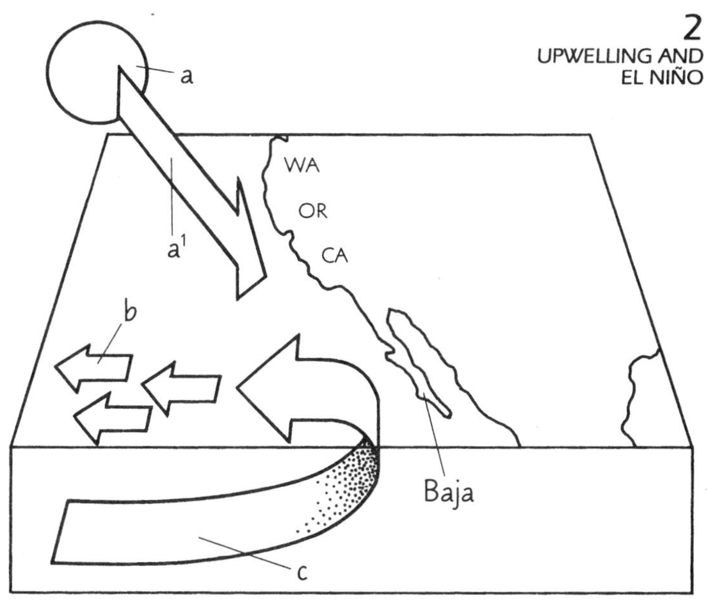

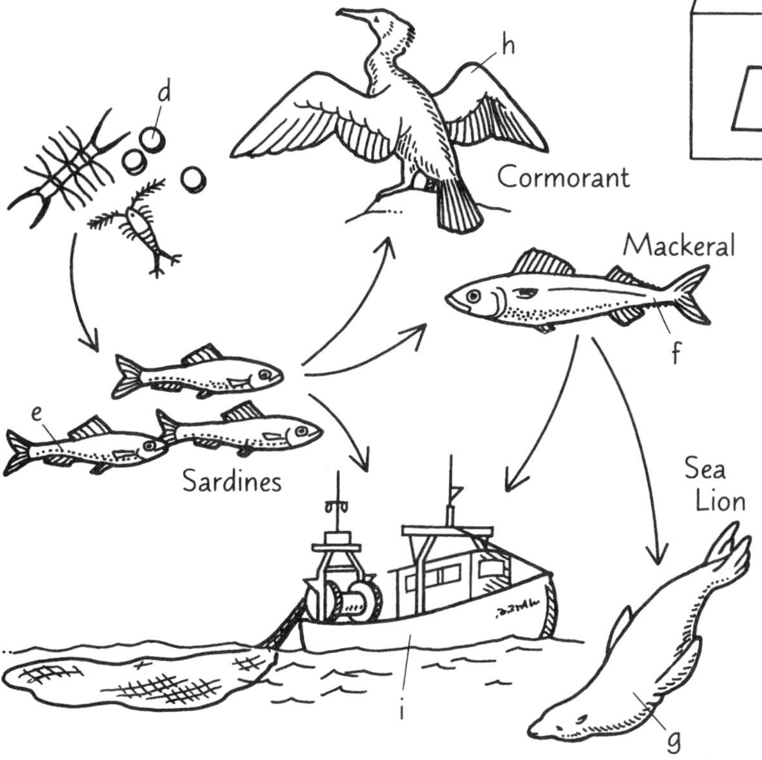

UPWELLING FOOD CHAIN*
- PLANKTON$_d$
- PLANKTON-FEEDING FISHES$_e$
- LARGER FISHES$_f$
- MARINE MAMMALS$_g$
- BIRDS$_h$
- MAN$_i$

EASTERLY TRADE WINDS$_j$
WARM SURFACE WATER$_k$
COLD WATER CURRENT$_l$

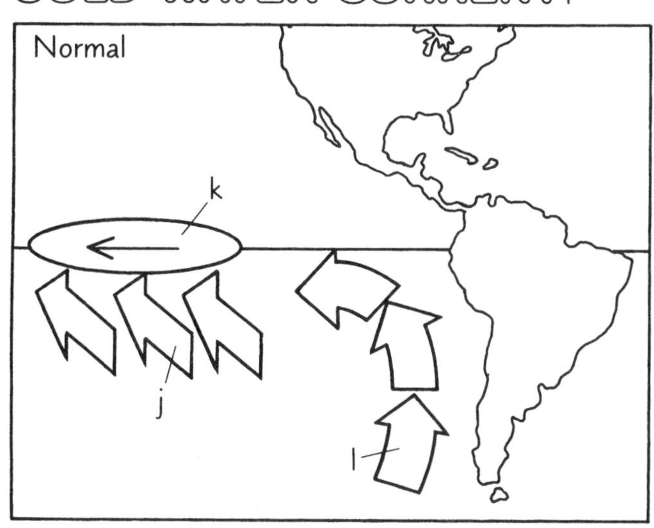

EL NIÑO CURRENT$_{k^1}$

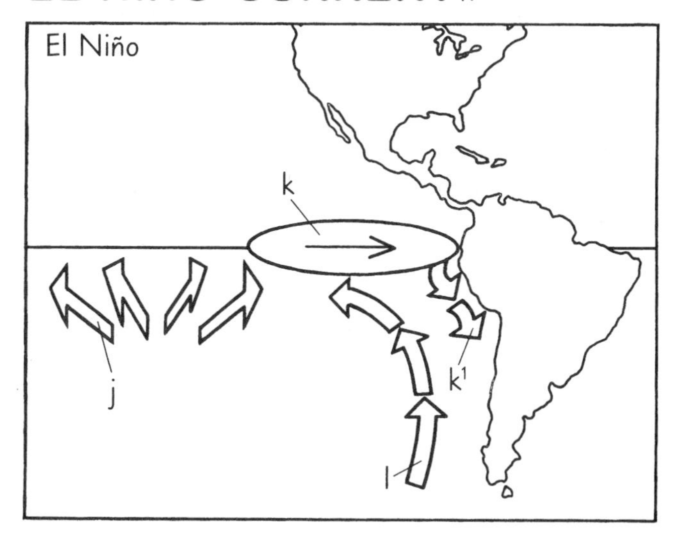

3
TIDES

As we have seen in the last two plates the ocean's surface is moved by wind and wave and is in constant motion, flowing in slow, oceanwide currents. There are also more locally experienced currents caused by tidal action. This plate will explore the nature and origin of tides.

Tidal variations are caused by the interplay of the earth, sun, and the moon. As the moon travels around the earth, and as they both, together, complete a larger path around the sun, their combined gravitational attraction along with other factors cause the alternate rising and falling of the surface of the ocean, called tides.

Color the diagram of the earth, the moon, and the sun at the upper left. Then color the three drawings of the earth at the upper right. The top drawing illustrates the moon's gravitational pull on the earth's surface water. The second drawing illustrates the tide caused by centrifugal force, and the third indicates the combined effects of the two forces.

Gravitation (gravity) is an attraction between two masses; it is proportional to the product of their masses and inversely proportional to the square of the distance between them. Thus, although the *moon* is much smaller than the *sun*, it is so much closer to the *earth* that it plays the major role in causing tidal variations.

Visualize an earth completely covered with a layer of water, spinning on its axis. As the moon orbits the earth, that part of the earth directly opposite the moon will be pulled by the moon's gravity, and the water will *bulge* away from the earth toward the moon.

The spinning motion of the earth-moon pair creates what is known as a *centrifugal force*. This force causes the water on the side of the earth farthest away from the moon to also *bulge* out, away from the earth's surface. The gravitational pull of the moon and centrifugal force modify one another, and, at any given time, a bulge of water will occur on the side of the earth facing the moon and on the side farthest away from the moon.

Because the earth revolves completely on its axis approximately (but not precisely) every 24 hours, every spot on the earth will experience both types of bulge within that 24-hour period. The bulges are called lunar tides; the peaks are the high tides, and the troughs (between the bulges) are the low tides.

In an isolated earth-moon system, each spot on the earth would experience four tides each lunar day (24 hours, 50 minutes); two high tides and two low tides of equal magnitude. But the sun has an important modifying effect.

Color the diagram in the center of the page that demonstrates the position of the earth, sun, and moon during the moon's phases, and illustrates the tidal bulges. Then color the illustration of tidal range and the map that shows tidal variations.

The moon orbits the earth once every 27.5 days (the lunar month), and in the course of its orbit, the moon is in a different position relative to the sun every day. At the time of the new moon and full moon, the sun, earth, and moon are in a direct line with one another, and the combined gravitational pull causes extra high and extra low tides. These tides are called *spring tides* and are depicted in the tidal range illustration to the left of the piling.

During the first-quarter and third-quarter phases of the moon, the three masses of earth, sun, and moon are not in a direct line. The sun's gravitational pull minimizes the effects of the moon's gravitational pull, and the tidal variations are of much smaller magnitude. These tides are called *neap tides* and are depicted in the tidal range illustration to the right of the piling.

There is considerable tidal variation on earth. Geographical position, the shape of the ocean basin, and a host of other local, global, and planetary factors act to modify tidal range and frequency. The tides at the end of a long inlet will be considerably amplified due to the large volume of water pressed into a very narrow space, whereas the tides on a mid-ocean island will hardly be noticed.

Similarly, the frequency of the tides varies from place to place. The coastline of northern Europe experiences two equal high and low tides every day. The coast of California also experiences two high and two low tides daily, but they are not equal. The Gulf of Mexico has only one high and one low tide each day.

TIDES

EARTH$_a$
ORBIT$_{a^1}$
ROTATION$_b$
CENTRIFUGAL FORCE$_{b^1}$ TIDE$_{b^2}$
MOON$_c$
ORBIT$_{c^1}$
GRAVITATIONAL PULL$_d$
TIDE$_{d^1}$

TIDAL BULGE$_{b^2+d^1}$

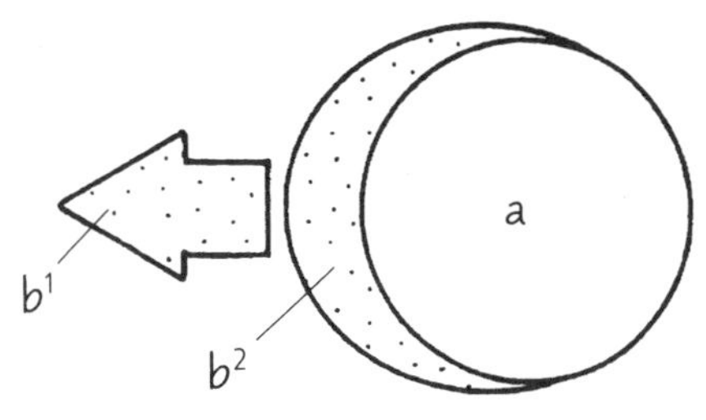

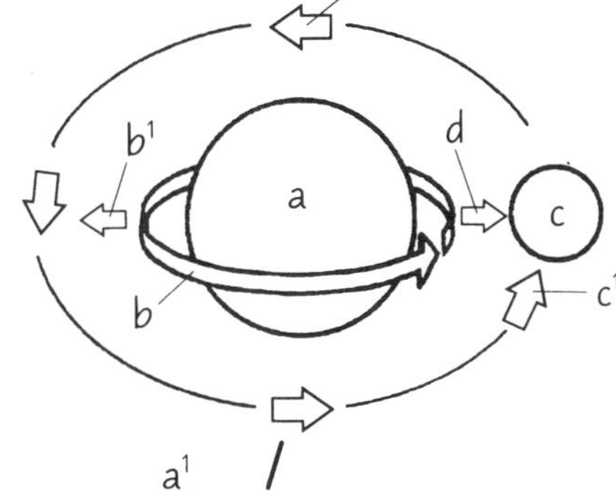

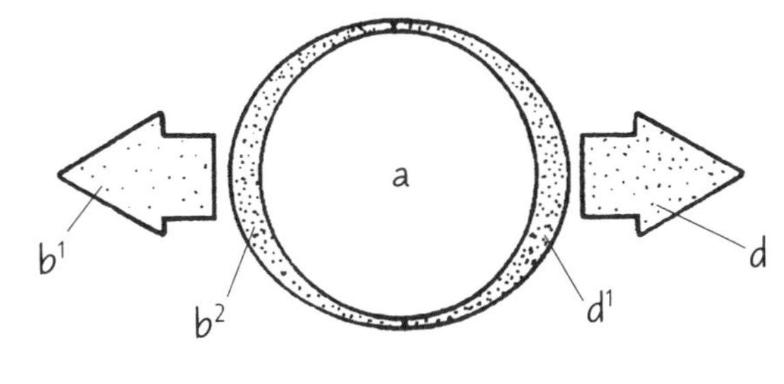

SUN$_e$
SPRING TIDE$_f$
NEAP TIDE$_g$

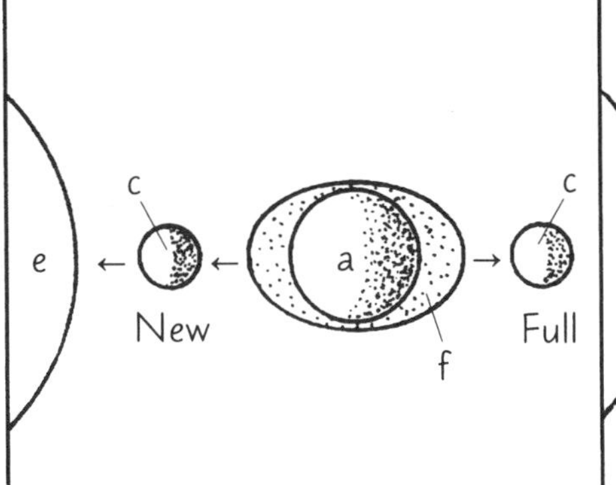

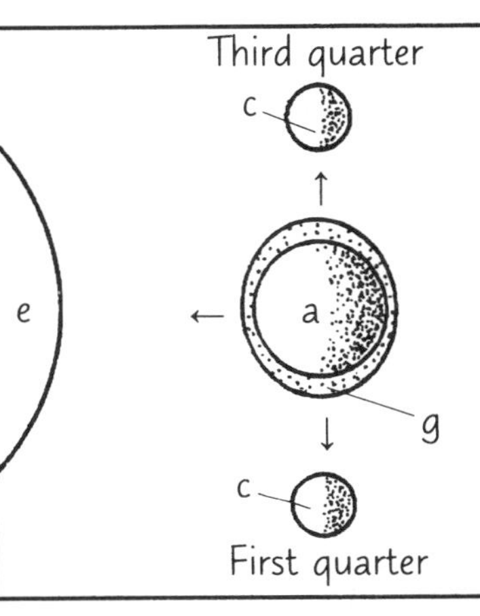

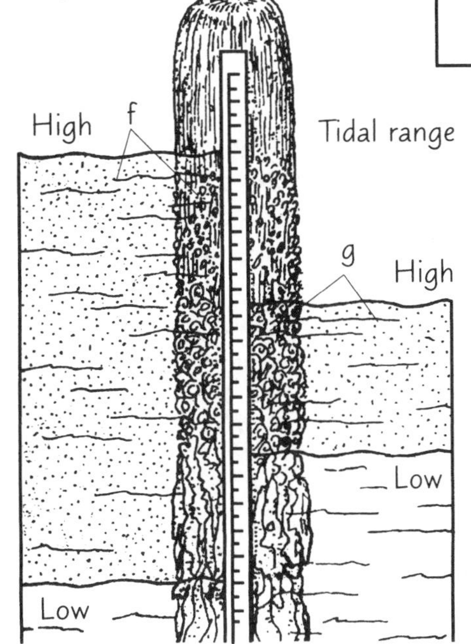

TIDAL VARIATIONS* LARGE$_h$ SMALL$_i$

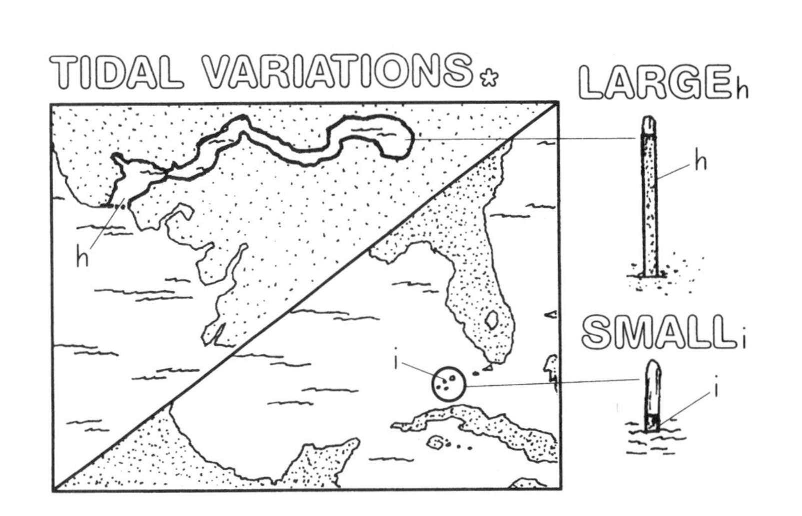

4
TIDAL ZONATION PATTERNS

The intertidal zone, also called the littoral zone, is the most readily accessible of marine habitats. The intertidal zone is that portion of the ocean shore that is periodically covered by the highest spring tides and exposed by the lowest spring tides. It is the meeting place of the marine and the terrestrial (land) environments and is exposed, at least a portion of it, to elements of both during each tidal cycle. During high tide, the water temperature is relatively even, although the intertidal inhabitants may be pounded by waves and wave-borne debris. When the tide is out (low tide), the intertidal zone is exposed to extremes in temperature, variations in light, and to fresh water; snow and ice may occur in the intertidal zones in some areas.

The degree to which a given intertidal area is influenced by the marine or terrestrial conditions is governed by the amount of time the area is exposed to the air and the frequency with which such exposures occur. Conditions are more terrestrial in the upper intertidal and correspondingly more marine in the lower intertidal. The wide range of physical conditions encountered in the intertidal environment influences where different kinds of organisms can live, and contributes to the creation of intertidal zonation patterns that correspond to particular tidal levels.

The intertidal zonation phenomenon is most visible at low tide on protected shores, where obvious horizontal stripes or zones are often apparent. These zones reflect the distinct texture or color of the spatially dominant organisms present. There is a remarkable similarity in rocky intertidal zonation patterns around the world. After studying world intertidal zonation for thirty years, Anne and T.A. Stephenson devised a general scheme by which to describe this zonation. Using their scheme, we will look at the zonation pattern found at one rocky intertidal area (in Coos Bay, Oregon), with the understanding that this scheme could be applied almost anywhere in the world, although it would include different organisms.

A representative dominant resident of each tidal zone is pictured on the right side of the plate. As each zone is introduced in the text, color the resident and its corresponding zone shades of the same color. If you wish to use realistic colors, color (a) light green, (b) gray, (c) light gray, (d) medium green, and (e) dark green.

The *supralittoral zone* is the area above the high tide mark that receives both wave splash and sea-water mist. Here live terrestrial organisms, such as lichens, that can tolerate some sea water, and marine animals that are becoming less dependent on, or less tolerant of, the ocean than those living lower in the intertidal. (An example is the large isopod in Plate 35.) The small *green alga* (seaweed), is found in the supralittoral zone: fresh water seeping down the cliff face and sea water splashing upwards provide conditions uniquely suited to this plant.

Below the supralittoral zone is the *supralittoral fringe* or "splash zone." This is the upper level of the high tide zone and receives a regular splash of waves when the tide is in. Here, the marine *periwinkle* snail is found. It can tolerate long periods of exposure to air and needs only an occasional wetting.

The lower limit of the supralittoral fringe is marked by the beginning of a barnacle zone. This area is called the *midlittoral zone* and encompasses the majority of the intertidal area. It extends down to the upper limits of the habitat of the large brown algae, which characterize the next zones. The midlittoral zone supports a great variety of marine animals, including *barnacles* and mussels (Plate 5).

The *brown alga*, *Alaria*, marks the *infralittoral fringe*, which includes the lowest level exposed by extreme spring tides. This area is often occupied by the brown alga, *Laminaria*, which extends from the infralittoral fringe into the *infralittoral zone*, or subtidal area, marking the beginning of the marine environment that is below the tides. Although *Alaria* and *Laminaria* are in the brown algae division (Plate 21), they are green in color. Sponges, sea urchins, and abalone are also found in the infralittoral zone and fringe.

Low spring tides reveal most of the shore and are the best times to venture into the intertidal zone.

TIDAL ZONATION PATTERNS

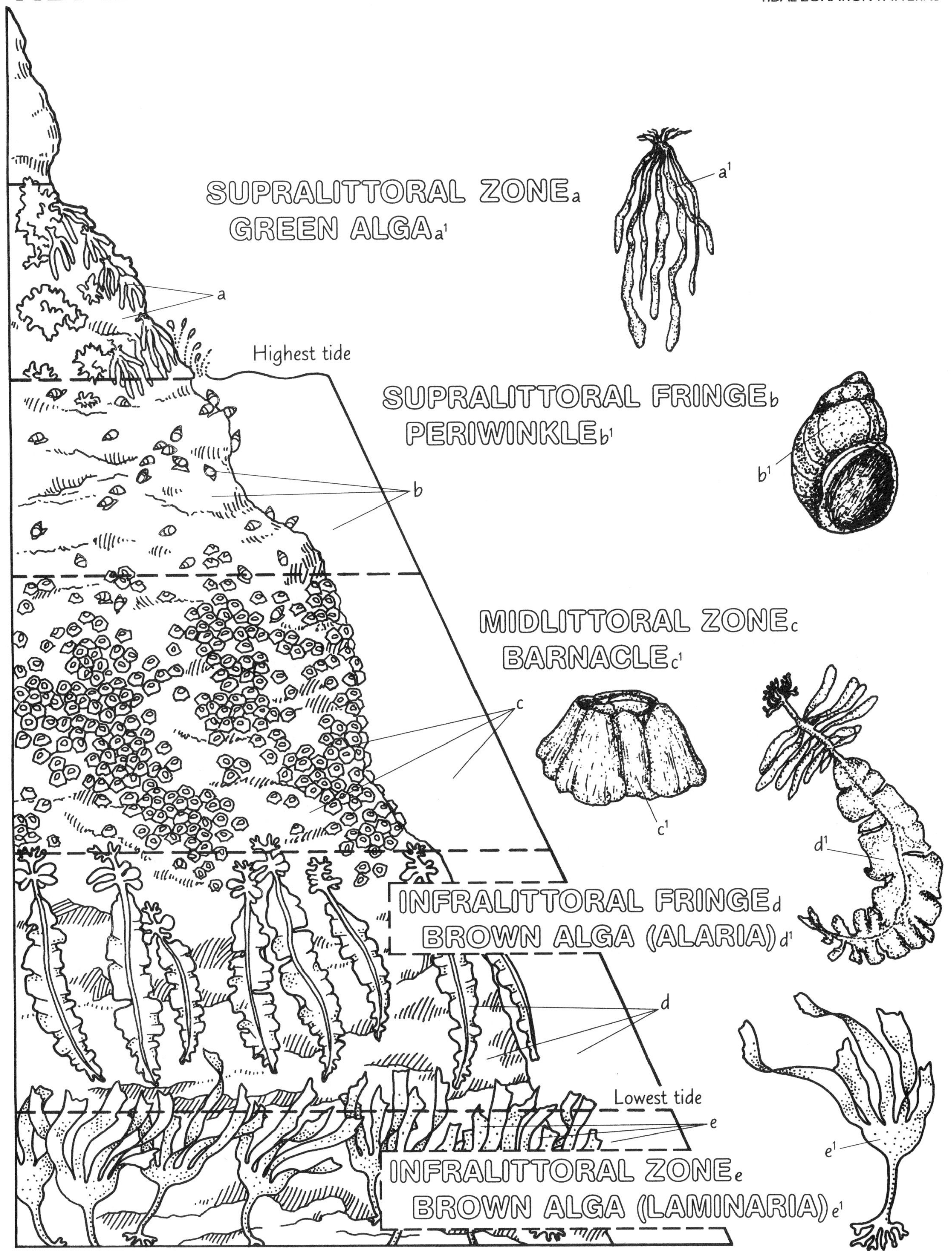

5
ROCKY SHORES

The diverse twists and turns of the earth's geological history are often visible in the scoured rocks of the shoreline, where the land meets the ocean. Some rocky intertidal zones are relatively flat with little surface relief, while others are steep or irregular with boulders, ledges, overhangs, and tidal (surge) channels. Irregular and varied surfaces offer a great deal of potential living space for organisms.

All open-coast rocky intertidal areas are subjected to the battering of waves as well as periodic exposure to the air. Using this plate, we will investigate how some of the more common rocky intertidal animals of the west coast of North America are adapted for survival in this rigorous habitat.

Begin your coloring with the limpets near the top of the rocks, and color each animal as it is mentioned in the text. The rocky substratum and the unidentified plants should not be colored. Note that the tide is low in this illustration and that the approximate levels of high and low tide are indicated. The bottom third of the page, from the water line down, may be colored over with a light blue or light green.

Limpets are commonly found in the high intertidal zone. These small, gray-brown gastropods are able to hold onto the substratum with a large muscular foot. They seek out depressions and ledges that shield them from direct sunlight; this helps them to avoid being dried out during low tide. Limpets also secrete a sticky mucus that seals their shells to the rocky surface and prevents the evaporation of water from their soft bodies. Some species use their radulas (a filelike structure; Plate 106) to excavate a depression in the rock. These limpets fit into the hole perfectly during low tide. When high tide returns, they move about on the rocks, grazing on algae, and then return "home" to their self-made, shallow "fox hole."

The green-lined *shore crab* has a flattened body that allows it to slide under and between rocks to avoid exposure. This nimble scavenger can live high in the intertidal zone. It is most active at night and at high tide.

Mussels are commonly found in the middle intertidal zone, often in large numbers. This bivalve mollusc (Plate 30) anchors itself to the rocks or to other mussels. By keeping its dark blue shell tightly closed, the mussel is able to resist desiccation (dehydration) during low tide. A large grouping of mussels, called a mussel clump, provides living space for many other small marine creatures.

Another organism found abundantly in the high and middle intertidal region is the *aggregating anemone* (Plate 96). This small anemone (2.5 cm, 1 in across) forms masses often comprising hundreds of individuals. The crowding reduces water loss caused by exposure. The gray-green bodies of these anemones are commonly covered with bits of shell and rock, which further protects them from direct exposure.

An algae-eater of the middle intertidal zone is the *mossy chiton* (ky-ton). This mollusc's shell is divided into eight separate gray plates, or valves, that move against one another (articulate) allowing the chiton to conform to the irregular rocky surface. The chiton uses its foot to maintain a strong grip against the wash of the waves. When the environment is calm, the mossy chiton forages on small algae, and remains on the rock's surface during both high and low tide.

In the lower intertidal zone, there are often sharp ledges and overhangs where light cannot penetrate and the growth of algae is prevented. Bright red, orange, and green *sponges* and other delicate animals such as the straw-colored ostrich-plume *hydroid* attach themselves to these protected underledges.

The purple *sea urchin* lives in the lower intertidal zone and below, in the subtidal zone. It survives the pounding of waves by excavating a pitlike depression in the rock with its spines. The pits remain moist at low tide and protect the urchins from the direct crashing of waves and wave-borne debris at high tide.

ROCKY SHORES

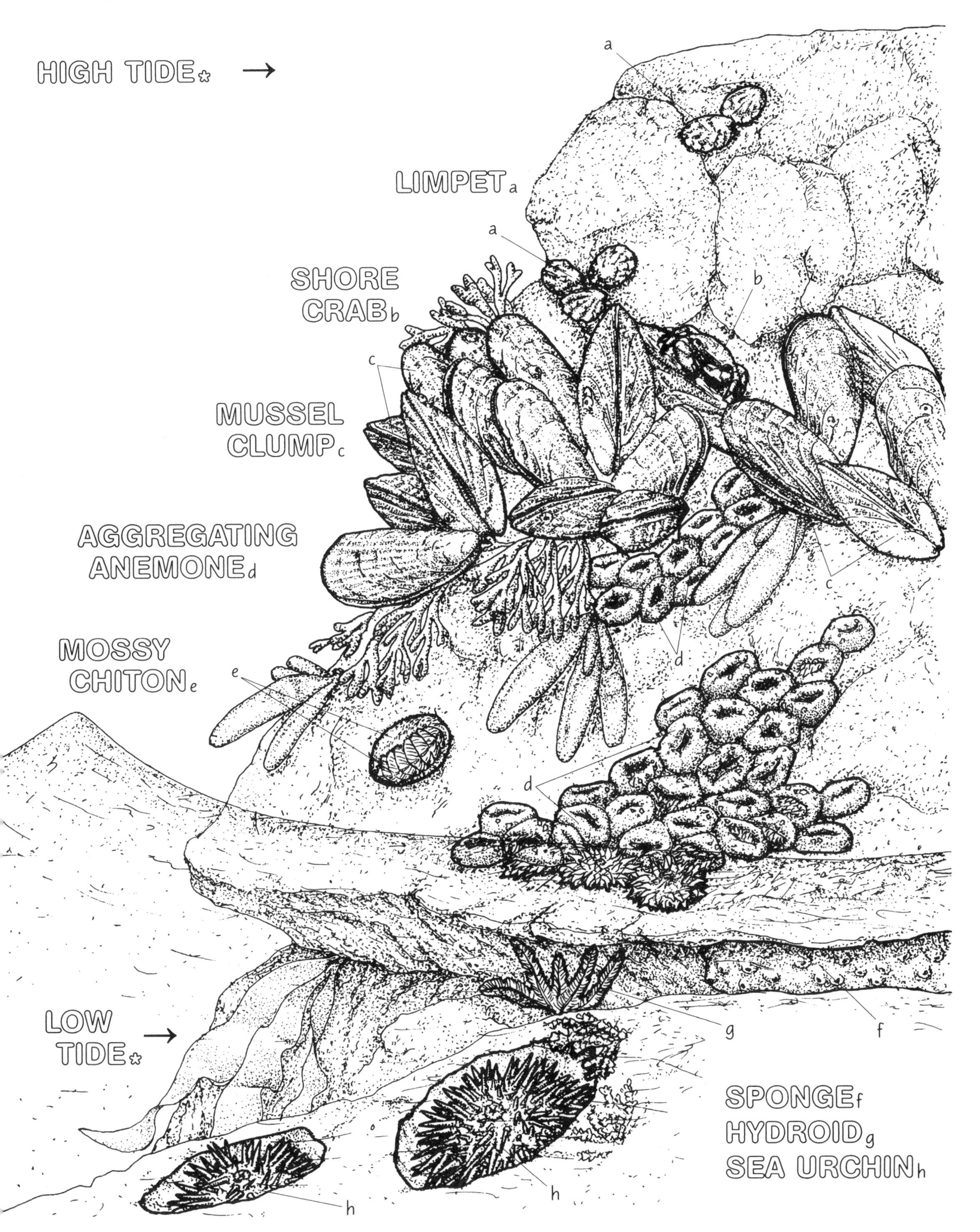

6
TIDE POOL

Tide pools are microcosms of the marine life found below the tide line, and are readily accessible when the tide is low. They are formed where depressions in the rock hold reservoirs of sea water. These pools are colonized by sessile (attached, or stationary) plants and animals, and also by small motile forms that live permanently in the pool or seek refuge there when the tide recedes.

Although these pools are found at every level between the high- and low-tide lines, it is in the pools at the lower levels, where there is less fluctuation in physical conditions, that the most abundant and diverse life forms reside. The organisms discussed here represent the inhabitants of a typical low-intertidal pool on the central California coast.

Color each organism as it is mentioned in the text. You may wish to use the organism's natural color where suggested.

The background coloration in tide pools is created by the substratum and the attached plants and animals residing there. In the well-lighted areas of the pool are the pastel pinks of the *encrusting* and *articulated coralline algae*, the bright green of *surf grass*, and the subtle reds and purples of the delicate *red algae*. In the shadowed areas are the bright reds and yellows of encrusting *sponges* and the vibrant orange of the small *solitary corals*. Also tucked in among the shadows is the *giant green sea anemone*, waiting for unwary prey or dead material washed within its grasp.

Moving through this colorful seascape is a multitude of very small crustaceans, molluscs, and worms (not shown). The larger animals include the *sculpin* that darts about briskly and then settles motionless on the bottom, fading into the background by virtue of its camouflaging mottled gray-brown coloration. These fishes (5–7.5 cm, 2–3 in) venture considerable distances from the tide pool when the tide is in, and often return to the same pool when the tide recedes.

Another well-camouflaged traveler in the pool is the *broken-back shrimp* which is capable of rapid movement by the sudden flexure of its powerful abdomen. These crustaceans (2.5 cm, 1 in) display a range of colors from solid greens to colorful mottled combinations. They are sometimes found in considerable numbers, but are often difficult to see because of their camouflage.

The *hermit crab* is fairly easy to spot in the tide pool. This crab occupies an empty sea snail shell, then exchanges it for a larger shell as it increases in size. If disturbed, the crab rapidly disappears into its sanctuary; when unmolested, the hermit crab scavenges for food along the bottom of the tide pool.

Bearing a shell of its own creation, the *dunce cap limpet* may be seen feeding on encrusting coralline algae. The dunce cap limpet is one of the few animals able to consume this unpalatable alga, which is even found growing on its own shell, giving it a pale pink color.

The small (5–7.5 cm, 2–3 in) six-rayed *sea star* may be found patrolling the bottom and sides of the pool. This dull red or green generalist carnivore feeds on a variety of the small attached life forms in the pool.

If the tide pool has any loose rocks on the bottom where small amounts of sediment have collected, delicate *brittle stars* and burrow openings of *polychaete worms* can be found. If sufficient sediment has collected on the bottom, one may also discover brick-red *rock crabs* that have burrowed in, with only their stalked eyes and antennae visible above the sand.

TIDE POOL

ATTACHED*
CORALLINE ALGAE (a)
ENCRUSTING a^1
ARTICULATED a^2
SURF GRASS b
RED ALGAE c
SPONGE d
SOLITARY CORAL e
GIANT GREEN SEA ANEMONE f

MOTILE*
TIDEPOOL SCULPIN g
BROKEN-BACK SHRIMP h
HERMIT CRAB i
DUNCE CAP LIMPET j
SEA STAR k
BRITTLE STAR l
POLYCHAETE WORM m
ROCK CRAB n

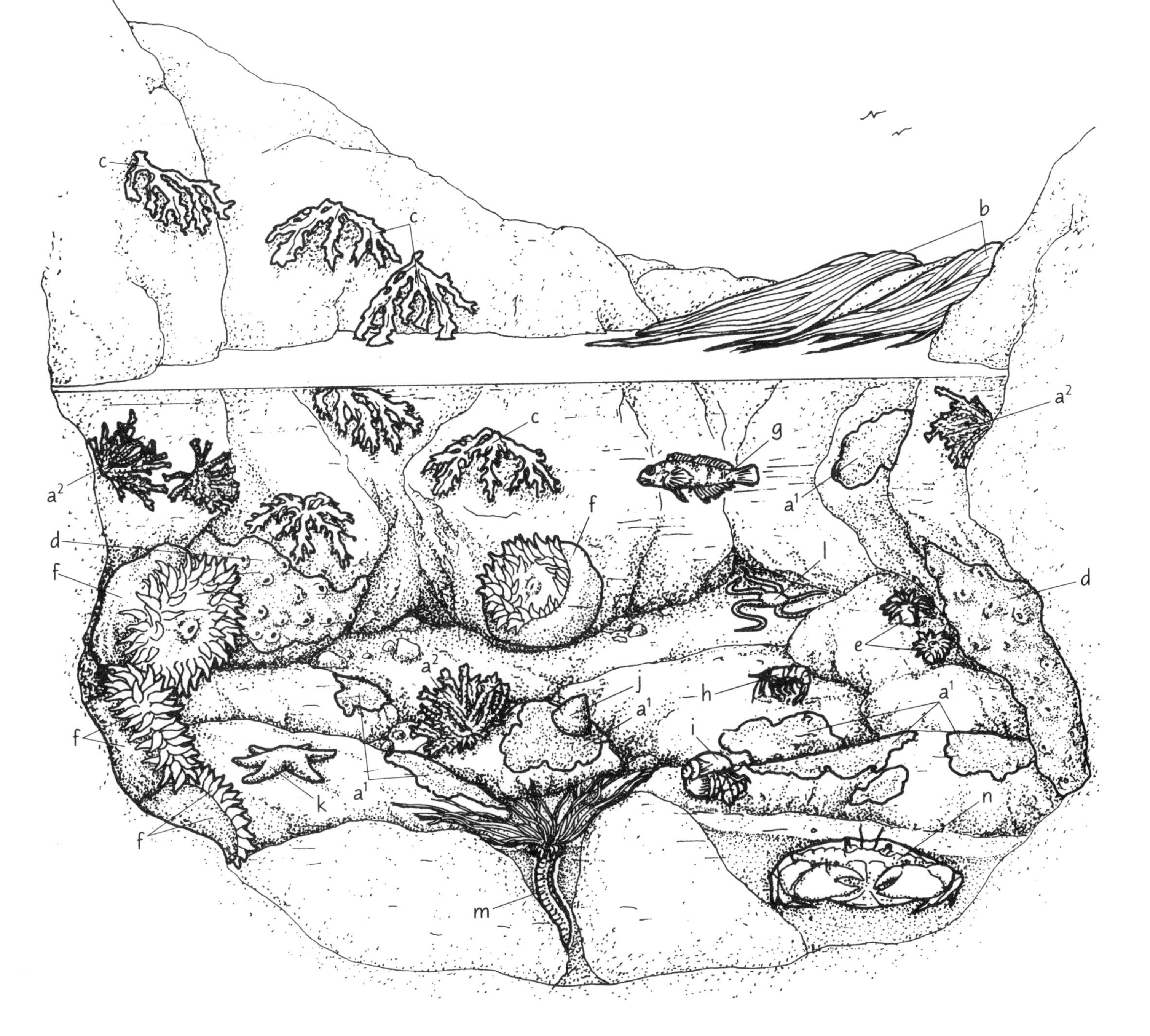

7
SALT MARSH

The previous two plates characterize marine habitats with hard, rocky substrates. While these habitats are very interesting because of the high diversity of organisms they support, the majority of intertidal habitats have soft substrata.

Along many coastlines are areas well protected from the direct onslaught of waves. These areas are designated coastal wetlands and include coastal lagoons and embayments, estuaries, and tidal sloughs. Many of these environments are supplied with fresh water from rivers or coastal streams. In the quiet waters of these wetland environments, fine sediments and organic detritus carried by rivers and tidal currents are deposited. A muddy substrate forms which, from mid-tide level and higher, harbors one of the most conspicuous and important habitats found in coastal wetlands—the salt marsh. Below the salt marsh, from mid-tide down, open tidal flats occur. These are treated in the following plate.

Reserve a light green color for the cord grass, blue for the tidal channel, and brown for the mud flat. Begin by coloring the single cord grass plant in the center of the page, including the roots. Color the areas of cord grass in the picture of the salt marsh. At the top of the page, color the enlargement of the blade of cord grass showing salt discharge. Color the arrow indicating detritus falling into the tidal channel area. Next, color the channel and mud flat and each of its inhabitants as they are mentioned in the text. Except where noted, most of these animals are gray or mud colored. Finally, color the arrow for animal waste.

Salt marshes stretch over millions of acres along the broad coastal plain of the southeastern coast of the United States and along the Gulf of Mexico. They also fringe the edges of the major estuaries on the northeast and west coasts of the United States. At the heart of the marsh, from the mid-tide to the high-tide lines, is a rugged flowering plant known as *cord grass*. Cord grass is one of the few flowering plants that can survive immersion in salt water: excess salt coming into the plant is accumulated and discharged through the leaves.

Cord grass sends its roots deep into the rich marsh mud, tapping the nutrients derived from decaying organic matter and sending out underground stems from which new plants sprout. Cord grass can soon monopolize an area; the spreading plants trap more sediment and build up the substrate, expanding the marsh seaward. In the Georgia salt marshes, cord grass can produce an annual average mass of 200 tons per hectare (80 tons per acre), making it one of the most productive habitats in the world.

Only about 10 percent of the cord grass that grows is consumed directly by animals in the salt marsh. Most of it dies, breaks apart, and is carried into the *tidal channels* by the ebb and flow of the tides. Here the plant material is decomposed by bacteria and fungi and becomes *detritus*. This detritus forms the basis of the food chain that fuels many major fisheries of the east coast. Much of the detritus is transported directly out of the marsh with the ebb tide and finds its way into near-shore food chains.

Among the cord grass roots and along the tidal channels are *mussels*, which feed by filtering out the detritus and minute plants suspended in the water. Detritus is utilized by other filter feeders, such as *oysters* and *clams* inhabiting adjacent *mud flats*, as well as the young *menhaden* fish. The salt marsh forms the hatching ground and "nursery" for the first eight months of the menhaden's life before it goes to sea to join the largest fishery on the east coast.

Small, green *grass shrimp* are abundant in the tidal channels and eat the detritus found there. These shrimp, in turn, fall prey to *flounder* and young *striped bass*, who follow the tide into the salt marsh to feed. The *blue crab* also lurks in the tidal channels waiting for prey.

When the tide recedes, the tidal channels and the lower mud flats are left exposed. Hordes of *fiddler crabs* emerge from their mud *burrows* to sift the rich mud for plant detritus. During the mating season, large male crabs stand by the entrances to their burrows and wave their outsized claws in an attempt to attract a female partner.

All marine animals in the channel and the salt marsh deposit their *wastes* on the marsh bottom, where they are decomposed by bacteria and recycled into the cord grass through its roots.

All around the salt marsh the sounds of shorebirds and migrating ducks and geese may be heard. The salt marsh serves a vital role in linking the land with the sea. The marsh's productivity and its role as a nursery area figure importantly in the success or failure of near-shore fisheries.

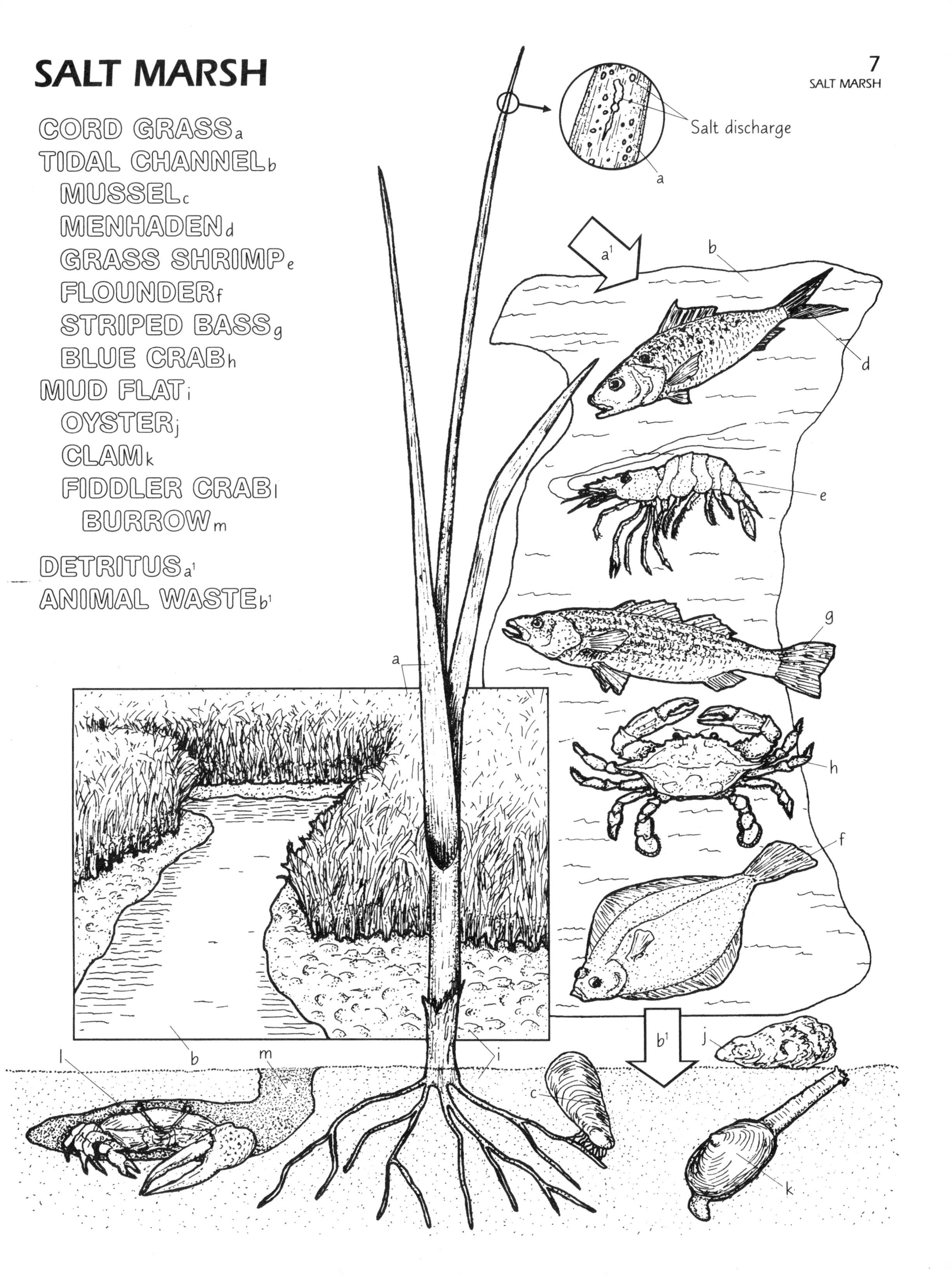
SALT MARSH
7
SALT MARSH
CORD GRASS a
TIDAL CHANNEL b
MUSSEL c
MENHADEN d
GRASS SHRIMP e
FLOUNDER f
STRIPED BASS g
BLUE CRAB h
MUD FLAT i
OYSTER j
CLAM k
FIDDLER CRAB l
BURROW m
DETRITUS a1
ANIMAL WASTE b1
Salt discharge
a
a1
b
d
e
g
h
f
a
b
l
m
i
c
b1
j
k

8
TIDAL FLATS

Most marine bottom environments are mud or sand. These soft bottoms directly result from water movement. Faster water contains more energy and carries larger suspended sediment particles. As water slows, it loses energy and the larger, heavier particles fall to the bottom, forming a sandy sediment. Sandy sediments are found on exposed beaches and those estuaries and coastal embayments with strong tidal currents or river flow (Plate 9). In quiet-water marine habitats, only the smallest, lightest particles have remained in suspension and settle to the bottom. These particles, known as silts and clays, are deposited as marine muds.

Organic particles are small and light and have a settling pattern similar to clay particles. Availability of organic detritus directly influences the character of the substrate and its potential inhabitants. Marine muds contain a large amount of organic detritus, a potential food source for animals. The mud flat fauna is dominated by deposit feeders (Plates 27, 28, 30). On sandy bottoms, the organic detritus remains suspended, and the fauna is dominated by filter feeders (Plates 29, 30, 105).

Color the large and small illustration of each organism as it is discussed.

Formed by gradual sedimentation, tidal flats show little physical relief. Surface signs of burrowing and feeding activities of resident organisms can be quite extensive. The pink to dark green *lug worm* ingests sediment and digests any available organic matter. The remainder is defecated onto the surface in characteristic fecal mounds, giving away the lug worm's presence below (Plate 27).

The bivalve molluscs known as clams (Plates 29, 30) are very much at home in protected, tidal flat habitats. The relatively undisturbed substrate allows them to sit vertically in semi-permanent burrows. Clams have posterior, elongated siphons which stretch to reach the water for filter feeding while the body of the clam remains safely burrowed. Siphons are especially effective for clams that live where there is little water circulation below a very shallow surface sediment layer. Beneath this layer, oxygen is depleted by decomposing bacteria, and the substrate turns dark and has the odor of rotten eggs. Often the shells of these clams will be stained black, as is frequently encountered in the large *gaper* or horseneck *clam* of the west coast of the United States. The large siphons of the gaper clam can not be fully retracted, so the posterior end of the shell gapes open to allow the clam to close its shells around the rest of its body. Gaper clam shells may reach over 20 cm (8 in), the clam can weigh as much as 2.8 kg (6 lb) and can burrow over a meter (3 ft) into the sediment.

The *bent-nosed clam* (Plate 30) lies burrowed on its left side with the bent posterior end of the grayish-white shell facing upwards. The yellow siphons of this clam are long, thin, and separate from one another. The incurrent siphon probes about the surface for deposited organic matter and vacuums the food into the clam's mantle cavity. Here it is trapped on the surface of the gills in the same manner as a filter-feeding clam (Plate 29). The excurrent siphon remains buried, out of the way of the busy incurrent siphon.

The dull brown *mud snail* is a native of the east coast of the United States. These common 2.5 cm (1 in) long snails pepper estuarine mud flats and salt marsh channels. Mud snails actively sweep elongated snouts across the mud and pick up deposited organic detritus.

Fascinating sandy mud flat crustaceans, *ghost shrimps* are active burrowers and their excavations can extend a meter or more beneath the surface. As the pale bluish-white shrimp burrows through the sediment, it sifts out and eats buried organic particles. The shrimps' continuous excavations can turn over the upper layer of a local tidal flat area several times a year.

Solitary *moon snails* (Plate 79) are commonly seen on sandy mud flats, usually indicating their burrowing presence as a bulge in the sand or just the top of their light brown shell showing. When one of these large predatory snails is unearthed at low tide, it often has its huge foot extended and wrapped around a bivalve prey.

The sand is subjected to stronger water currents and sometimes, especially in coastal embayments, wave action. For the clams that live here, this means that a burrow may be disrupted occasionally or buried. The *basket cockle* of the west coast of the United States (Plates 29, 30) has a robust, heavily ribbed shell mottled with white and brown that can reach 10 cm (4 in) in diameter. The cockle's siphons are very short, barely protruding beyond the posterior end of the gaped valves. The foot is large, stretching out over twice the length of the shell when fully extended and allows the clam to reburrow quickly if exposed. The cockle lives at such a shallow depth that it is easily excavated by the predatory sea star, the *short-spined pisaster*. However, the cockle has a remarkable escape response up its sleeve (shell?) (Plate 103).

The pink short-spined pisaster is found from sand flats and the rocky intertidal zone to pier pilings, and well offshore on subtidal sand and mud bottoms. It feeds primarily on bivalve molluscs. Short-spined pisasters find and extract burrowed clams by extending special elongated tube feet into the substrate. The tube feet have suckers at their tips which attach to the clam and pull it to the surface.

TIDAL FLATS

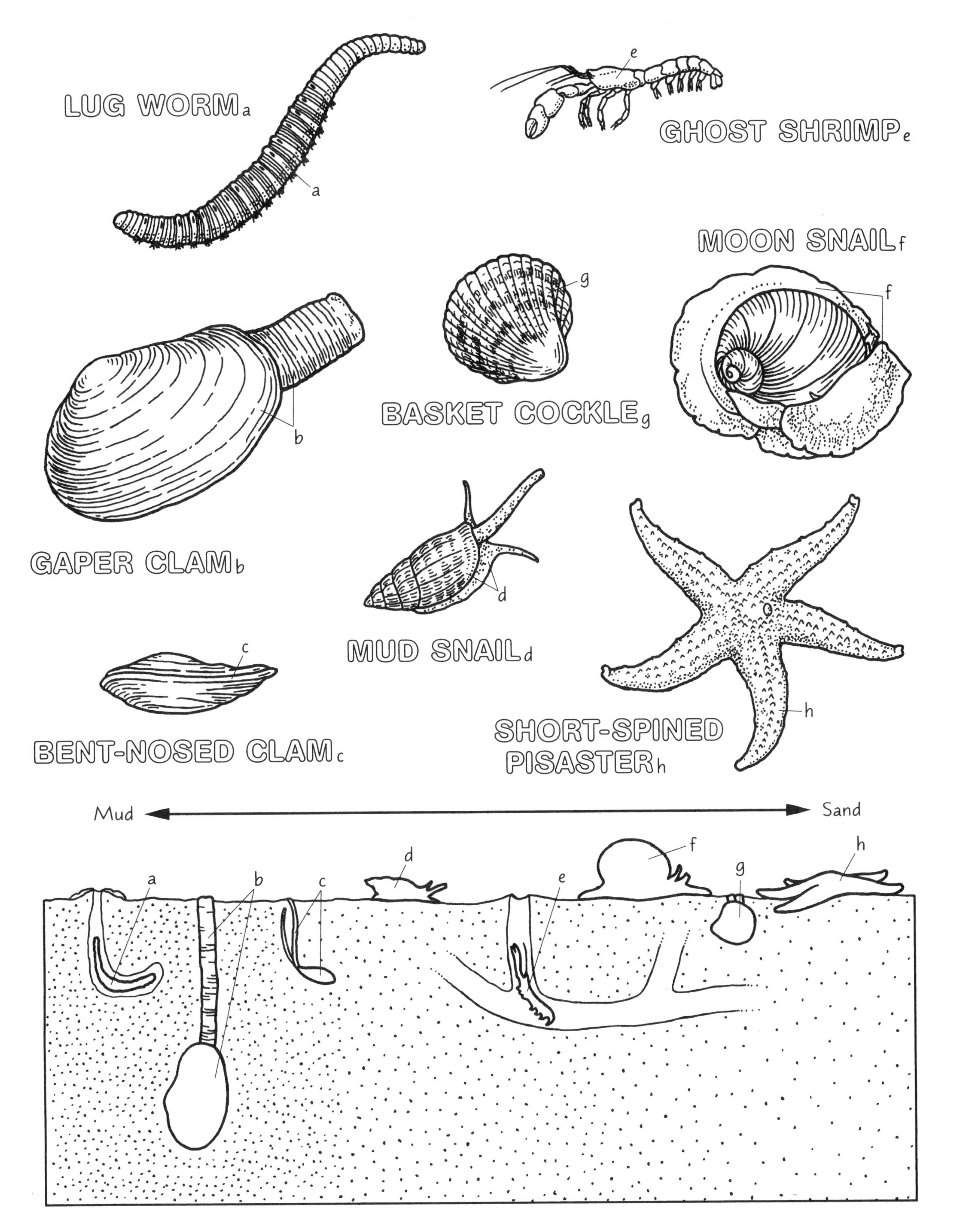
LUG WORM a
GHOST SHRIMP e
MOON SNAIL f
BASKET COCKLE g
GAPER CLAM b
MUD SNAIL d
BENT-NOSED CLAM c
SHORT-SPINED PISASTER h
Mud
Sand
a
b
c
d
e
f
g
h

9
SANDY BEACH

Sandy beaches, a familiar sight along open coastlines, form by the accumulation of sand particles which have been carried and deposited by waves. The particles that make up these sandy beaches either are eroded from the land's surface and carried into the sea by rivers and the wind, or are weathered from near-shore habitats by the pounding waves. A typical mainland beach is composed of small particles of the minerals quartz and feldspar. The beaches of tropical islands are sometimes composed of eroded coral and shell, and the black sands of certain Hawaiian beaches came from the erosion of lava flows.

Once a beach forms, it changes continuously; through the seasons, the waves constantly rework the sand and reshape the beach.

Color the illustrations of the sandy beach, shown in spring, summer, fall, and winter. The curved arrows in the fall illustration represent the wave-driven removal of sand from the berm.

During spring and summer, gentle *waves* deposit *sand* onto the *beach platform*, forming a broad sandy slope or *berm*. The large waves of the first fall storm begin to remove sand from the beach and deposit it offshore in ridges, called sandbars. The winter beach may have nothing remaining but the rocky beach platform (the eroded edge of the coastline) and *cobblestones* too large for the winter waves to carry off.

The beach is a rigorous environment, and organisms that live here must adapt to shifting sand and waves. Successful species are able to ride the waves and to burrow deep into the sand, or are able to reside just above the tide line. Sandy shores support relatively few species, which can occur in great numbers of individuals due to reduced competition between species. The waves bring a steady source of detritus and larger pieces of organic debris, such as loose seaweed and dead fish. These materials provide a dependable food source for the marine organisms of the sandy shore.

Color the large and small illustration of each sandy-beach organism as it is mentioned.

The dark gray *sand crab* uses the waves in moving up and down the shore. It feeds by burrowing its posterior end into the sand and then unfurling its long antennae into the overlying water to filter food from the wave backwash (Plate 36).

The small (2.5 cm, 1 in) wedge-shaped, buff-color *bean clam* is a rapid burrower and rides the waves, staying in the area of greatest water movement, where suspended food is most abundant.

The *razor clam* is found along the beaches of the Pacific Northwest near the low-tide line. Razor clams thrive on surf-zone diatoms (microscopic plants) that bloom throughout spring and summer. Anyone who has pursued these delectable yellow-shelled clams can tell you of their burrowing speed.

Like the razor clam, the *bristle worm*, a scavenging polychaete, remains in the lower zone of the beach and relies on rapid burrowing to keep from being washed away by the tide.

Beach hoppers live in burrows above the high-tide line and emerge at night to feed on deposited drift algae. The predaceous *rove beetle* also lives above the high-tide line. It comes down onto the shore at night in pursuit of the unwary beach hopper.

Another nocturnal carnivore is the pale white *ghost crab* of tropical and semitropical sand beaches. These crabs are highly tolerant to air exposure and live in burrows above the high-tide line.

The gun-metal blue *swimming crab* and the silvery barred *surfperch* ride in with the tide to feed on sand crabs and other burrowing animals.

At low tide winged predators arrive to forage on the beach. *Sanderlings* (dark gray wings and white body) follow the waves' retreat to feed on sand crabs and worms.

SANDY BEACH

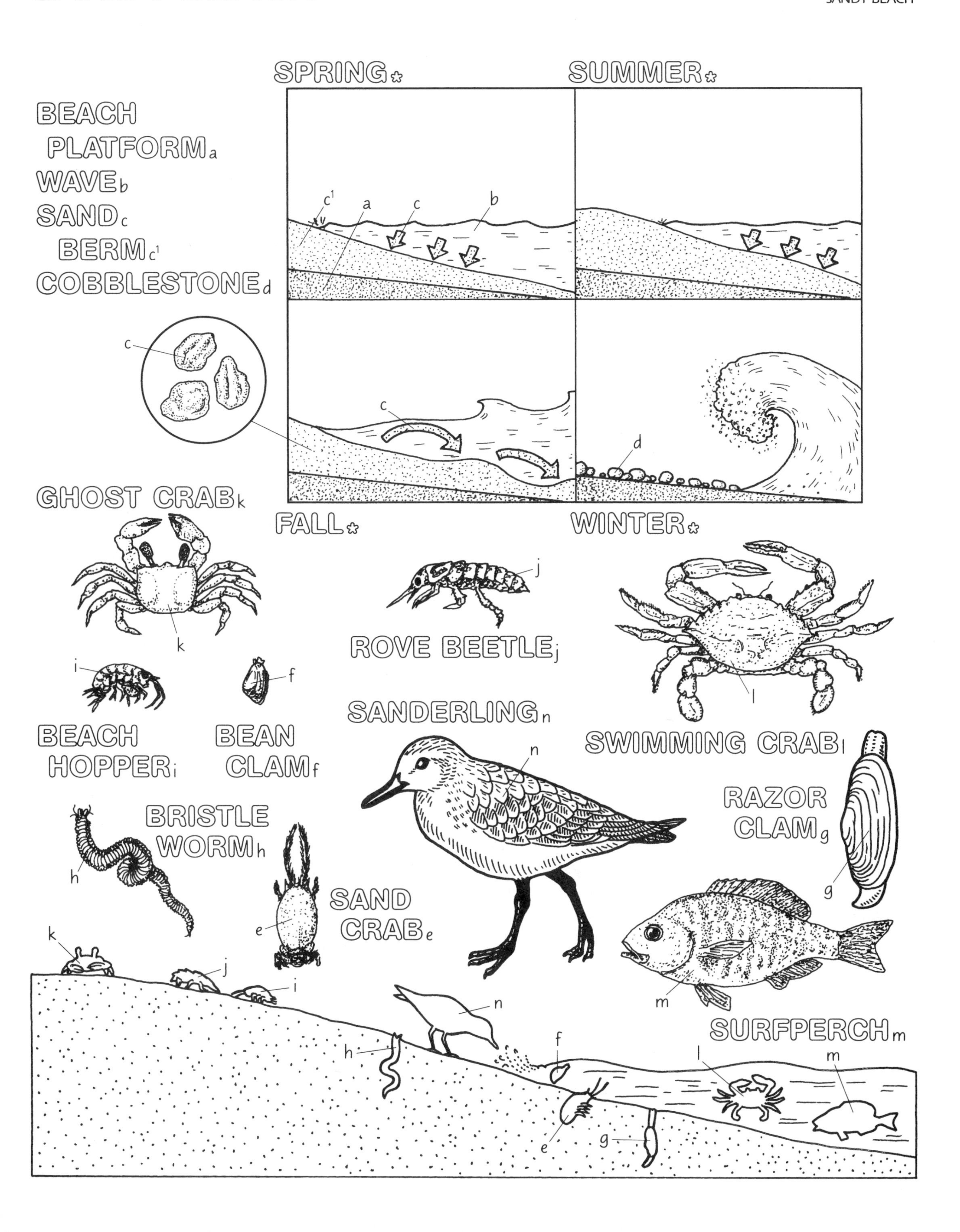
SPRING*
SUMMER*
FALL*
WINTER*
BEACH
PLATFORM a
WAVE b
SAND c
BERM c1
COBBLESTONE d
GHOST CRAB k
ROVE BEETLE j
BEACH HOPPER i
BEAN CLAM f
SANDERLING n
SWIMMING CRAB l
RAZOR CLAM g
BRISTLE WORM h
SAND CRAB e
SURFPERCH m

10
SUBTIDAL SOFT BOTTOMS

The previous six plates have dealt with intertidal habitats. Subtidal habitats of the near-shore, from the lowest intertidal zone to the edge of the continental shelf, will be treated in this and the next three plates. The distribution of soft bottom types in subtidal habitats is controlled by the same factors discussed for intertidal soft bottoms: the degree of water movement and the character of the materials in suspension determine the type and composition of the bottom.

Soft bottoms on the continental shelf are composed mainly of inorganic materials washed from the land by rivers or carried away by wind. The shallow bottom closest to shore experiences the strongest wave action and water currents. Here sandy bottoms predominate. At greater depths or in sheltered areas, water movement is reduced, and finer particles are deposited, forming mud. A gradient of soft bottom results, ranging from coarse sand near shore to muddy bottoms offshore over the continental shelf. As the water becomes calmer offshore and suspended organic detritus settles to the bottom, the fauna shifts from filter feeders to deposit feeders similar to the faunal patterns in tidal flats.

Color the small and large illustration of each animal as it is mentioned in the text. The animals are grouped as epifauna, living on the surface, and infauna, which live in the substratum.

Residing below the low-tide line along the central California coast is the large (15 cm, 6 in) *Pismo clam.* Pismos live in clam beds and feed on the large amount of suspended detritus in the surf zone. They depend on their heavy shells to keep them in place. Just beyond the surf zone, beds of the filter-feeding epifaunal *Pacific sand dollar* (Plate 105) may be found. These beds may be several meters wide and stretch for kilometers offshore of exposed sandy beaches, with the dark purple sand dollars numbering in the millions. Amongst the sand dollars, the scavenging *elbow crab* and *hermit crab* scuttle about looking for food trapped in the bed. Predators such as the sandy-gray *sand star* feed on the sand dollars, and the buff-brown *moon snail* feed on the numerous, burrowed filter-feeding clams, like the bright white *sea cockle*. Many species of flatfish occur, including the camouflaged *sanddab* which blends into the sand bottom and feeds on small worms and crustaceans. Larger fish include the *angel shark* which conceals its flattened body by partially covering itself with sand when not actively foraging for burrowed prey. Other flat elasmobranch fishes like skates and rays (fishes that lack true bones, Plate 52) also hunt for food on these soft bottoms, often excavating deep pits with their broad pectoral fins to uncover clams and other prey.

In deeper, calmer water, large aggregations of pale gray *brittle stars* may be found swarming on the bottom, feeding on deposited material, or burrowed into the sediment, with only their arms protruding through the mud. Another spiny-skinned animal found here is the dark gray *heart urchin*, or sea porcupine, which ingests sediments as it burrows horizontally just below the surface, digesting the organic material contained therein.

Worms are among the most abundant infaunal animals. Their sleek, elongated bodies are perfectly adapted for efficient burrowing. Many types of worms live in these soft bottom sediments. Some ingest the sediment as they burrow through it; others live in vertical tubes buried in the substrate and feed on detrital material deposited on the surface; still others filter suspended materials from the water above with elaborate tentacles while their bodies rest safety protected in a burrow (Plate 28). Shown here is the shimmy worm, *Nepthys*, a robust (5–10 cm, 2–4 in) silver-gray, burrowing *polychaete worm. Nepthys* is a carnivorous scavenger that can re-burrow very rapidly if uncovered.

The soft bottoms of the continental shelf are very monotonous, homogenous habitats that offer limited places for plants and animals to live. These habitats host a low diversity of organisms compared to the many different kinds of plants and animals that can occur in hard substrate habitats. However, individual members of the soft sediment fauna can be extremely abundant, like the sand dollar beds mentioned here, and can play important roles in marine food webs.

SUBTIDAL SOFT BOTTOMS

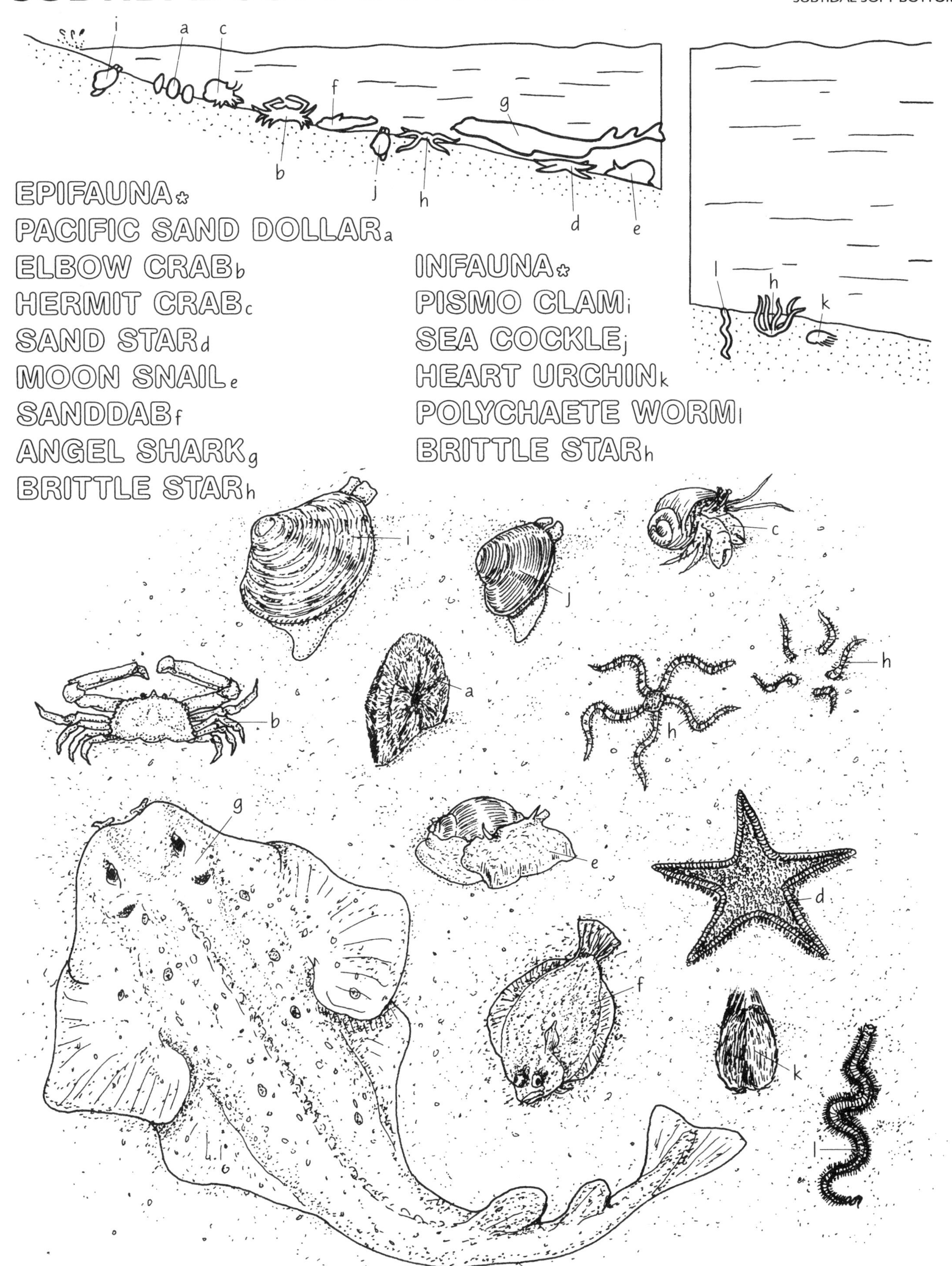

11
KELP BED

The kelp bed is a most productive and interesting cold-water marine habitat. Kelp beds are near-shore areas dominated by the presence of very large brown algae, called kelps. These plants require cold sea water and a solid rocky bottom on which to attach. They also require significant water movement to insure a constant supply of dissolved nutrients to fuel their photosynthetic processes. Kelp beds commonly grow in about 20 meters (65 ft) of water; if the water is exceptionally clear, they may grow at depths up to 30 meters (100 ft). The kelp bed structure and some of its more conspicuous inhabitants are discussed here.

Color each organism as it is introduced in the text. Note that the invertebrates and fishes in the foreground are drawn on an exaggerated scale relative to the large kelps for purposes of illustration. The sea lions and sea otter are drawn in the background to suggest the extent of the kelp bed.

Kelp beds occur in cold temperate water of the Pacific and Atlantic oceans. In some areas the dominant kelp species are relatively short, a few meters in length, and they make up kelp "beds." Elsewhere, large dominant species form kelp "forests." Off the coast of southern and central California, the dominant kelp is the *giant kelp*. Stipes (Plate 21) of the giant kelp can grow to lengths of 30 meters (100 ft) from their holdfast anchor on the rocks to the surface of the water. Its blades spread to form a thick canopy that soaks up the sun's energy for photosynthesis. These giant plants are kept afloat by the bulb-shaped air bladders (pneumatocysts) at the base of each blade.

Beneath the overhead canopy created by the giant kelp, the smaller *palm kelp* grows. This species has a thick elastic stipe that bends with the water movement, but is strong enough to hold the plant upright above the bottom. The palm kelp uses the light that filters through the overhead canopy for its photosynthesis.

On the rocky bottom is a turflike layer of small *red algae*. If the area is very densely shaded, it may contain a variety of attached invertebrates: *sponges*, sea *anemones*, sea squirts, and barnacles (not shown). Among these attached creatures live millions of smaller, motile animals. Brittle stars, gastropods, amphipods, and isopods abound.

The herbivorous (plant-eating) *sea urchins*, *sea hares*, and *abalones* can often be found on the bottom, where they take advantage of the large amounts of plant material produced in the kelp forest. The omnivorous (plant-and animal-eating) *sea bat* is variable in color; red, orange, brown, yellow, or green, and is a conspicuous kelp forest resident. The large pink to purple carnivorous *sunflower star* eats the prickly sea urchin, other sea stars, and a variety of other invertebrates.

The waters below the kelp canopy are rich in fish life. The *sheephead*, boldly colored in dark gray with a contrasting salmon-pink midsection and white lower jaw, comes to feed on larger invertebrates that live among the kelp stipes. The kelp forest is also home to several species of *rockfishes*, which feed on other fishes in the forest and a variety of invertebrates. Individual rockfish species occupy relatively discrete subhabitats within the kelp forest, thus avoiding direct competition with each other.

Two marine mammals that frequent kelp forests are the *sea lion* and the *sea otter*. The sea lion stalks fish and sometimes plays between the large kelp stipes. The sea otter may spend almost its entire life in the kelp forest and has a very important role in this habitat (Plate 113).

KELP BED

SEA OTTER n
SEA LION m
GIANT KELP a
SHEEPHEAD k
ROCKFISH l
PALM KELP b
RED ALGAE c
SEA HARE g
URCHIN f
SEA BAT i
ANEMONE e
SPONGE d
ABALONE h
SUNFLOWER STAR j

12 TYPES OF CORAL REEFS

Coral reefs are spectacular marine habitats. The unique hard bottom substrate is composed of the calcium carbonate skeletons of living and dead corals.

The underlying platform of the reef is formed by the massive skeletons of large coral colonies that have lived on the reef, and in death have tumbled to its base. The spaces between these large coral skeletons are filled by a rain of calcareous material including: the skeletal parts of smaller coral species and other coral reef inhabitants, the excavation products of boring animals that attack the corals, and the excreta of parrotfish (Plate 111) and other coral-grazing species. This jumble is lithified into a solid substratum by the precipitation of calcium carbonate dissolved in the sea water. Living coral colonies grow on the surface of this reef platform and maintain the reef's presence against the onslaught of wave action and boring organisms.

Corals are simple animals belonging to the phylum Cnidaria that also includes jellyfishes and sea anemones. The individual coral animal, called a coral polyp, looks very much like a sea anemone and feeds on zooplankton in the water (Plate 23). The coral polyp secretes a calcium carbonate cup into which it can retreat. As polyps grow, new coral cups are laid down over the old ones, and the coral colony grows outward and upward. Only the surface of the coral colony has live coral polyps.

Reef-building corals have very specific environmental requirements. They will grow only in clear, shallow sea water that has an average temperature of at least 20°C (68°F), which limits coral reefs to tropical regions. Modern coral reefs are found between 30° North and 30° South latitudes in the Caribbean Sea, and the Indian and tropical Pacific oceans.

Corals maintain single-celled plants known as zooxanthellae (modified dinoflagellates, Plate 19) in their tissues. The coral-plant relationship is a mutually beneficial (symbiotic) interaction (Plate 91). The plants have a safe place to live, and utilize the coral's metabolic waste products (nitrates and phosphates) as nutrients. The corals receive a substantial portion of their nutritional energy from the photosynthetic products of the zooxanthellae. Without this energy input, corals would be unable to secrete the amount of calcium carbonate necessary to form and maintain the coral reefs. Clear, shallow water is vital for adequate light exposure to the zooxanthellae and, thus, for survival of the reef.

Color the coral reef profile using light colors.

All coral reefs have a similar basic profile. On the seaward side, the coral reef has a sloping *reef front* that rises up from the depths. The reef front is not a solid wall, but is interrupted by channels that dissipate wave action and allow water and sediment to flow around the reef. Above the reef front is the *reef crest* that intercepts the onslaught of waves and reaches to the surface or just below it. Behind the reef crest is the protected, gradually sloping *reef flat* area that ends at the shore on fringing reefs and slopes into the lagoon on barrier and atoll reefs. The reef flat may be narrow or quite extensive with deep channels and a variety of substrates. These three reef areas experience very different exposures to light and wave action and harbor species of coral reef corals specifically adapted to these conditions.

Color the fringing reef, the barrier reef, noting the complex pattern of reef and lagoon areas, and the atoll reef as each is discussed. The three reef types are not drawn on the same scale.

The most common type of coral reef is the *fringing reef.* This reef projects seaward directly from the *shore.* The reef flat area may have coral rubble with small corals as well as *sea grass beds* where sediments have accumulated.

Offshore of large tropical land masses where a shallow continental shelf occurs, a barrier reef may be formed. The Great Barrier Reef of northeastern Australia and the barrier reef off the east coast of Belize are the two main examples of barrier reefs. These reefs occur well offshore of the land mass and the *lagoon* between the land and the reef can be many miles wide. Barrier reefs are actually made up of many smaller *reef elements,* forming a most complex marine habitat. Barrier reefs support a great biodiversity and have served as the evolutionary points of origin for many forms of marine life.

Atoll reefs begin as fringing reefs growing adjacent to steep oceanic islands. The island is actually the top of a submarine *volcano* that over geologic time has reached the surface due to volcanic activity or a drop in sea level. If the sea level rises and the change in depth is gradual, the coral reef may keep up with the slow increase in depth by growing upward. As the volcano sinks deeper, the shape of the reef may take on a circular or horseshoe configuration with a lagoon in its center. Extensive and complex *patch reefs* may be found on the lagoon floor. Reef flat areas of the main reef may become exposed as low-lying *islands* fringing the central lagoon. Coral sediment accumulates and plants become established from drifting seeds, forming the typical coral atoll island of the tropical Pacific Ocean.

TYPES OF CORAL REEFS

CORAL REEF PROFILE*
REEF FRONT a
REEF CREST b
REEF FLAT c

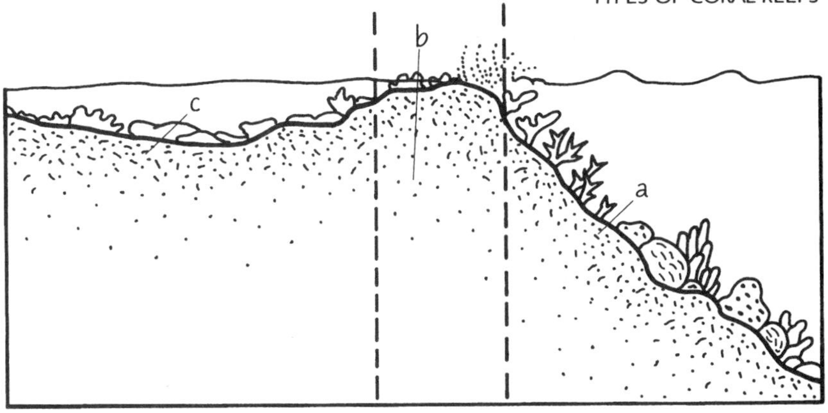

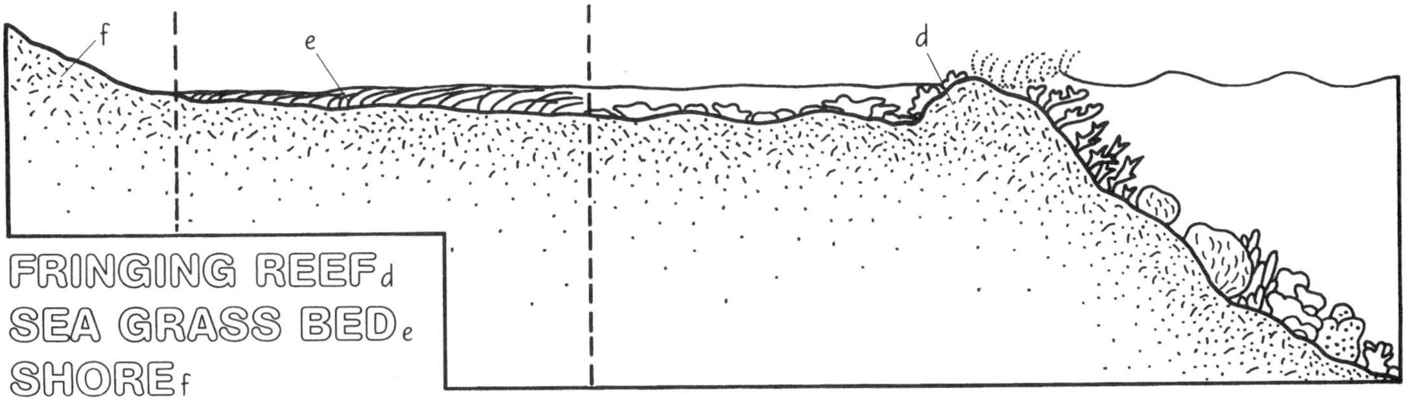

FRINGING REEF d
SEA GRASS BED e
SHORE f

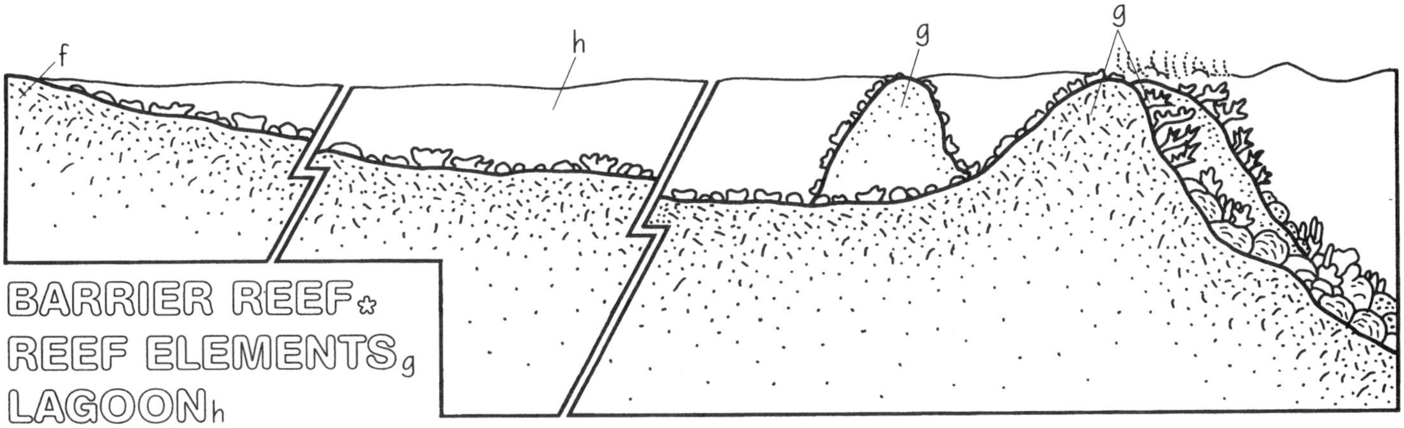

BARRIER REEF*
REEF ELEMENTS g
LAGOON h

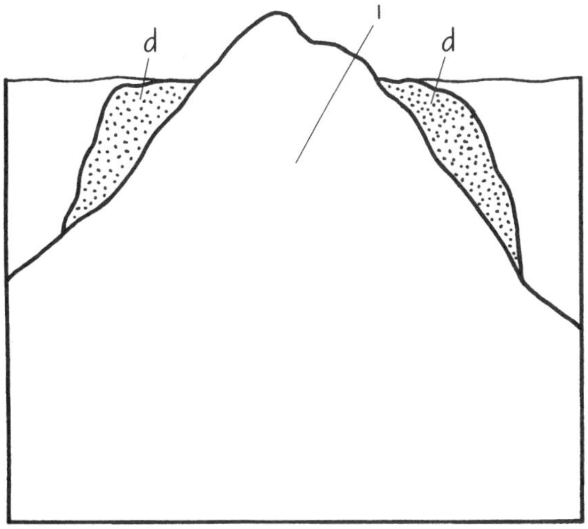

ATOLL REEF*
VOLCANO i
PATCH REEF j
ISLAND k

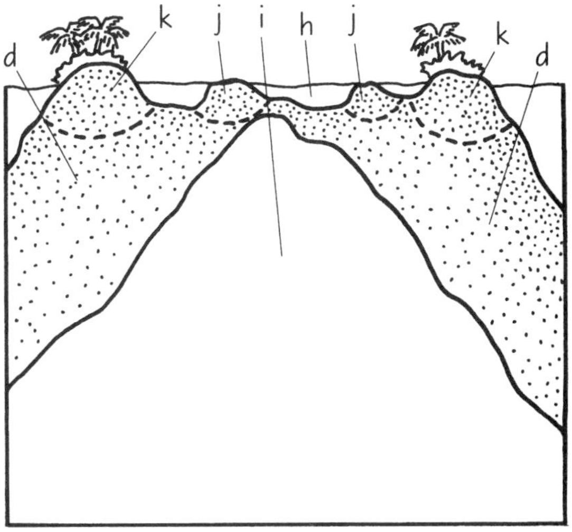

13 CORAL REEF RESIDENTS

The previous plate revealed that coral reef corals are dependent on the photosynthetic products of algal symbionts for their successful growth. Many coral species have growth forms designed to expose these symbionts to maximum light, much like land plants. Gathering light, coupled with withstanding wave action and competing for attachment space has resulted in a remarkable variety of coral shapes and sizes. Added to this assembly are other animals that stretch out into the water from a narrow point of attachment, such as the cylindrical, erect, purple *sponges* and extensively branching, red and orange *sea fans*. Collectively these organisms create a labyrinth of small spaces, caves, and overhangs for mobile animals to inhabit. This plate introduces some of the variety found on a coral reef. For the purposes of illustration, coral species from a wide range of depth are represented.

Begin by coloring the upper picture as each organism is mentioned in the text. The upper picture represents a view of the coral reef during the day. The bottom picture represents the same setting at night; you may wish to color gray over the water and the unlabeled residents.

On the reef crest where light penetration and water movement are greatest, the large, light tan *elkhorn coral* predominates. Each coral formation is an entire colony that may be many meters across, the product of many years of coral growth. Elkhorn coral grows very rapidly, up to 15 cm (6 in) of linear growth a year. Because of its exposed location, elkhorn coral is frequently damaged by storm waves. Whole colonies are toppled or large branches are broken off. However, it is one of the few coral species known that can regenerate a new colony from a broken branch, and therefore is able to recover from disturbance relatively quickly.

On the reef front, massive corals like the mound-forming *star coral* and *brain corals* are found. Compared to the elkhorn coral, these gray and brown colored corals have a relatively small surface area exposed to light and therefore depend less on symbiotic algae and rely more on the zooplankton-gathering of their large individual polyps (Plate 91). Mound-forming corals grow much more slowly than elkhorn corals. In death, their massive skeletons are very important contributors to the formation of the coral reef platform (Plate 12). Because the large polyps of these species make a big target for fishes that pick off coral tentacles, the polyps remain closed during the day and open only in the dark. This may seem counterproductive, but many of the reef zooplankton, also potential prey of the day-feeding fishes, come out only at night.

At the base of the reef front, an area of decreased light penetration, the pale white *plate coral* grows, spreading out flat to maximize exposure to the remaining light.

The coral reef supports both a day and night shift of active animals. During the day, brightly colored orange *groupers*, yellow *butterflyfish*, dusky gray *damselfish*, electric-blue *parrotfish*, and others swim busily about the reef. Brightly striped *cleaner shrimps* are found waving their long white antennae, waiting for the fishes to visit their "cleaning stations" (Plate 92).

As night approaches, the daytime feeders take refuge in holes or crevices in the reef; the parrotfish secretes a protective thin mucous coating around its body. Out from their daytime refuge come the orange big-eyed *squirrelfishes* and the brightly striped yellow *grunts* to begin their nocturnal foraging. The brown and green *moray eels* also come out to feed.

Some large invertebrates also emerge at night: the colorful *feather stars* climb to a suitable perch and unfurl their many arms to begin filter feeding. The green *spiny lobster* crawls out from under a deep crevice or overhang and scavenges about the reef for food. The long-spined *sea urchin*, visible by day only as a phalanx of formidable black spines protruding from a crevice, moves with surprising speed towards the small algae that it eats.

As daylight appears, the nocturnal animals return to their daytime retreats; the coral polyps once again retract; and the coral reef recommences its daytime activity.

CORAL REEF RESIDENTS

SPONGE$_{a}$
SEA FAN$_{b}$
ELKHORN CORAL$_{c}$
STAR CORAL$_{d}$
BRAIN CORAL$_{e}$
PLATE CORAL$_{f}$

DAY SHIFT*
GROUPER$_{g}$
BUTTERFLYFISH$_{h}$
DAMSELFISH$_{i}$
PARROTFISH$_{j}$
CLEANER SHRIMP$_{k}$

NIGHT SHIFT*
SQUIRRELFISH$_{l}$
GRUNT$_{m}$
MORAY EEL$_{n}$
FEATHER STAR$_{o}$
SPINY LOBSTER$_{p}$
SEA URCHIN$_{q}$

14
PHOTIC ZONE

The pelagic zone is the most extensive of the marine environments; it encompasses the volume of water from just above the ocean floor to the surface, and it extends from shore to mid-ocean. The pelagic environment is divided according to its proximity to the continents and depth of water. This plate is concerned with the upper layer of the pelagic zone, where light penetrates through the water, a region called the photic zone.

Light penetration is necessary for the growth of plants and for the presence of animals that must see in order to feed. The water temperature has a great influence on organisms' growth rate and metabolism, and water movement plays an important role in the pelagic zone where organisms either drift with the current or swim within it.

Begin by coloring the three drifting animals belonging to the neuston group as they are discussed in the text.

Neuston are those organisms that float on the surface of the water. Although neuston are moved by both *wind* and surface *currents*, the wind is the most important mover of these animals. One of the most common coastal water drifters is called the *by-the-wind sailor*. The by-the-wind sailor (a cnidarian) is actually a whole colony of individual organisms that live, feed, and float together. As a colony, they set *sail* to the wind, producing a thin bright blue membrane of stiff material, oriented diagonally across the colony's elliptical body. As wind carries the colony across the surface, one individual in the center of the colony feeds, while others function in defense or reproduction.

Another colonial cnidarian is the *Portuguese man-of-war*, which utilizes a *gas-filled sac* as both a float and a sail, while long prehensile tentacles dangle into the water to catch prey.

The *violet snail* is also a member of the neuston group. The snail secretes *bubbles* that keep it afloat. This animal feeds on the by-the-wind sailor.

Now color the enlarged plankton which are moved by the current.

Plankton includes the drifting organisms that are found at most depths in the water column. The single-celled *phytoplankton* are minute plants and are successful only in the uppermost section of the pelagic zone where there is plenty of light for photosynthesis. Phytoplankton are traditionally represented by diatoms and by dinoflagellates (illustrated here, see also Plate 19), larger cells that are captured in standard plankton nets. Biological oceanographers now recognize that much of the photosynthesis that occurs in the photic zone is accomplished by much smaller phytoplankton that pass through plankton nets. Not illustrated here, these small, flagellated, photosynthetic cells are members of the protozoa: single-celled, eukaryotic organisms that include flagellates, and heterotrophic ciliates and pseudopodia-bearing sarcodinids. These protozoans along with bacteria constitute a "microbial loop," organisms that were previously overlooked because of their small size, but are now thought to play a pivotal role in photic zone food chains.

Zooplankton (animal plankton), such as *copepods* and *euphausiids*, feed on the abundant phytoplankton. Two larger, sarcodinid protozoan groups with skeletons are also included in the zooplankton. *Radiolarians* with siliceous skeletons and *foraminiferans* with calcareous skeletons extend protoplasmic strands called pseudopodia to trap other plankton. The predatory *arrow worm*, or chaetognath, is known as the "tiger of the zooplankton" for its voracious appetite. These small animals dart about and devour prey as large or larger than themselves. Zooplankton play an important role in the food chain and are the main source of nourishment for the young of many fish species, as well as much larger animals that strain the water for food.

Color the nekton which are self-propelled.

Nekton are those pelagic dwellers that can swim in directions independent of the ocean's currents. This category includes the earth's largest mammal, the *blue whale* which can reach 27 meters (90 ft) in length. The blue whale together with several other whale species, is a filter feeder, whose principal food source is the zooplankton, especially euphausiids, or "krill." The nekton include many fast and efficient swimmers, such as the *squids*, which utilize their own variety of jet propulsion. *Albacore* prey on the smaller *herring*. The sharks are a highly successful and predatory group, represented here by the *blue shark*.

The phytoplankton growing in the sunlit upper layers of the photic zone provide the basis of virtually all life in the entire pelagic zone. The productivity of the phytoplankton thus governs the number of organisms that can survive throughout the vast reaches of the pelagic zone. This explains why, with increasing depth or distance from land and its input of nutrients, fewer creatures are found.

PHOTIC ZONE

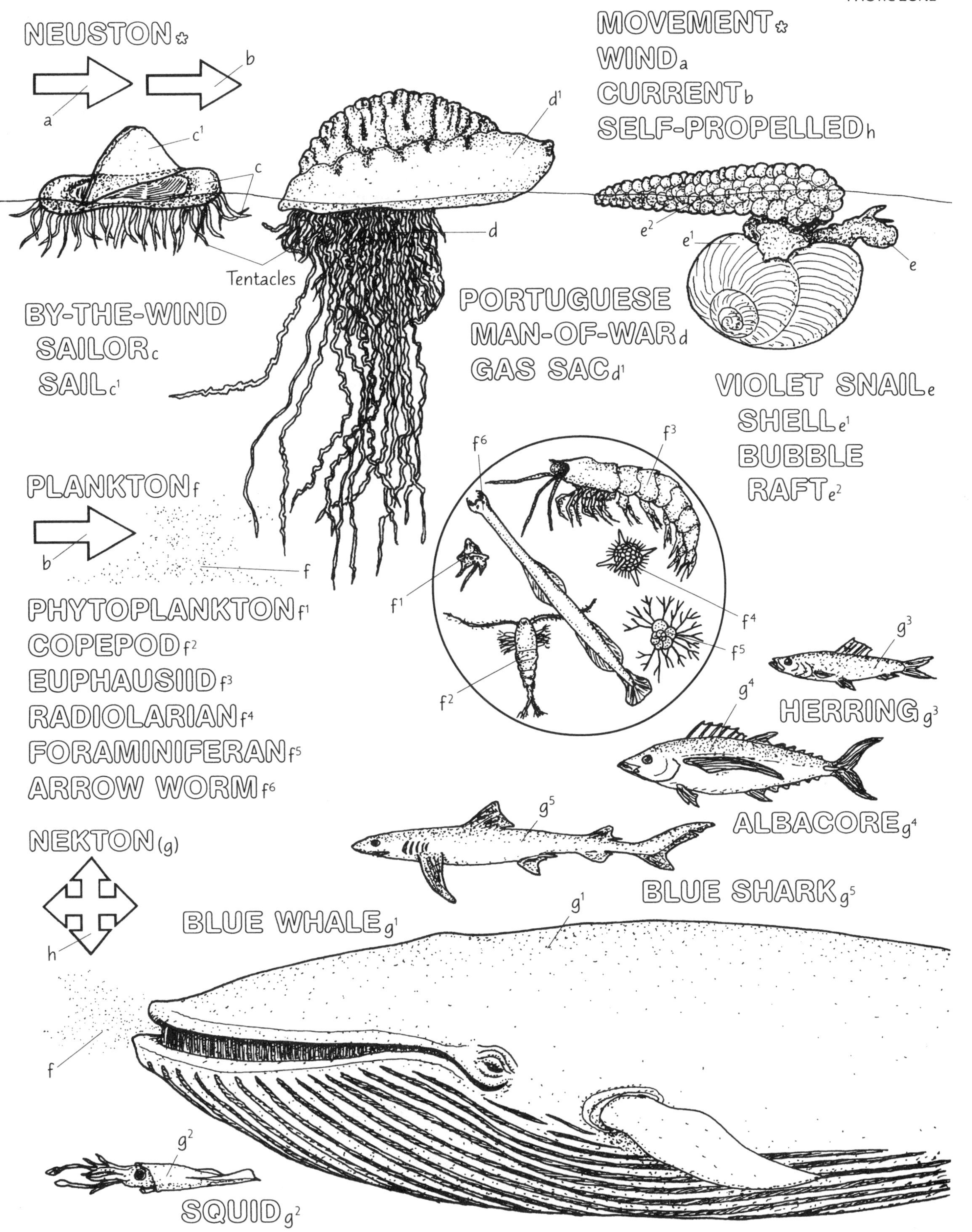
NEUSTON*
MOVEMENT*
WIND a
CURRENT b
SELF-PROPELLED h
a
b
c¹
c
d¹
d
Tentacles
e²
e¹
e
BY-THE-WIND
SAILOR c
SAIL c¹
PORTUGUESE
MAN-OF-WAR d
GAS SAC d¹
VIOLET SNAIL e
SHELL e¹
BUBBLE
RAFT e²
PLANKTON f
b
f
f⁶
f³
f¹
f⁴
f⁵
f²
PHYTOPLANKTON f¹
COPEPOD f²
EUPHAUSIID f³
RADIOLARIAN f⁴
FORAMINIFERAN f⁵
ARROW WORM f⁶
g³
HERRING g³
g⁴
ALBACORE g⁴
g⁵
NEKTON (g)
BLUE SHARK g⁵
h
BLUE WHALE g¹
g¹
f
g²
SQUID g²

15
MIDWATER REALM

The upper, lighted photic zone penetrates to 300 meters (1000 ft) in the clearest of sea water. Beginning just below the photic zone, and continuing downward to just above the ocean's bottom, is the mysterious midwater realm. The average depth of the ocean is 3.8 kilometers (2.4 miles), thus leaving the tremendous volume of the midwater realm to constitute the largest habitat on earth.

The midwater realm is divided into two depth zones. The first, called the disphotic, twilight, or mesopelagic zone occurs just beneath the photic zone and may extend to a depth of 700 to 1000 meters (2300–3300 ft), depending on water clarity, location, and a number of other factors. Here the light intensity is low and not sufficient to permit photosynthesis, but does allow some visual cues to resident animals capable of discerning and responding to visual stimuli. This mesopelagic zone is occupied by a large number of species and individuals as it is close to the rich planktonic food source of the photic zone. Many residents such as copepods, euphausiids, shrimp, and small fishes (Plates 14, 48) migrate vertically to the photic zone at night to feed. The fishes tend to be black and the crustaceans red, colors difficult to see at these depths. These organisms have large eyes and many have light organs which produce "animal light" or bioluminescence (Plates 69, 70).

Below the mesopelagic zone is the perpetually dark bathypelagic zone which extends downward to just above the deep-sea bottom zone, discussed in the next plate. In the bathypelagic zone, food is scarce and there are fewer species and individuals. Animals tend to be translucent or white, with reduced eyes and fewer light organs (Plate 48).

In the past 50 years, deep-sea research has greatly advanced. The development of new sampling technology, (Plates 114, 115) which includes manned submersibles and unmanned Remotely Operated Vehicles (ROVs), has allowed a view of a midwater community occupied by wondrous creatures. Among the most intriguing are the gelatinous marine animals that were never seen alive and rarely made it to the surface intact with traditional plankton net sampling.

Color the calycophoran siphonophore. This transparent colonial animal has several distinct parts.

The siphonophore is closely related to the Portuguese man-of-war seen in the previous plate. Siphonophores are large, polymorphic cnidarian colonies composed of modified polypoid and medusoid members (Plates 23, 24). There are several types of siphonophores in the midwater realm; shown here is a calycophoran siphonophore. At the anterior is a rocket-shaped swimming bell or *nectophore*, equivalent to the bell of a medusa, which contracts rapidly, quickly propelling the animal through the water. Feeding members are located on a *stem* below the nectophore. They string out long *tentacles* studded with stinging organelles called nematocysts (Plate 23) in very precise, species-specific fishing configurations. The tentacles can extend out for several meters, and form a lace curtain of death for zooplankton that swim into it.

Color the ctenophore, which is also transparent in life. Color the ctene rows a light, vibrant color and the oral lobes with a contrasting color. You may lightly color the spaces between the outlined areas.

Other exotic gelatinous animals of both midwater zones are the ctenophores. Mainly known from shallow water species, midwater ctenophores look like miniature space stations. The ctene plates (fused cilia) beat rhythmically along the length of the eight *ctene rows* and reflect back the imaging lights of ROVs in prismatic flashes of color (Plate 69). The transparent ctenophores are carnivores that catch prey on sticky tentacles or trap them in large *oral lobes* surrounding the mouth as seen in the ctenophore shown here.

Next color the planktonic polychaete worm, *Tomopteris*.

Polychaete worms (Plates 27, 28) are most frequently encountered in bottom habitats. However, there are several families of polychaetes that are strictly planktonic and are found throughout the midwater realm. The translucent white *Tomopteris* uses long, rigid, swept-back *antennae* as stabilizers as it swims, and large segmental appendages, called *parapodia,* provide ample paddles to propel the worm through the water. As is the case in all planktonic polychaetes, *Tomopteris* lacks the chitinous setae (stiff bristles) that support the parapodia in bottom-dwelling forms. The absence of setae is considered a weight-reduction adaptation.

Color the cock-eyed squid deep red.

The cock-eyed squid, *Histioteuthis heteropsis*, shows a unique adaptation for living in the twilight of the mesopelagic zone. Sunken into one side of its head is a small *blue eye*, while on the other side a *yellow eye* sticks out from the head and is more than twice the diameter of the blue eye. The most logical explanation for this is that the large eye is adapted for seeing in shallow water and the small eye for deep water. The cock-eyed squid's deep red, 20 cm (8 in) *body* is covered with rows of small white light organs.

MIDWATER REALM

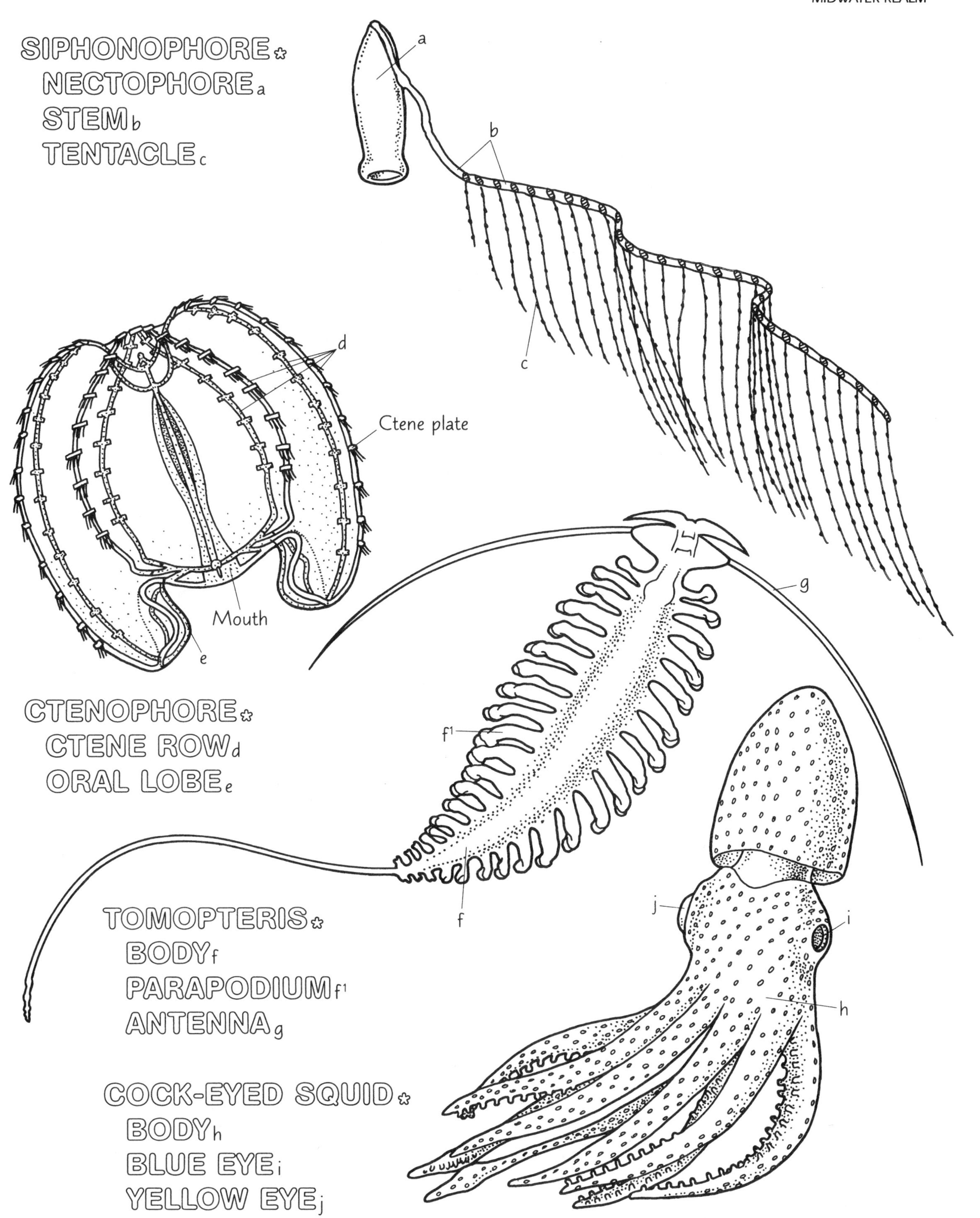

16 DEEP-SEA BOTTOM

The benthic or bottom zone of the deep sea is a cold, dark, generally uniform habitat. No light other than bioluminescence is found here. The bottom sediments reflect the biological and physical processes happening in the water above. Sediments made up of phytoplankton and zooplankton skeletons or precipitated chemical compounds stretch for thousands of square kilometers of bottom. The average temperature of the deep ocean is 4°C (39°F). In addition to low temperature, the extreme pressure generated by the overlying sea water severely slows the metabolism of bottom dwellers. As a result, they grow much more slowly and generally to a smaller size compared to their shallow water counterparts. Conversely, many live considerably longer, decades as compared to years.

Most food reaching the deep-sea bottom originated from the lighted photic zone above. As food falls to the bottom, it passes through the midwater communities and rarely settles in a form that the bottom organisms can eat directly. Relatively refractory organic detritus, like fecal pellets and molted exoskeletons from zooplankton are first attacked by bottom bacteria, and the bottom animals feed primarily on this bacterial protoplasm attached to the deposited detritus. Occasionally large food items, like the bodies of large fishes and marine mammals reach the bottom and are devoured surprisingly quickly. The physical and biological realities of the deep-sea bottom present a very severe and demanding environment yet life persists and the diversity of form is truly confounding. Fresh insights into this diversity are provided by the new tools that are now able to reach an estimated 98 percent of the deep ocean zones (Plates 114, 115). Some of the deep-sea bottom fishes are presented in Plate 48. This plate explores the invertebrates of the deep-sea bottom.

Color the brittle stars. Some are buried with only their arms protruding.

In this unlit world of fine detrital food stuffs, echinoderms are prominent (Plates 40, 41). *Brittle stars* with thin, spiny limbs carpet vast stretches of bottom, often so close together their arms intertwine. These dull gray animals are mainly detrital feeders, taking their meals one small particle at a time. However, scientists in a submersible have videotaped a species of brittle star that holds its arms into the water column lasso-style and catches small fishes and squid that swim too close. When trapped, the prey is quickly pounced upon and dismembered by a host of nearby brittle stars.

Next color the sea cucumbers grazing on the bottom. The feeding tentacles are modified tube feet and should be given a shade of the same color as the body tube feet.

Another fascinating group of deep-sea echinoderms are the sea cucumbers. These elongated cousins of sea urchins and starfish (Plates 39, 40, 41) are found in a variety of bizarre forms. Shown here is a species that lifts its gray, sausage-shaped *body* off the soft substrate with large, leg-like *tube feet*. The cucumber sweeps its mouth and feeding *tentacles* back and forth across the substrate, picking up organic detritus. Herds of these cucumbers roam across the bottom.

Color the bait container and the time-lapse camera and strobe light suspended above it. Next color the deep-sea fishes and the amphipods which are light gray to off-white.

In an ingenious experiment, scientists from Scripps Institute of Oceanography lowered five-gallon *containers* baited with fresh fish to the bottom. A time-lapse *camera* was suspended above to take periodic pictures of the fate of the *bait*. To the amazement of the scientists, the containers were quickly discovered and their contents devoured by a variety of deep-sea *fishes*, suggesting that at least some of the fishes living here survive by a constant search over wide areas for windfalls of food. Even more surprising were the scavenging *amphipods* that arrived to feast on the bait. Amphipods are common, small crustaceans in many shallow water habitats, usually a few centimeters or less in size. The amphipods captured by the camera were 25 cm (10 in) long. How do the largest known members of this group survive in one of the most food-limited marine habitats? They are an example of abyssal gigantism, extreme size achieved by very slow-growing, long-lived species adapted to the rigors of the deep-sea bottom.

Color the vampire squid.

One of the eeriest forms encountered in the deep sea is the dark red vampire squid, *Vampyroteuthis infernalis.* Seen both in the water column and on the bottom, this small animal (5–7 cm, 2–3 in) has two small *fins* on the sides of its *body* which it uses as paddles. It has eight short, webbed arms tipped with bioluminescent organs and lined with stout spines. In videotape sequences filmed by an ROV, the squid has been observed rolling back its arms so the spines face outward, perhaps forming a spiked net to ensnare prey as it drifts along or as a defensive posture. The large *eyes* of the vampire squid are deep blue and close down like the diaphragm of a camera, yet another strange behavior in this sinister looking deep-sea creature.

DEEP-SEA BOTTOM

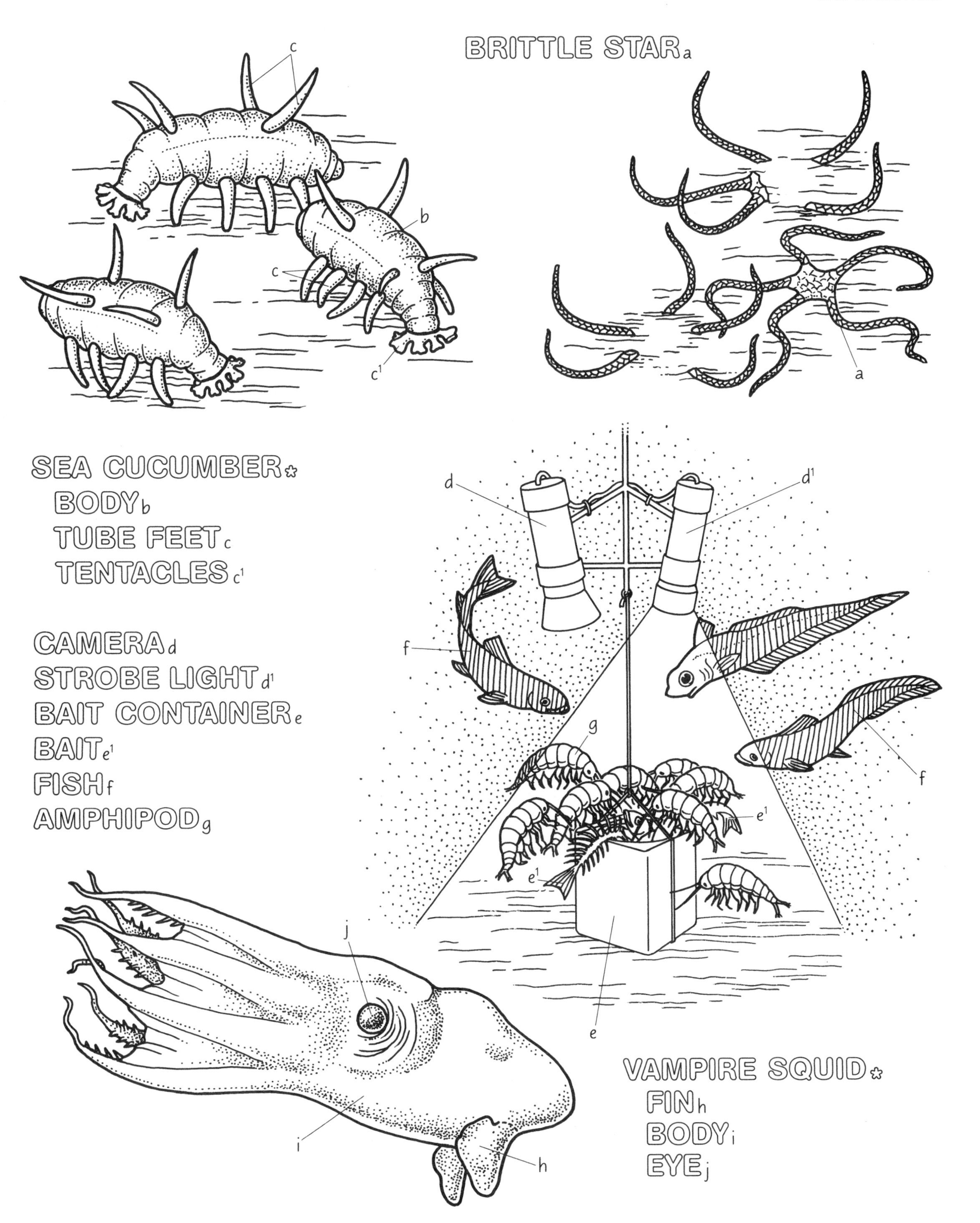

17
DEEP-SEA HYDROTHERMAL VENTS

The earth's surface is a thin crust of rocky plates that float on a molten core. These floating tectonic plates include continental and oceanic plates.

Color the diagram of the rift zone along the oceanic ridge. Next color the research submersible *Alvin*. Color each member of the vent community as it is discussed.

Rift zones occur along *mid-oceanic ridges* on the deep-sea floor of the oceans' basins, where volcanic activity, deep in the earth, pushes up molten rock (*magma*). This new rock material added to the edges of the oceanic plates results in sea-floor *spreading*. The heavy oceanic plates grow and spread away from the rift zone, and push against the lighter continental plates. Along this margin, they slide underneath, to be consumed by the molten core.

In 1977, marine geoscientists used the deep-diving research submarine *Alvin* to explore a rift zone 368 km (229 miles) northeast of the Galapagos Islands. The scientists expected the rift zone to be a volcanic wasteland, instead they were surprised by a thriving bottom community 2700 meters (almost 9000 ft) deep that existed entirely independent of the productivity of the ocean's lighted surface.

Rift zone communities have now been studied in some detail. Sea water seeps through fissures created in the spreading sea floor and contacts the molten rocks below. Here the water is super-heated and rises through chambers or vents back up to the ocean floor. As the super-heated sea water rises, it leaches materials from the rocks and sediments. The heated water contains high concentrations of these materials, especially hydrogen sulfide. Hydrogen sulfide is a high-energy chemical compound. Anaerobic bacteria can extract this chemical energy by adding oxygen (oxidation) and trapping the released energy to run their metabolism. This process, called chemosynthesis, uses a chemical energy source to drive the synthesis of carbon compounds (not light energy, as with photosynthesis). Hydrogen sulfide is highly poisonous to oxygen-breathing organisms. It binds with a critical chemical component of the respiratory pathway and shuts down respiration. However, over evolutionary time several marine animal groups have joined in symbiosis with sulfide bacteria to reap the energy benefit of hydrothermal vents.

The most conspicuous symbiont member of the hydrothermal vent community is the large Galapagos tube worm, *Riftia*. This worm reaches a meter (3 ft) in length and resides in white, chitinous tubes over 3 meters (9 ft) long. A large tentacular plume projects about 30 cm (12 in) beyond the tube. The tentacles, attached at right angles to a stiff base and layered one upon the other, receive a rich blood supply from the worm's circulatory system, giving them a bright red color. *Riftia's* blood contains two different blood proteins. One is hemoglobin, which is used to bind oxygen for the worm's respiration. The second blood protein binds hydrogen sulfide in such a way that it cannot harm the worm. The bound hydrogen sulfide is carried by the worm's circulatory system to a special organ, the trophosome, which is packed with symbiotic sulfide bacteria. The bacteria metabolize the hydrogen sulfide and part of the energy captured by the bacteria is released to the worm. The worm is totally dependent on this symbiotic relationship because it has completely lost its digestive system.

Two bivalve molluscs, a *clam*, *Calyptogena*, and a *mussel*, *Bathymodiolus,* also utilize symbiotic bacteria which they harbor in their gills. Growth studies of these bivalves reveal the high-energy nature of the hydrothermal vent environment. These molluscs grow just as fast or faster than bivalves in shallow-water communities and dramatically faster than other deep-sea animals, which are severely limited by the small amount of available food.

Some members of the rift community feed directly on *bacterial mats* which grow on the bottom. Others feed by filtering suspended bacteria from the water, as is the case with the large tube-dwelling polychaete called the *Pompeii worm*. There are also *crabs* and *shrimp* that are either predators or scavengers.

The water surrounding vent communities has a distinct thermal profile. The *heated water* may emerge as a gentle stream of 25°C (77°F) water, or jet out of chimney-shaped vents called *smokers*, where the water temperature reaches 300–400°C (600–750°F). The "smoke" is created when concentrated minerals rising with the heated water are precipitated by mixing with the cold bottom water. Near the mouths of the vents, where the *Riftia* occur, the water is quickly cooled by the surrounding sea to 16°C (61°F). Comparing the thermal profile to the distribution pattern of the vent community members reveals that each organism has a preferred thermal niche.

Hydrothermal vent communities are now known to be common along many ocean ridges. Apparently due to the highly volatile nature of the volcanically-influenced rift zones, the vent communities are fairly short-lived. Scientists have discovered dead communities along rift areas in which the hot, sulfurous water no longer welled out of the sea floor and the food chain collapsed. They estimate a community may persist for only 20 to 75 years.

DEEP-SEA HYDROTHERMAL VENTS

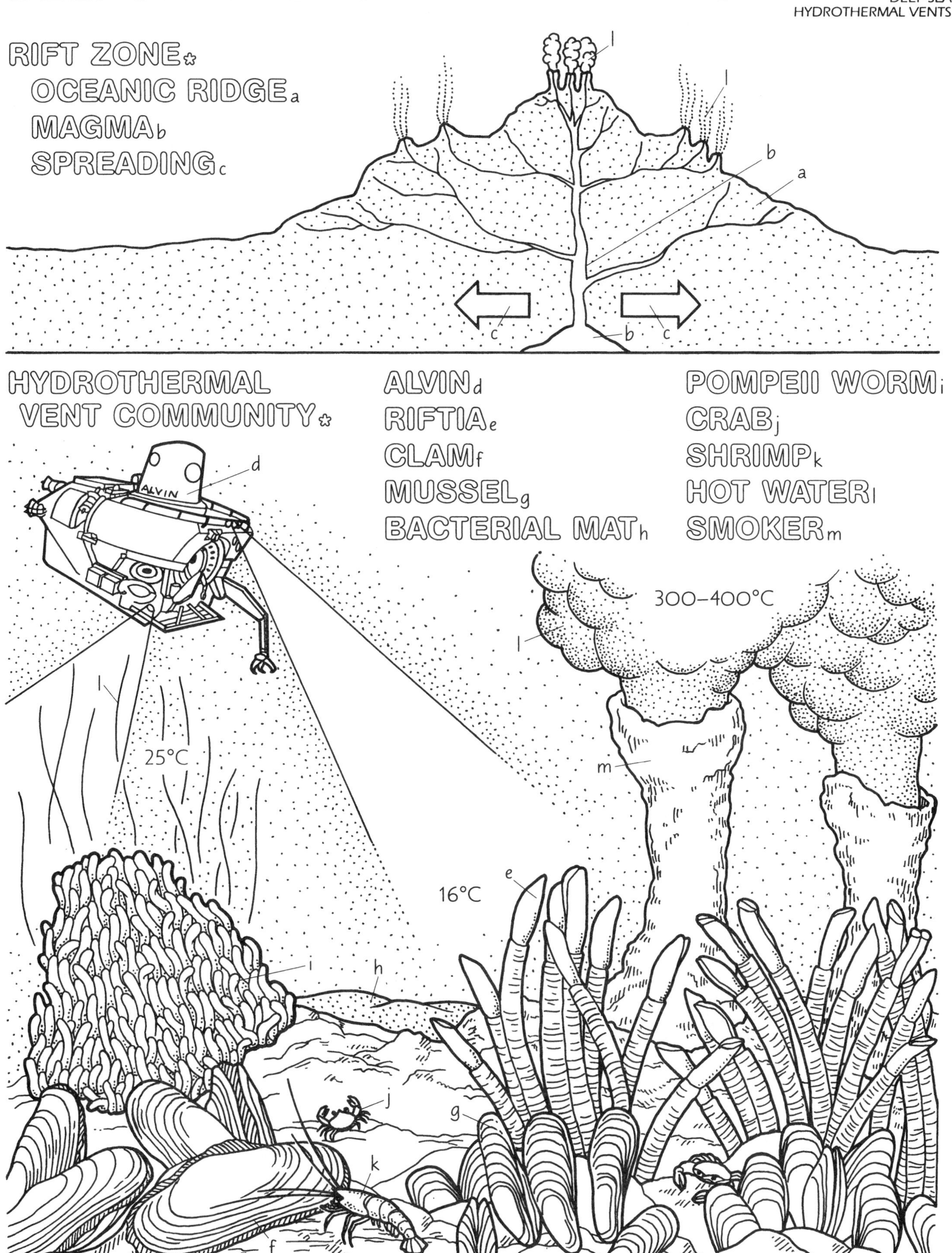

18
MARINE FLOWERING PLANTS

Most flowering plants become metabolically stressed if exposed to sea water, as they are incapable of dealing with the high salt content. This metabolic stress ultimately results in the death of the plant.

A few species of land plants have, however, successfully invaded and adapted to sea-water environments, flourishing in a variety of marine habitats. Unlike algae, land plants can grow only where they can absorb nutrients through their roots. These higher plants also need nearly direct sunlight, so they can only tolerate relatively shallow waters (1–30 meters, 3–100 ft).

Color the red mangrove, starting with the prop roots at the lower right. Notice the junction of the trunk and roots at the high-tide mark. Color the entire leaf mass green. Color the falling seed and new root protruding from it, and notice the four young leaves emerging from the seed.

One of the largest flowering plants to flourish in the marine environment is the red mangrove. These plants occur in tropical and semitropical regions from Florida to South America and reach tree size (1–3 plus meters, 3–10 plus ft). The mangrove grows best in the mud bottoms of estuaries, coastal lagoons, and near the mouths of large rivers, where silt is deposited. The mangrove remains stationary in the unstable mud bottom by sinking a mass of large, arching *prop roots* deep into the mud. The prop roots divide off from the *trunk* at the high-tide level; the roots proliferate and trap more sediment, raising the substratum level. This, in turn, creates more space where the mangrove can grow, and, in this way, the mangrove colony expands toward the sea.

Mangrove roots are the habitat for a myriad of attached algae and animals such as sea squirts, sponges, and sea anemones. The lattice-work of roots also offers protected nursery areas to young reef fishes and spiny lobsters. The mangrove *leaves* fall into the water and decompose, providing a detrital food source for many animals residing near the mangrove, and others further out to sea.

Seeds of the red mangrove sprout and begin to grow before they fall from the tree. A long (36 cm, 14 in), slender root emerges from the seed, and several leaves may also grow. If the seed falls at low tide, it may poke, dart-like, into the soft mud near the parent plant, and continue to grow. Seeds that fall during high tide float upright in the water and may be carried far from the parent plant. When the tide recedes, if the seed has been left on the mudflat with the root facing downward, it will be "drilled" into the mud by the jostling of small waves, thus beginning a new colony of mangroves.

Now color the turtle grass and the surf grass. Note that the grasses' stems receive shades of the same color as the mangrove's trunk, and the blades receive shades of the same color as the mangrove's leaves. These structures are homologous (same origin) and have the same function in these higher plants.

The flowering plants known as sea grasses can grow completely submerged. Various sea grass species occur from the intertidal zone to depths of 30 meters (100 ft). Turtle grass is found in eastern Florida, along the Gulf of Mexico, and in the Caribbean, where it thrives on soft bottoms of mud, shell, or sand. Once established, turtle grass proliferates by extending underground *stems* (rhizomes) from which leafy *blades* grow. Turtle grass forms extensive shallow-water meadows, which are frequented by numerous juvenile reef fishes. At night, parrotfish and sea turtles come into these meadows to feed. The stems and *roots* of turtle grass bind the substratum, which permits the colonization of burrowing animals, such as sipunculid worms and sea cucumbers (Plates 25, 41).

Unlike most sea grasses, which require quiet waters, surf grass grows in the wave-swept rocky intertidal zone of temperate waters. Surf grass seeds possess two stiff, bristled, pointed projections. If the seed lodges on an alga, it can germinate and send out tenacious roots to grab hold of the substratum. The surf grass roots trap sediment, and a patch of long, thin, bright green blades may develop in a tide pool or at the low-tide line. This creates an environment that shelters many species of worms and other small creatures.

MARINE FLOWERING PLANTS

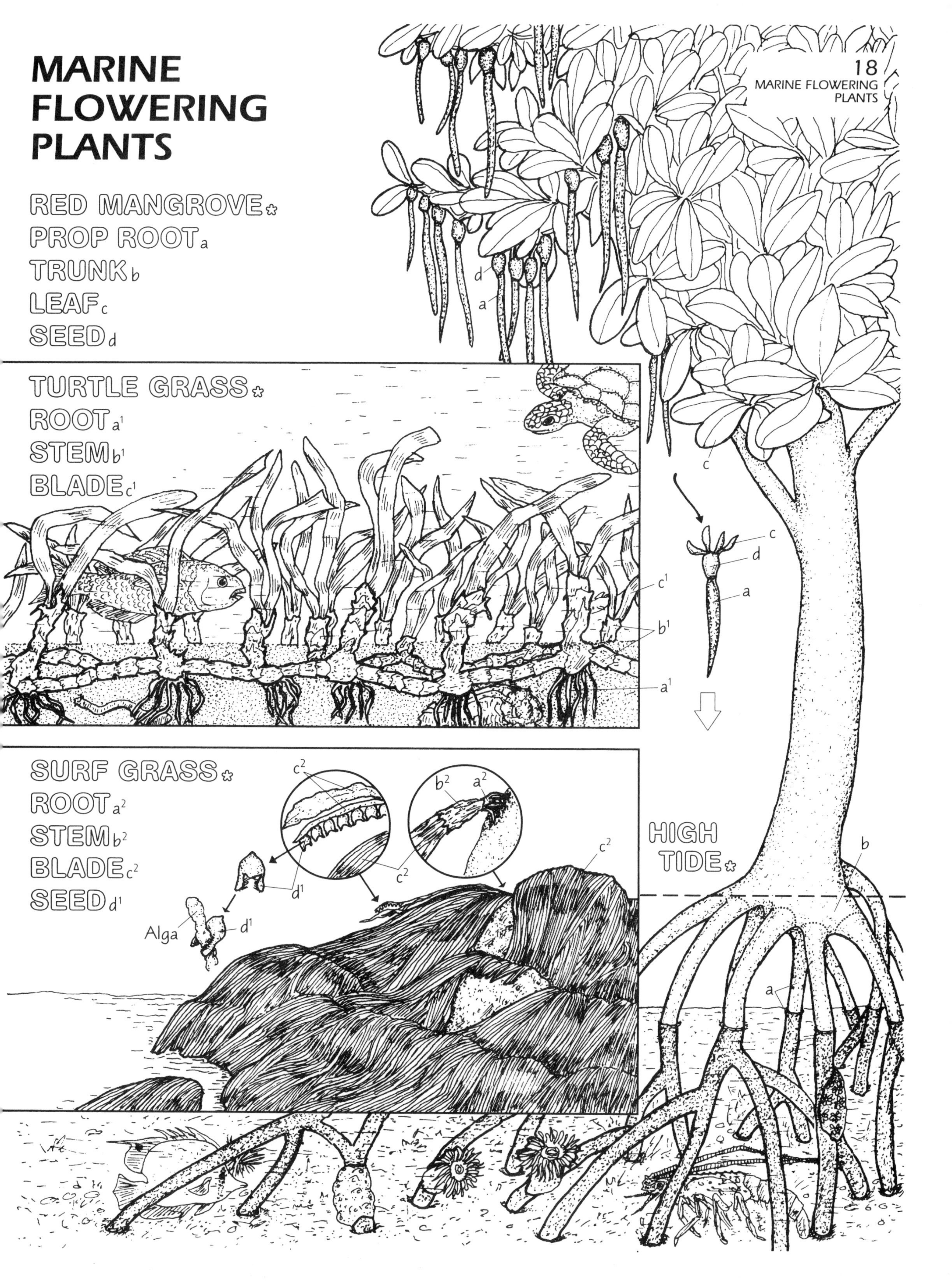

19
PHYTOPLANKTON: DIVERSITY AND STRUCTURE

Beyond the shallow depths where marine flowering plants and the large seaweeds grow, the sunlit layer of water is dominated by the single-celled plants known as phytoplankton. Phytoplankton includes a variety of plant forms, all of which are autotrophic: they capture energy from sunlight, and require nutrients (phosphates, nitrates, etc.) and carbon dioxide (CO_2) for photosynthesis. A single liter of rich coastal sea water may contain dozens of different species of phytoplankton, and possibly as many as 10 to 20 million individual one-celled plants. Some species are small, flagellated forms, much too tiny to be captured in the finest-mesh plankton net. The larger, more common phytoplankton species are diatoms and dinoflagellates, which are found abundantly in temperate waters. These species can grow up to 1 millimeter across, but most are much smaller.

Color the diatoms in the upper left. Color each diatom structure as it is discussed. Use a dark color for the pores.

Diatoms are found in both marine and freshwater habitats. Marine diatoms are of two basic types: the elongated forms (Pennales or pennate diatoms), such as *Pleurosigma*, which are usually found in very shallow areas; and the round or wheel-shaped forms, such as *Coscinodiscus* (Centrales or centric diatoms). In the case of *Coscinodiscus*, you see that the diatom consists of a two-part *frustule* which is made of silica and appears like a glass jewel when viewed under the microscope. On the top of the frustule, an elaborate pattern of *pores* radiates out from the center; the pores help reduce the weight of the floating diatom and allow diffusion of materials into and out of the cell. Viewed from the side, the shape of the frustule can be seen: the upper half, or *epitheca*, fits over the smaller bottom half, the *hypotheca*. Inside the frustule is the *nucleus* of the diatom, which contains the genetic material, and the *chloroplasts*, or photosynthetic organelles.

In addition to diatoms that are found in the plankton, some diatoms grow attached to hard substrata or on the surface of soft substrata. Many pennate and centric diatom species grow attached to hard surfaces. When attached forms divide they often fail to separate, forming diatom chains or colonies. Some pennate species grow into flat, branching filamentous colonies that may be confused with small seaweeds. These colonies can coat intertidal rocks in a slippery brown mass during certain times of the year, and form an important food source for intertidal herbivores like chitons and limpets. Other species of pennate diatoms grow on the surface of mud flats, and can form colonies that coat the mud's surface with a conspicuous brown-green sheen during spring and early summer. Short chains of centric diatoms often occur attached to animate objects like the covering of hydroid colonies (Plate 23), where they grow outward from the point of attachment like a beaded necklace.

Next, color the various species of the diatom genus *Chaetocerus*. The generic name is abbreviated to *C*. Note the differences in the setae.

Diatoms cannot move in the water column under their own power, but have developed adaptations that keep them afloat. Within the widely distributed diatom genus *Chaetocerus*, a variety of adaptations are visible. The individual *Chaetocerus* cell has two pairs of thin spines or *setae* projecting from either end. These setae fuse with those of other cells to form long chains, thereby increasing the buoyancy of the chained group. As you see from the illustration, the length and shape of the setae vary with different species. *Chaetocerus decipiens* is found in cool, dense water, and needs only relatively short setae to stay afloat. *Chaetocerus* species living in warmer, less dense water have developed long setae that provide more resistance to sinking. *Chaetocerus denticulatus* has secondary spines on the setae to keep it from sinking in warm water.

Now, color the dinoflagellates in the upper right corner.

Unlike the diatoms, dinoflagellates are able to swim and move up and down in the water column. They have long *flagella* that are used for locomotion. The whiplike flagella are located in grooves — the longitudinal *sulcus* and the transverse *cingulum*. Dinoflagellates have a multilayered covering of cell material. In the armored (thecate) dinoflagellates, such as *Peridinium*, the cell is encased in an expandable, overlapping layer of cellulose *plates*; this is absent in naked, or unarmored dinoflagellates, such as *Gymnodinium*. Many dinoflagellates are known to be bioluminescent (light-producing), and this group also includes the organisms that cause the sea water to turn red during the so-called "red tides" (Plate 72). A group of dinoflagellates known as zooxanthellae live within the cells of a number of invertebrate hosts in a mutually beneficial symbiotic relationship (Plates 12, 91).

DIVERSITY AND STRUCTURE

DIATOMS *
PLEUROSIGMA $_{a}$
COSCINODISCUS $_{(b)}$
FRUSTULE $_{b^1}$
PORE $_{c}$
EPITHECA $_{d}$
HYPOTHECA $_{e}$
NUCLEUS $_{f}$
CHLOROPLAST $_{g}$

a
b[1]
c
b[1]
d
e
g
f
g

DINOFLAGELLATES *
PERIDINIUM $_{(j)}$
FLAGELLUM $_{k}$
SULCUS $_{l}$
CINGULUM $_{m}$
THECAL PLATES $_{j^1}$

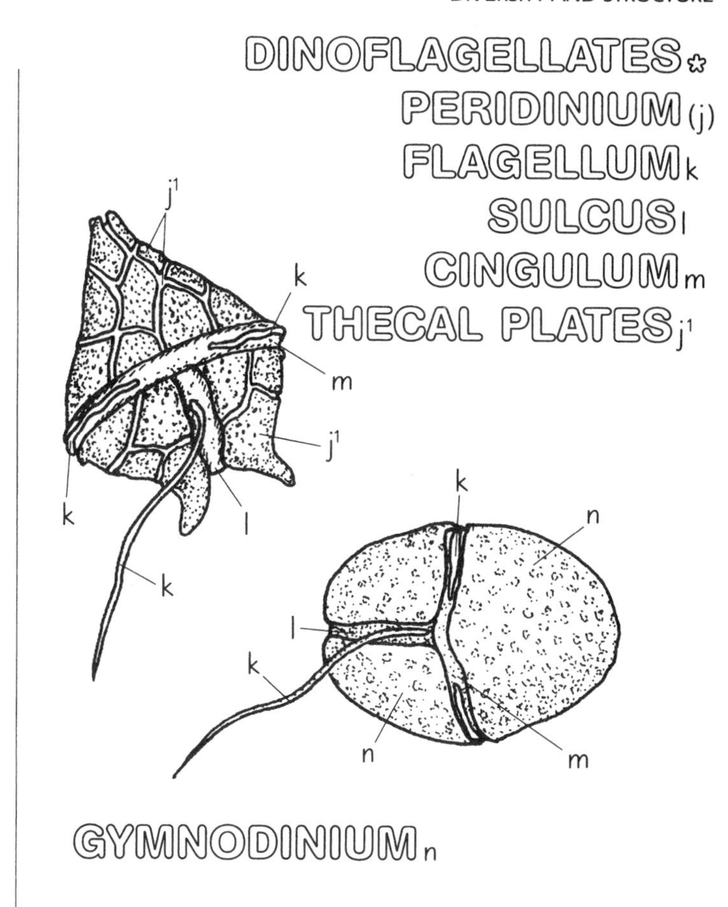

GYMNODINIUM $_{n}$

SPECIES OF CHAETOCERUS $_{(h)}$
SETA $_{i}$

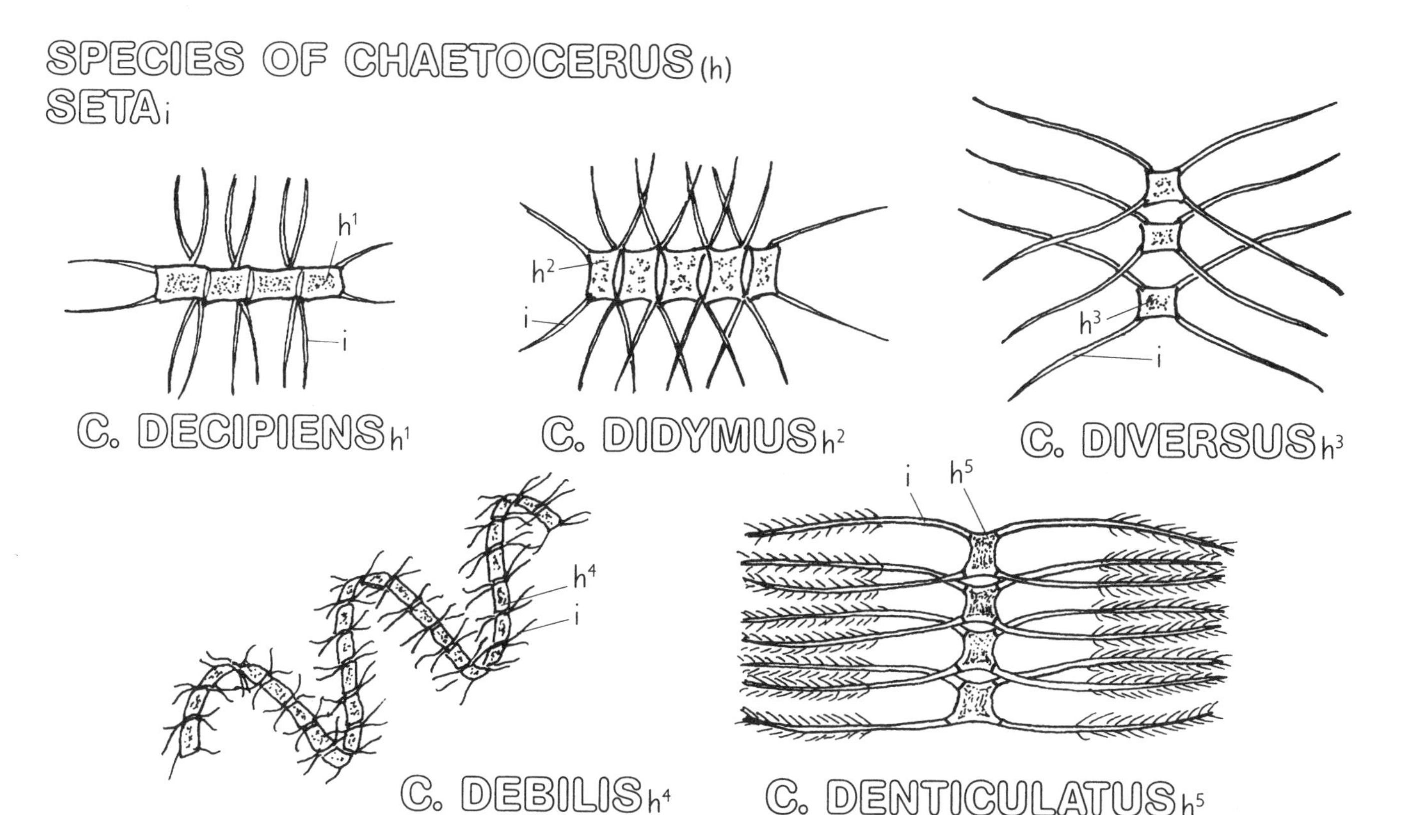

20 SEAWEED ADAPTATIONS: RED AND GREEN ALGAE

The multicellular seaweeds, or macroalgae, have evolved some very important adaptations allowing them to survive in the near-shore shallow marine environments. Macroalgae are primarily limited to areas of shallow water and rocky shores; the algae need light for photosynthesis, and solid substrata for attachment sites. But the near-shore habitat poses a challenge to the algae; wave action, desiccation during low tide, and grazing by numerous herbivores are all potential threats to these plants.

Begin by coloring the sea lettuce. The illustration at the upper right shows various algae attached to a rocky surface; the other illustrations are larger drawings of the algae. As each type is mentioned in the text, color both representations. Choose greens for the green algae, pink for the coralline algae, and reds and purples for the red algae. *Smithora* is attached to green surf grass.

The green algae are abundant in freshwater environments, but are represented by relatively few marine species. One green alga of the marine intertidal zone is the thin-bladed (two cells thick) *sea lettuce*. The delicate sea lettuce loses its moisture during low tide and becomes quite crispy, yet remains alive. When the tide rises, it again absorbs water. Some species of sea lettuce occur on soft substrata in calm water habitats where the alga attaches to a shell, rock, or other object and may grow over a meter (3 ft) in length. The green alga *Cladophora* grows in small, thick tufts in the middle intertidal zone. These tufts are composed of thousands of tiny multi-branched filaments that serve to trap sand. Together the filaments and trapped sand hold precious water used for survival during low-tide exposure.

The red alga known as the *salt sac* is a medium-sized, 2.5–5 cm (1–2 in) diameter alga of the middle intertidal zones, often found in large patches. The salt sac is aptly named, for in its hollow core it holds a reservoir of sea water to keep it from drying out at low tide. These water sacs are home to a particular type of tiny copepod (crustacean).

The grazing activities of invertebrate herbivores can severely injure an alga, and several of the alga species have evolved ways to combat grazing. The *coralline red algae* secrete calcium carbonate (lime) in their cell walls, making them a tough, crusty plant. Most herbivores avoid eating the coralline red algae, with the notable exceptions of the dunce cap limpet (Plate 6) and the lined chiton (Plate 107).

Two general types of coralline red algae are present in the marine intertidal zone. The *encrusting corallines* occur in shady areas of tide pools and cover the rocks in bumpy-textured pink sheets. The jointed or *articulated coralline algae* also occur in the lower intertidal zone. They are small (5–7.5 cm, 2–3 in), erect, branched plants that have calcified sections interlaced with flexible joints. The overall calcified nature of the articulated corallines offers resistance to the pounding waves, with enough flexibility to bend with the water movement.

Many red algae are highly branched, with beautiful and intricate patterns of growth. This growth pattern increases the light-gathering surface of the plant for photosynthesis, and also makes it more difficult for small grazers to attach themselves to the branched structure. One such alga is the *pepper dulce*. As its common name implies, the plant has a biting, peppery taste; the plant may possibly sequester noxious chemicals in its tissues, making it still more unappealing to herbivores.

The intertidal area is often crowded with plant life, and there are few open spaces to which an alga can attach. Some algae attach to other plants or animals. One prominent example is the red alga *Smithora*, which is found only attached to eel grass and to surf grass. *Smithora* has small (1–2 cm, 0.4–0.8 in) reddish-purple blades that may entirely cover individual surf grass blades (as shown). Biologists believe that this arrangement may be mutually beneficial. *Smithora* has a place to attach and grazing herbivores prefer to feed on the fleshly, more prominent alga, leaving the surf grass alone.

RED AND GREEN ALGAE

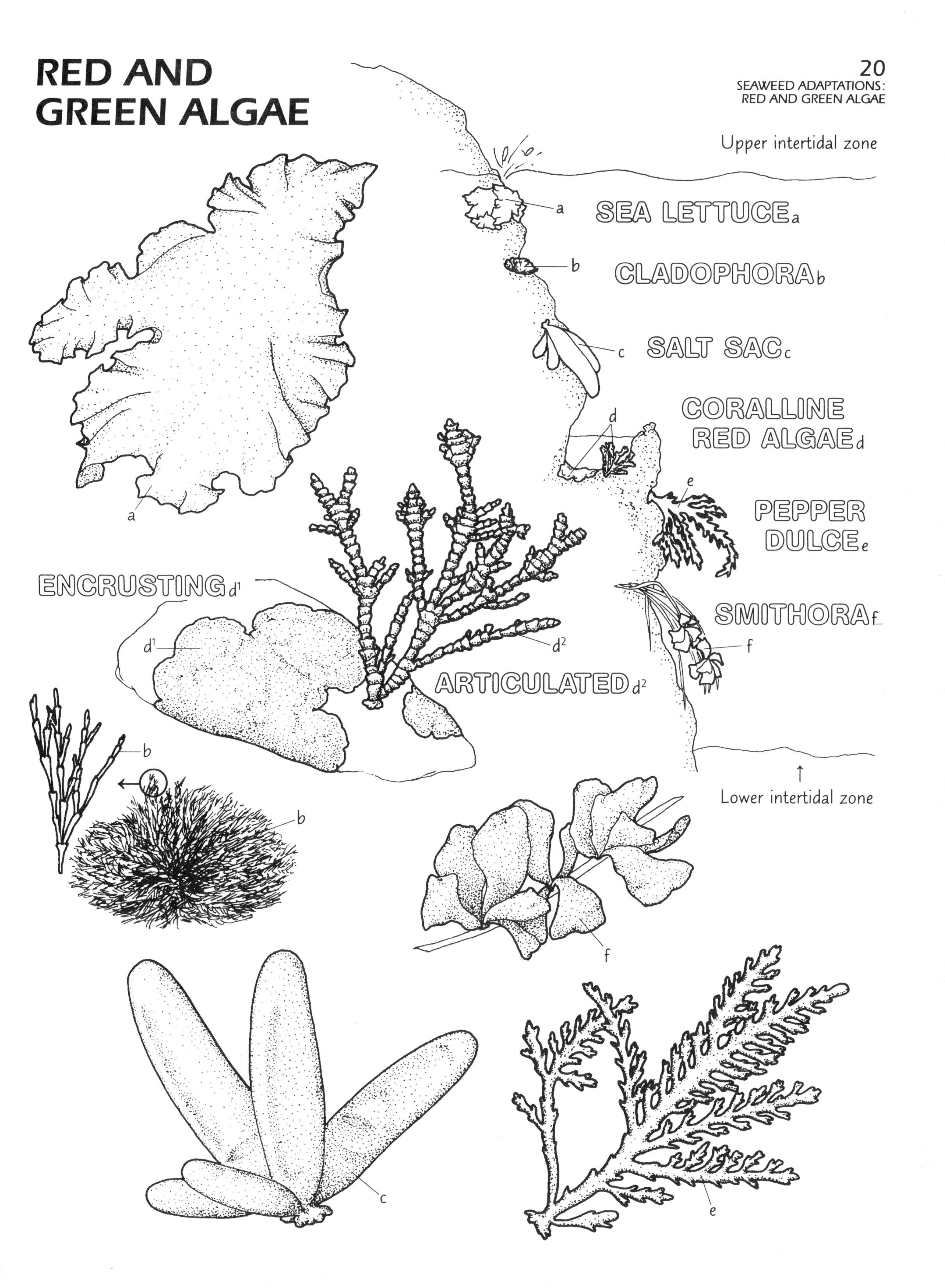

21
SEAWEED ADAPTATIONS: BROWN ALGAE

The brown algae are subjected to the same environmental stresses that exist for the red and green algae: damage from wave shock, desiccation, grazing by herbivores, and competition for available attachment space and for light for photosynthesis. Some brown algae are much larger than other types of algae, which is often a survival advantage. Some species have developed several other survival mechanisms as well.

Color the parts of the rockweed at the upper left. Color each brown alga as it is treated in the text. Notice the differences in relative proportions of the holdfast, stipe, and blade among these plants. The feather-boa kelp has especially long stipes and small blades.

Rockweed is a common brown alga growing in the high and middle zones of the rocky intertidal worldwide. These zones are a potentially very stressful area, being most frequently exposed to air at low tide. The rockweed is successful here because it can tolerate considerable desiccation, such an event being retarded by thick cell walls and high concentrations of polysaccharides (sugars). Rockweed does not grow structures that are readily distinguishable as stipes or blades; the entire plant is called a *thallus*. At high tide, small air bladders (not shown) along the sides of the thallus cause the alga to float up from the bottom where there is better exposure to light. The swollen tips of the thallus are reproductive structures called *receptacles*.

The large brown algae known as kelp are common subtidally and in the low intertidal zone, where the wave action is most severe. The feather-boa kelp has long (10 meters, 33 ft), supple, strap-like *stipes* that wash back and forth with the movement of the water. The stipes sometimes snap off during storms, leaving the large *holdfast* behind. The holdfast resembles a mound of intertwined roots and anchors the kelp. It can persist over a winter, and in the spring, new stipes will sprout.

Another wave-zone kelp is the oar weed, a type of *Laminaria* (Plate 4). It, too, has a stout holdfast, but also limber, hollow stipes. The resiliency of the stipes serves to keep its deeply incised *blades* upright for better exposure to sunlight.

Lessoniopsis is a kelp found in the most exposed, wave-battered areas of the low intertidal. Its huge holdfast, resembling a thick woody trunk, enables it to survive in this environment. *Lessoniopsis* is a perennial alga that may persist for many years. Each year it adds another layer to its "trunk," which can grow to 20 cm (8 in) thick. The stipe of *Lessoniopsis* is highly branched and sports long (1 meter, 3 ft), thin blades. A large plant may have over 500 blades.

The bull kelp is one of the "giant kelps." It is an annual alga (lives for a single year) (Plate 73), and occupies the shallow subtidal zone to depths of 30 meters (100 ft). The bull kelp possesses an elongated, rapidly growing stipe (10 cm or 4 inches per day under ideal conditions). The stipe ends in a large, gas-filled *air bladder* (pneumatocyst), which lifts the plant to the water's surface. Large blades grow from the pneumatocyst and spread out and take advantage of the sunlit water. In the fall and winter, storm waves dislodge the bull kelps and wash them onto the shore in a tangled mass.

Humans have put the rapidly growing giant kelps to use. In southern California, large barges with paddlewheel mowers harvest the kelp beds. The kelp is rendered to yield algin—a stabilizer and emulsifier used in many products, including paint, ice cream, and cosmetics.

BROWN ALGAE

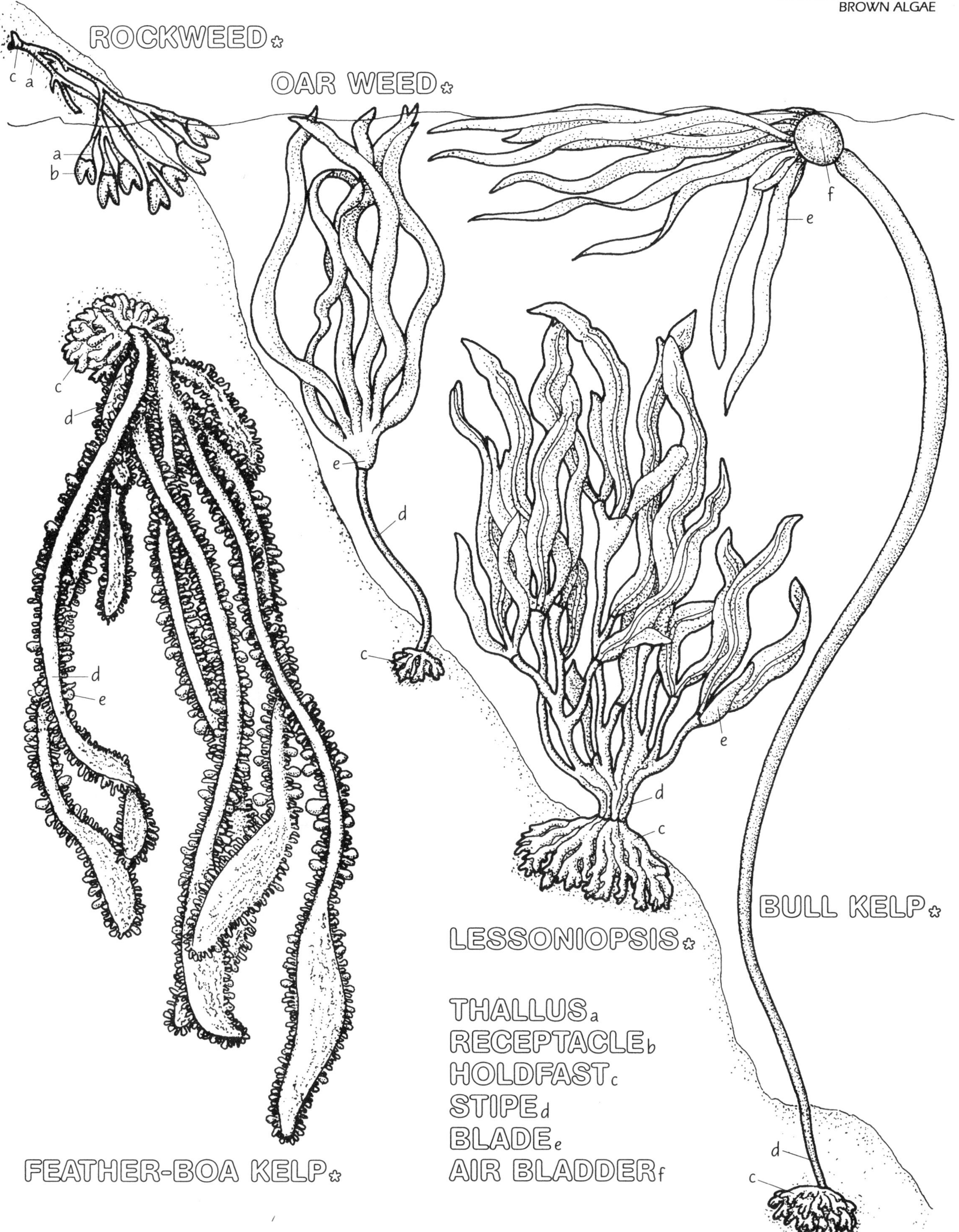
ROCKWEED*
OAR WEED*
BULL KELP*
LESSONIOPSIS*
FEATHER-BOA KELP*
THALLUS a
RECEPTACLE b
HOLDFAST c
STIPE d
BLADE e
AIR BLADDER f

22 MARINE SPONGES: SPONGE MORPHOLOGY

Sponges (phylum Porifera) are considered to be among the simplest living organisms. Almost all of the 5000 species are marine, and all operate as filter feeders, collecting small particles of food from the sea water flowing through their bodies.

Begin by coloring the cutaway diagram in the upper right corner, which shows the circulation of water through a generalized sponge. Color the body wall and the ostia. The size of the ostia has been exaggerated in the drawing; they are actually microscopic in a living sponge. Next, color the filter chambers, the atrium, and the osculum. Color the arrows which indicate water flow.

The *body wall* of the sponge is perforated by many small pores or ostia (singular *ostium*), through which water enters the sponge. In simple sponges the water flows into the *atrium* and finally out a larger opening called the *osculum*. In more complex sponges with folded body walls, shown here, the water flows through the ostia into small *filter chambers*, and finally out the osculum. Sponges rely on this flow of water for feeding, gas exchange, excretion, and often, reproduction.

Color the enlargement of the single filter chamber in the circle. Next color the enlargement of the collar cell, noting how the water flows through the collar. Finally, color the enlargement of the spicules.

The heart of the sponge's water system is a unique cell called a *collar cell* or choanocyte. In simple sponges, the collar cells line the inside of the atrium, and in more advanced sponges, they are found lining the filter chambers. Each collar cell possesses a single whiplike flagellum that beats in a rhythmic fashion. The independent rhythmic beating of many flagella creates a positive pressure within the atrium of the sponge, forcing water out the large osculum and pulling water in through the ostia. As the water moves through the sponge, it passes through the fencelike *collar* of the collar cells. The collar consists of bundles of fused cilia with spaces in between. Small particulate matter becomes trapped on the collar where it can be engulfed by the *body* of the collar cell or picked off by motile amoeboid cells and then digested in food vacuoles. Amoeboid cells may also engulf larger particles that get caught at the entrance to a filter chamber or an ostium on the sponge's surface.

Sponges vary in size from tiny lumpish forms to massive vaselike structures. Small skeletal elements called *spicules* are embedded in the body wall of the sponge and support its structure. In most sponges, the spicules are scattered individually in the body wall, as in this illustration. In the glass sponge, however, and some others, the spicules are organized into an elaborate, latticework skeleton. Instead of, or in addition to the spicules (siliceous or calcareous), some sponges have fibers of a protein called spongin.

The morphology of sponges varies from very simple tubular forms, to those with a filter chamber system (as shown), to complex systems involving more infolding of the body wall and the proliferation of smaller and more numerous filter chambers. The increased number of filter chambers allows more water to be filtered through the sponge; a 10 cm^3 (0.6 in^3) sponge is capable of filtering 20 liters (5.3 gal) of water in 24 hours.

Color the four different types of sponges. The body of each sponge can be colored the same color as the diagram above or the natural color, as given in the text.

The form of sponges is influenced greatly by the available space, type of substratum, and the strength of the water movement. Most sponges are attached to hard substrata in relatively shallow water.

The purple *encrusting sponge* is found low in the rocky intertidal zone, often in large patches. This sponge may grow to 2.5 cm (1 in) thick, and its oscula are quite large. The encrusting sponge does not grow tall in areas of heavy wave action that would quickly tear and destroy it; in quiet waters the oscula are raised on elevated volcano-shaped projections of the body wall.

Not all encrusting sponges attach to inanimate substrata. The smooth pink *pecten sponge* is found on the shells of scallops — a mutually beneficial relationship. In exchange for the substratum, the sponge covers the scallop with its porous, yielding body, offering some protection from sea star predators.

In the quiet water of subtidal habitats, such as the coral reef, large sponge forms flourish. The azure blue *tubular sponge* grows very tall.

The *boring sponge* burrows into the shells of abalones, oysters, and other molluscs. This yellow sponge lives in the tunnels it chemically etches out of the shell. Its tunneling can be extensive, severely weakening the shell. Some species of boring sponge attack corals and are responsible for much erosion of coral reefs.

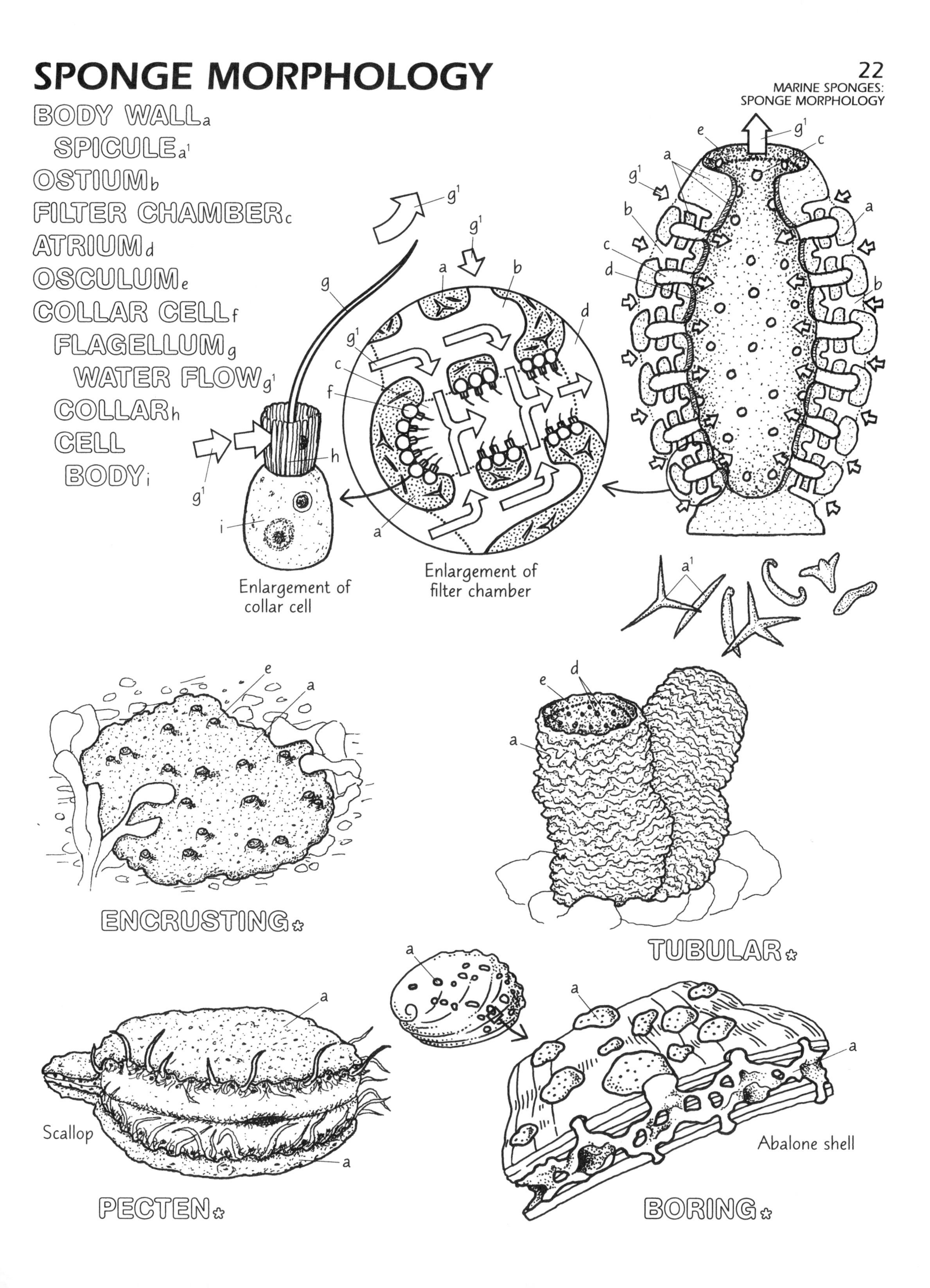
SPONGE MORPHOLOGY
22
MARINE SPONGES:
SPONGE MORPHOLOGY
BODY WALL a
SPICULE a1
OSTIUM b
FILTER CHAMBER c
ATRIUM d
OSCULUM e
COLLAR CELL f
FLAGELLUM g
WATER FLOW g1
COLLAR h
CELL
BODY i
Enlargement of
collar cell
Enlargement of
filter chamber
ENCRUSTING
TUBULAR
Scallop
PECTEN
Abalone shell
BORING

23
CNIDARIAN DIVERSITY: POLYPS

The phylum Cnidaria is a group of animals with a very simple, functional body structure. The digestive tract of the cnidarians lacks a second, or anal, opening and consists only of a mouth and a saclike cavity (the coelenteron or gastrovascular cavity).

Color the drawings which compare the structure of the polyp and the medusa, using a light color for the body.

There are two basic cnidarian body forms: the free-swimming medusa and the sessile polyp. Both types have a radially symmetrical *body* organization (body units arranged in a circle, Plate 39) with the *coelenteron* located in the center. The basic difference between the polyp and medusa is this: the medusa floats free of the substratum, with its *mouth* and *tentacles* facing downward; the polyp is attached by its *pedal disc* to the substratum, with its mouth and tentacles facing upward.

Now color the enlargement of the nematocysts.

The cnidarian mouth is surrounded by a ring of tentacles where highly specialized cells are located. These cells contain *nematocysts*—small stinging, whiplike structures that are discharged from the cells in response to outside chemical or mechanical stimuli or direct nerve stimulus.

When potential prey make contact with the tentacles of the polyp, the nematocyst-bearing cells are stimulated, causing the nematocyst to rapidly uncoil and, in some cases, penetrate the victim. Many nematocysts contain a venomous liquid that subdues the prey; some types of nematocysts are barbed or sticky, and some types actually wrap around the prey. When the prey is subdued, the tentacles maneuver it into the mouth of the polyp, and it is digested in the coelenteron. Undigested parts are regurgitated back out of the mouth.

Both polyp and medusa are known as passive predators: the polyp waits for prey to wander into its deadly tentacles, and the medusa trails its tentacles in the water as it floats along, catching its food.

Now color the different polyp types, beginning with the sea anemones, then the coral polyp, and finally, the hydroid. The non-living covering around the *Obelia* colony is transparent and should not be colored.

The familiar sea anemone of rocky shores is a large, single polyp. Its body is basically cylindrical in shape, with an *oral disc* at the top end, and a pedal disc anchoring the anemone to a solid substratum at the other end. Some species use special sticky nematocysts, combined with mucous secretions, to ensure a tight seal against the substratum. Sea anemones are generally attached to a solid surface, although some species prefer a burrowing existence in sand or mud, and others attach themselves to the shells of other animals. Anemones vary in size from a few centimeters in diameter to animals whose oral discs are 30 cm (12 in) or more in diameter.

The squat anemone in the illustration is a giant green sea anemone of Pacific coast tidepools. This anemone grows to 25 cm (10 in) or more in diameter, and is capable of compressing its pale green body to just a few centimeters in height. It feeds on nearly any organisms that are washed or swim unaware into its green tentacles.

The long-columned, white-plumed sea anemone in the illustration may reach a height of 30 cm (12 in), and is usually found subtidally in water 20 meters (65 ft) or deeper. Its numerous fine tentacles reach into the current to capture small organisms carried by the moving water.

The coral polyp is similar in structure to the sea anemone but is usually much smaller (less than 1 cm in diameter). The individual polyps of a coral colony are connected by a continuous layer of body tissue. Each coral polyp possesses a calcium carbonate *skeleton* into which the entire polyp can contract. The cup-shaped skeletons are secreted by the polyp's epidermis, and are the basic structural units that form the tropical coral reefs. Reef corals are found in colonies; these can grow to massive sizes in a remarkable variety of colors, shapes, and forms (Plate 12).

The polyps of marine hydroids occur generally in colonies as well, forming branched structures that may be attached to various substrata. An individual polyp is usually quite small, less than a centimeter, and is specialized for a particular function: feeding or reproduction. Illustrated here in the marine hydroid *Obelia* are feeding polyps, called gastrozooids, each of which have a mouth surrounded by tentacles, and reproductive polyps called *gonozooids*. Small medusae are produced asexually on the gonozooids and then swim away.

POLYPS

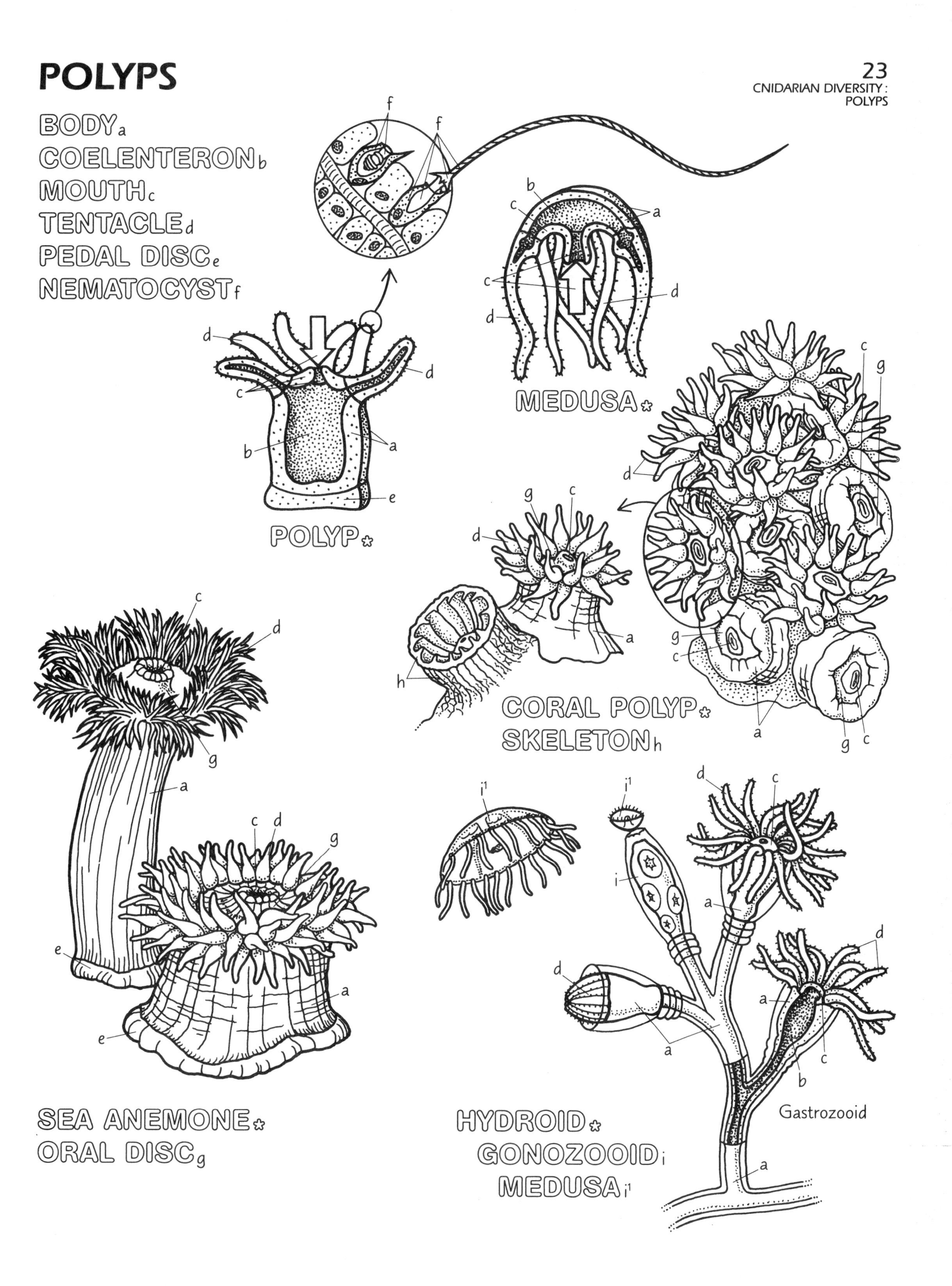
BODY a
COELENTERON b
MOUTH c
TENTACLE d
PEDAL DISC e
NEMATOCYST f
POLYP
MEDUSA
CORAL POLYP
SKELETON h
SEA ANEMONE
ORAL DISC g
HYDROID
GONOZOOID i
MEDUSA i1
Gastrozooid

24 CNIDARIAN DIVERSITY: MEDUSAE

The medusa represents the unattached, motile stage in cnidarians and is found swimming freely in the water column. Swimming is accomplished by muscular contraction of the dome-shaped umbrella, or bell, which forces water out and propels the medusa in the opposite direction. As the umbrellar muscles contract the bell, fibers of the protein collagen in the umbrella are compressed. When the muscles relax, the elastic recoil of the compressed collagen fibers causes the bell to expand and refill with water. Medusae remain largely at the mercy of the prevailing currents and therefore belong to the group of drifting pelagic organisms known as plankton (Plate 14).

Begin with *Polyorchis* at the top of the page, and color each medusa separately as it is mentioned in the text. Use a very light color for the umbrella. The tentacles have been removed from the front edge of *Polyorchis* to show the underside of the umbrella. In order to get the transparent effect of the umbrellar dome in *Polyorchis* and *Haliclystus*, color the inner structures first, and then apply the umbrella color. The umbrella of *Pelagia* is thick and not as transparent as those of the other medusae, so the underlying structures are not shown.

The medusa *Polyorchis* has a high-domed *umbrella* that is transparent and clearly reveals the organs within the umbrellar space. The elongate, *mouth*-bearing *manubrium* protrudes through the opening of the umbrella. It opens into the *stomach*, which itself opens into *radial canals*. Suspended from beneath these canals are elongated *gonads*, or reproductive organs. The long, extensible *tentacles* bear nematocysts that sting and capture zooplankton and other small marine animals. *Polyorchis* is a relatively common medusa of the bays and estuaries of the west coast of the United States and Canada; specimens with umbrellas over 5 cm (2 in) in height are not unusual.

The squat, milky translucent medusa *Aurelia* is a jellyfish with no manubrium, relatively short tentacles, and four elongated *oral arms* that surround the mouth opening. *Aurelia* is found, often abundantly, in coastal temperate water of both the Atlantic and Pacific oceans. Instead of entrapping prey with its tentacles, *Aurelia* is a suspension feeder. As it sinks through the water, *Aurelia* catches plankton in the mucus on the inside of its umbrella. The mucous-covered food is then carried to the margin of the umbrella and scraped off by the oral arms. The arms have ciliated grooves that carry food to the mouth. The umbrellar diameter of *Aurelia* reaches 15 cm (6 in).

Another typical jellyfish is *Pelagia*. This animal is quite striking with its purple and translucent gray umbrella often exceeding 75 cm (30 in) in diameter. *Pelagia* occurs in large numbers along the central and southern California coast. The four oral arms are very maneuverable. A large *Pelagia* trails its nematocyst-bearing oral arms as much as 2.5 meters (8 ft) below the umbrella, entrapping and subduing small organisms that swim into them. The oral arms contract to deliver prey to the mouth, which is located at the center of the umbrella. In southern California, many *Pelagia* are swept shoreward during the summer months and are the cause of a large number of the jellyfish stings experienced by bathers.

The small (2.5 cm, 1 in) jellyfish *Haliclystus* does not drift or swim freely, as do most medusae. Instead, *Haliclystus* attaches itself to surf grass in the rocky intertidal zone, and to eelgrass in quiet waters. Its small, stalked *attachment disc* protrudes from the center of the upturned umbrella. The gonads radiate out from the center and give the medusa the appearance of a webbed basket adorned with eight tentacle clusters. The mouth is at the center of the umbrella. *Haliclystus* feeds on small planktonic organisms and other small animals living on the plant to which the medusa is attached. Some species of *Haliclystus* are able to absorb pigment from the plant to which they attach, thus changing the color of their umbrella to match their substratum.

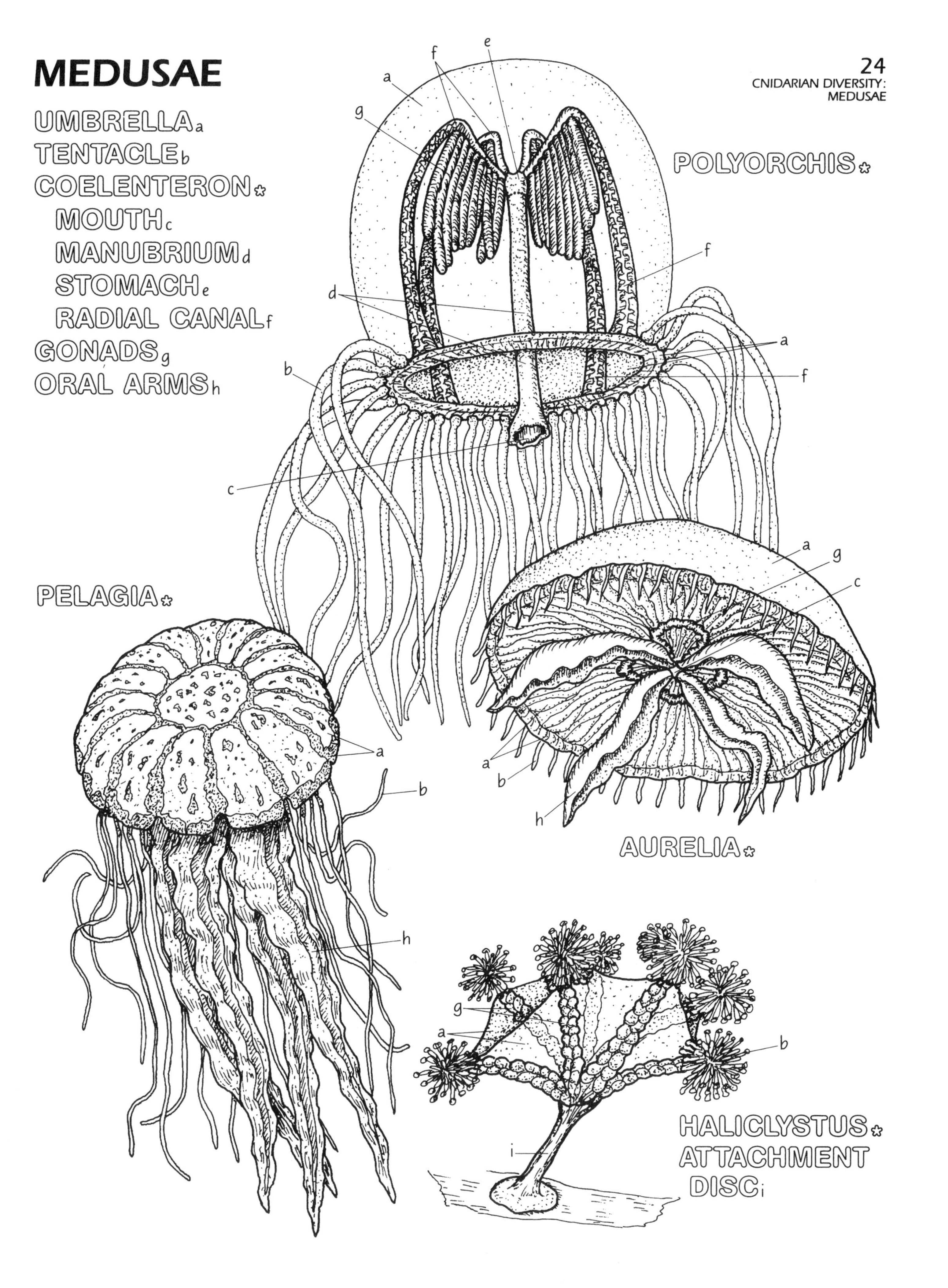
MEDUSAE
24
CNIDARIAN DIVERSITY:
MEDUSAE
UMBRELLA a
TENTACLE b
COELENTERON *
MOUTH c
MANUBRIUM d
STOMACH e
RADIAL CANAL f
GONADS g
ORAL ARMS h
POLYORCHIS *
PELAGIA *
AURELIA *
HALICLYSTUS *
ATTACHMENT DISC i
a
b
c
d
e
f
g
h
i

25
MARINE WORM DIVERSITY: COMMON WORMS

The marine worms include several groups of animals that occur in a great variety of sizes, shapes, and colors.

Color each worm as it is discussed in the text. Begin with the flatworm and notice that the pharynx, located on the under or ventral side, is visible from the top through the thin body.

The free-living flatworms range in size from almost microscopic to 60 cm (24 in) in length. The great majority of the 3000 known species are marine, and most of these are bottom dwellers, living in sand or mud under rocks and algae. The marine flatworm illustrated here depicts a generalized form (such as *Notoplana*), and is commonly found worldwide in the rocky intertidal zone. It lives on the underside of large boulders, and glides along the rock by means of the cilia on its ventral surface. *Notoplana* has two dark *eyespots* at the anterior of its brownish-gray *body*. The dark area located on the midline of the body is the only opening of the digestive tract. What appears to be a ruffled curtain is the retracted *pharynx*, which can be extruded for food gathering (shown in the illustration). The flatworm shown here is a nocturnal predator, feeding on small molluscs, crustaceans, and other invertebrates.

The ribbon worms (over 600 species) are generally thought to be related to flatworms. They are capable of tremendous lengthwise extension; a worm that measures 20 cm (8 in) when contracted can stretch to over a meter (3 ft)! The ribbon worm shown here lives in a parchment-like *tube* among the algae, mussels, and other organisms on low intertidal and subtidal rocks and pilings along the west coast of North America.

Unlike the flatworms, ribbon worms possess a complete digestive tract (both mouth and anus). Ribbon worms have a food gathering device called an eversible *proboscis*, which, unlike the eversible pharynx of polychaete worms (Plate 27), is not an extension of the digestive system. When not in use, the proboscis is retracted inside a cavity above the digestive tract. The proboscis can be everted (shot out) anteriorly to coil around the worm's prey. A very sticky mucus is secreted from the proboscis to aid in the capture of the prey. In some ribbon worm species, the proboscis may be longer than the rest of the worm, and may be equipped with a piercing barb, or stylet, and poison glands. Ribbon worms are generally nocturnal carnivores, feeding on other worms, molluscs, crustaceans, and small fish.

The peanut worms, or sipunculids, are a group of about 300 species. They live burrowed in mud or sand flats, in muddy crevices between rocks, in coral crevices, in abandoned shells of gastropods, or in the tubes of polychaete worms. Peanut worms range in size from 0.2 to 72 cm (0.08–28 in); the average length is about 10 cm (4 in). Their bodies consist of two basic sections: the rounded, bulbous *trunk* and the narrower *introvert*. The introvert is the anterior portion of the worm's body, and can be retracted into the trunk. The mouth is at the tip of the introvert and is surrounded by ciliated *tentacles* that are used in filter or deposit feeding. A peanut worm with a 5-cm (2-in) trunk can extend its introvert out 15 cm (6 in) in search of food, while it stays safely in a crevice.

The nematodes or round worms are the most common of the worm groups in both the terrestrial and marine realms. However, they are frequently overlooked because most are a few millimeters or less in length. Almost all nematodes look very similar. The worm is encased in a translucent, elastic *cuticle* which gives it a smooth, unornamented appearance. The *mouth*, surrounded by three blunt *lips*, opens at the anterior end of the worm. The posterior end tapers to a point. Nematodes are extremely successful parasites. The cuticle is impervious to noxious chemicals such as stomach enzymes, and many nematodes are intestinal parasites of vertebrates. They also parasitize plants, feeding on the sap and cell fluids. Many species are free-living, occurring by the millions in terrestrial soils as well as in marine muds and sands where they ingest the organic material contained in the substrate.

COMMON WORMS

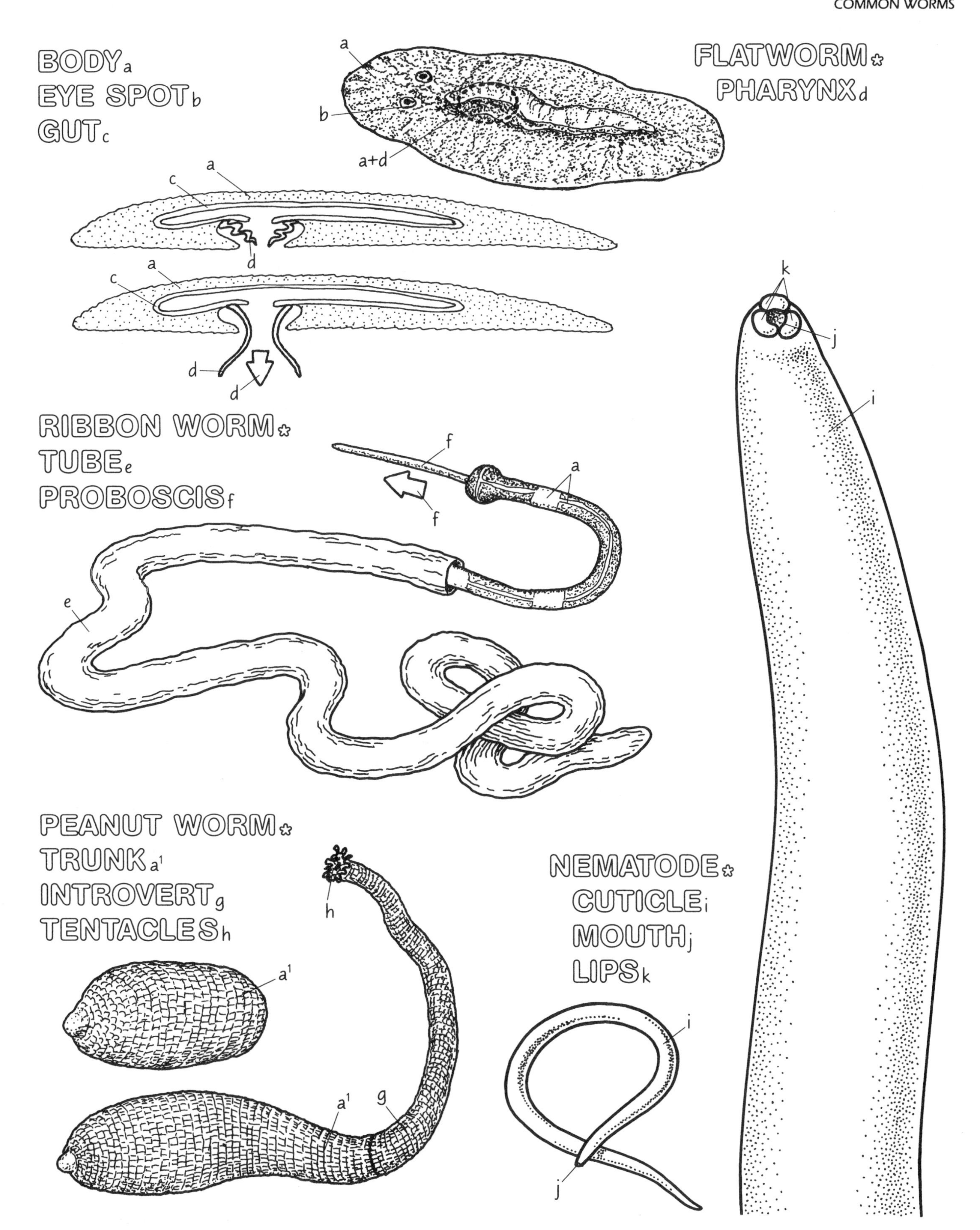

26
MARINE WORM DIVERSITY: INNKEEPER WORM

The fat innkeeper worm, *Urechis caupo*, is an interesting marine worm that inhabits semi-permanent burrows in sandy mud flats along the California coast. Normally about 20 cm (8 in) long, this round worm possesses a highly stretchable (distensible) proboscis and a pair of gold-colored, hooked bristles near its anterior end. A circle of similar bristles surrounds its anus.

Color the drawing of the innkeeper worm at the top of the plate a medium pink, which is its true color. Then color the diagram of peristaltic action. Here, the worm receives two colors, one for its body and one for the area of muscular contraction. Be sure to color the arrows that indicate the direction of the muscular contraction and water flow, using a light blue for the water. Now, color the worm in its mud flat habitat. Note that the sea water above the burrow and the sea water in the burrow receive shades of the same color, indicating that the burrow is flooded. Color the mud light brown. Color the proboscis of the worm in the burrow, then color over it with the light color chosen for the mucous net.

The innkeeper worm makes its *burrow* by digging with its *proboscis* and its anterior hooked *bristles*. The *anal bristles* and *body* movement carry the material backward and out of the burrow. Once the burrow is constructed, the innkeeper stays there unless disturbed.

The innkeeper worm pumps water through its burrow with waves of *muscular contraction* in a process called peristaltic action. These peristaltic waves create a moving "bottleneck" which travels down the worm's body, pushing *water* before it and eventually out the burrow. To feed, the innkeeper secretes a *mucous net* placed snug against the burrow walls near one opening. The worm backs down the burrow, secreting the net as it goes. When the net is between 5 and 20 cm (2–8 in) long, the innkeeper stops, positions itself in the burrow, and begins drawing water through the very fine mesh of the net, using peristaltic action. When the net is laden with food particles, the worm moves back up the burrow, eating both the net and small particles of food; large chunks are discarded. Thus, the innkeeper is a filter feeder whose filter is outside its own body.

Color the other marine animals in the innkeeper's burrow, noting their positions in the burrow. Also color the larger illustrations of these animals.

The large food particles discarded by the innkeeper worm do not go to waste. There are several marine animals that share the innkeeper's home and take advantage of both the extra availability of food and the security of the burrow. This situation, in which two or more dissimilar organisms live together, is another example of symbiosis (Plate 91).

The small (2 cm, 0.8 in) pale brown *pea crab* and the translucent white polychaete *worm* (4 cm, 1.5 in) share the burrow and fight over the food discards. The polychaete worm remains in contact with the innkeeper in order to gain an advantage over the quicker pea crab.

Two other frequent residents of the burrow are a *goby* and a small *clam*. The mottled gray-green goby uses the burrow as a home base and forages for its own food out on the mud flat at high tide. The "tap-in" clam (1.7 cm, 0.7 in) is the innkeeper's fourth guest. Instead of making its own shallow burrow at the surface of the mud flat where it might be washed away or eaten, the dull-white clam digs ("taps") into the wall of the innkeeper's burrow some distance below the opening. From here, the clam extends its short siphons into the water currents flowing through the large burrow, and from these currents it is able to siphon food.

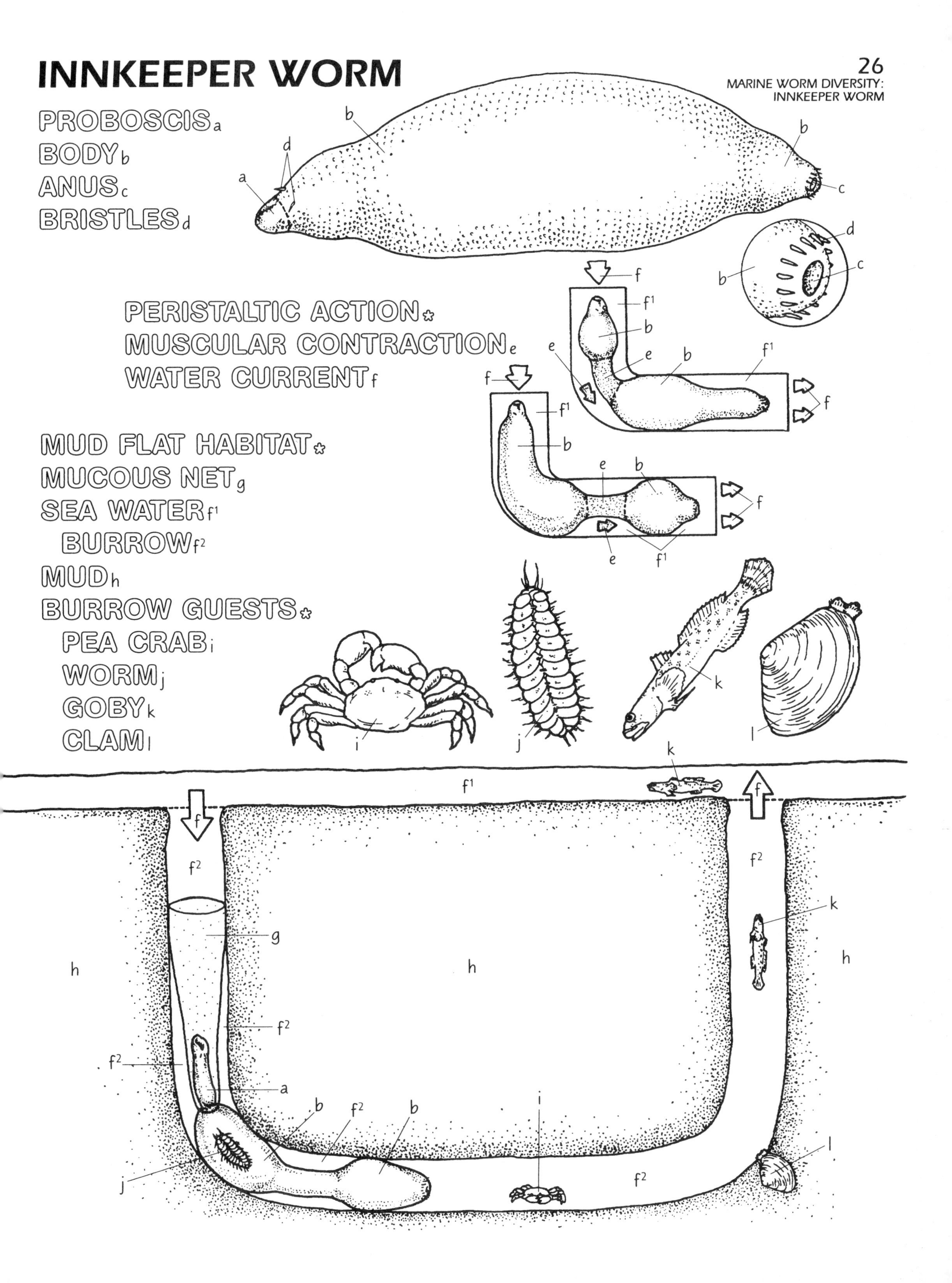
INNKEEPER WORM
26
MARINE WORM DIVERSITY: INNKEEPER WORM
PROBOSCIS a
BODY b
ANUS c
BRISTLES d
PERISTALTIC ACTION *
MUSCULAR CONTRACTION e
WATER CURRENT f
MUD FLAT HABITAT *
MUCOUS NET g
SEA WATER f1
BURROW f2
MUD h
BURROW GUESTS *
PEA CRAB i
WORM j
GOBY k
CLAM l

27
MARINE WORM DIVERSITY: POLYCHAETES

Of the many phyla of worms, the segmented or annelid worms are the most diverse and perhaps the most beautiful. This group includes the familiar earthworm, the leech, and, in the marine environment, the class of polychaetes, comprising over 5000 species.

The annelid body is divided by partitions into compartments (segments) that, in part, restrict the flow of body fluids. This segmentation enables the burrowing annelid to dig much more efficiently than the nonsegmented worm. In this plate, three polychaete worms will be introduced.

Color *Nereis* and the enlarged views of its head region at right. *Nereis* is often an iridescent blue-green.

The clam worm, *Nereis*, is a widely distributed genus and is minimally specialized. *Nereis* has what may be considered the typical polychaete body, consisting of repeated identical *body segments*, each with a pair of lateral paddle-shaped appendages called *parapodia*. The parapodia are flattened projections of the body and are equipped with rods called *setae*. The setae project through the parapodia and are connected to muscles that enable them to retract or extend. As *Nereis* crawls along, the setae aid in gripping the substratum.

The head of *Nereis* consists of two segments, a *prostomium* and a *peristomium*. The prostomium is positioned in front of the *mouth* and bears several sensory structures; these include the light-sensitive eyes, as well as the *antennae* and *palps*, which appear to be receptors for both chemical and tactile senses. The peristomium, just behind the prostomium, contains the mouth and three pairs of tentacular *cirri* that also act as tactile receptors. As *Nereis* moves through the environment, these sensory structures concentrated in the head area provide information.

The mouth of *Nereis* holds an eversible *proboscis*. The proboscis remains folded in on itself until contracting body wall musculature increases pressure on the body fluid, which, in turn, everts the proboscis. The proboscis is armed with *jaws* that swing open and then clamp shut as the body fluid pressure is reduced and the retractor muscles pull the proboscis back in.

Nereid polychaetes consume a variety of foods; some are carnivores, some omnivorous scavengers, and some are herbivores. Species of *Nereis* move about freely in many habitats, including the rocky intertidal zone, and burrow in mud and sand flats.

Now color *Glycera* and note its everted proboscis. The natural color of *Glycera* is a dark pink or light red. Also color the smaller drawing that shows the worm in its network of tunnels. Give the flooded galleries and the water above them a light blue color.

Glycera is a sand-flat-dwelling carnivore possessing a proboscis that is one-fifth its own body length and armed with four stout jaws, each with its own poison gland. *Glycera* constructs a nest of interconnecting burrows (*galleries*) in the substratum with many openings to the surface of the sand flat.

Glycera's prostomium is conical and adorned with four short antennae. It is sensitive to changes in water pressure, as for example when prey move above the nest. *Glycera* feeds on polychaetes and other invertebrates. Some species of *Glycera* reach a length of more than 50 cm (20 in).

Next color the lug worm, *Arenicola*, and its burrow environment. The arrows indicating the flow of oxygenated water into the burrow and through the sand should also be colored. The natural color of lug worms ranges from pink to dark green.

The lug worm, *Arenicola*, lives on muddy sand flats and is a deposit feeder, not a carnivore. It excavates an L-shaped *burrow* and lies with its head in the toe of the L. The lug worm engulfs the sand at the toe of the burrow, and passes the sand through its digestive tract, thus removing any organic matter. As the lug worm swallows the sand at the toe of the burrow, new sand falls in to replace it, forming a distinct depression on the surface of the sand flat above. *Arenicola's* burrow is easily identified by the pile of *fecal mounds* located at its opening. The lug worm ventilates its burrow using rhythmic peristaltic body contractions. *Oxygenated water* is pumped in from the surface through the burrow opening, flows over the *gills* on the body, and then disperses into the sand (*aerated sand*), thus aerating the upper layers of the sand flat. The worm's feeding and burrowing activities provide circulation and re-exposure of the buried sediments and the nutrients they contain.

POLYCHAETES

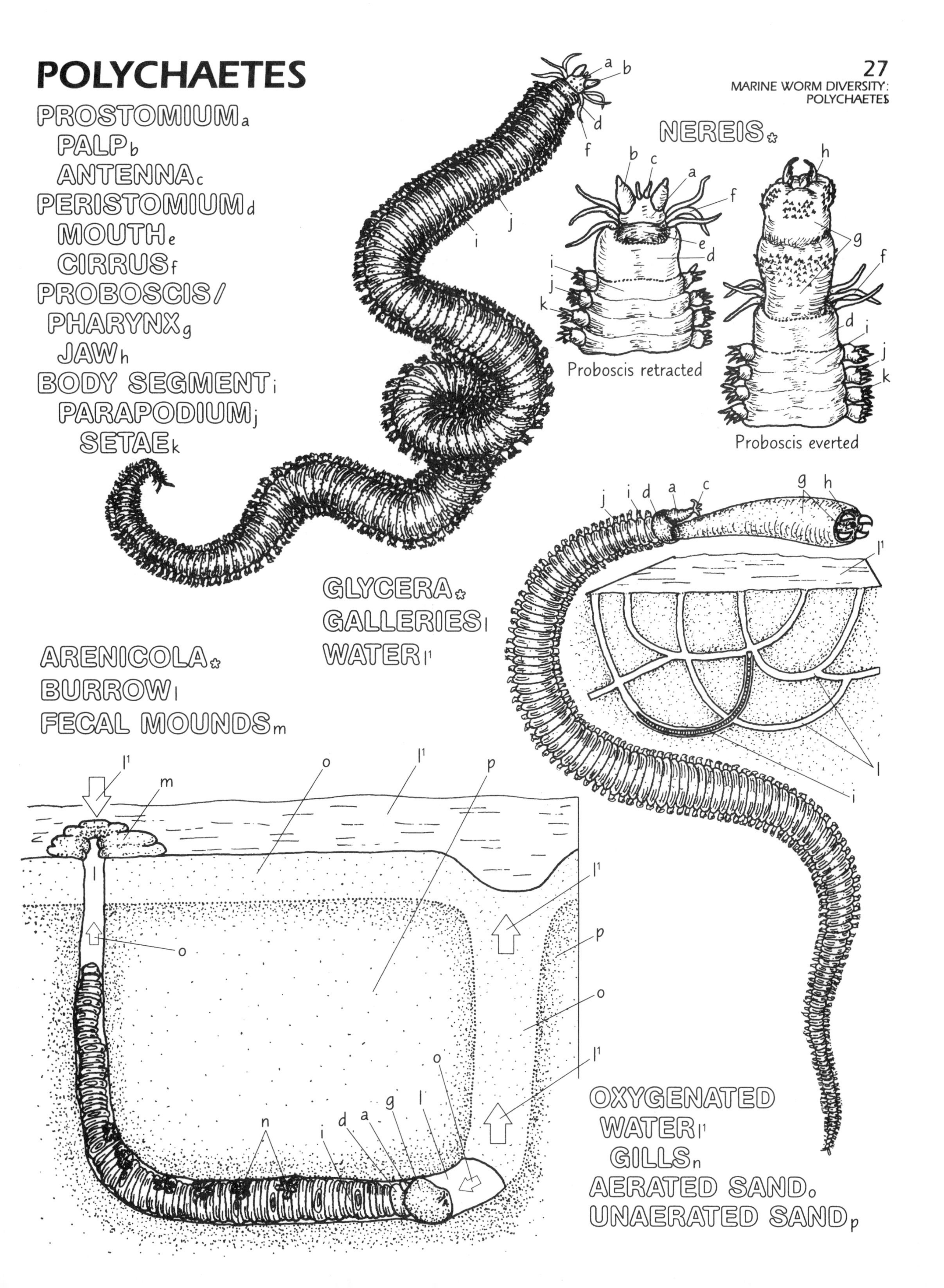

28
MARINE WORM DIVERSITY: TENTACLE-FEEDING POLYCHAETES

Free-living polychaetes, such as *Nereis*, actively seek out their food and shelter. Other types of polychaetes remain stationary, living in tubes, burrows, or crevices and obtain food by extending their tentacles out into the water. Two ways of attaining food with tentacles are exemplified by the fan worm *Sabella*, and the burrowing polychaete *Amphitrite*.

Color the fan worm in the upper left. You may wish to use a bright color for the mass of radioles, as they are very colorful reds and oranges in life. Next, color the enlargement of the tube and radioles. The tube is opaque, but for purposes of illustration it is shown as transparent so you can color the tan body segments of the worm as it resides in its tube.

Fan worms are conspicuous members of wharf piling and coral reef communities. Their beautiful fans, which resemble a flower, are actually a group of tentaclelike projections called *radioles*, extending from the prostomium (Plate 27). The radioles have lateral branches, or *pinnules*, covered with fine cilia. When the fan worm emerges from its parchmentlike *tube*, the radioles spread in a funnel-shaped crown around the *mouth*. The microscopic cilia beat in unison and create a current, bringing water up through the crown of radioles and out the top. *Food particles* carried in the current are trapped by cilia and moved down the pinnule to a food groove that runs the length of the radiole.

Now color the diagrammatic enlargement of the radiole and pinnules.

The ciliated *food groove* carries food toward the mouth; at the base of the radiole, the various food particles are sorted by size. Large particles are rejected; special ciliary tracks on the *palps* carry these back into the excurrent stream flowing up and out the center of the crown of radioles. The smallest particles are carried to the mouth and consumed. The medium particles are carried to a special *ventral sac*, where they are stored for use in tube building.

At the base of the radiole crown is a fleshy fold of tissue called the *collar,* which holds the fan worm securely at the top of its tube. Mucus, secreted from both the collar glands and the glandular *ventral plate,* is mixed with the medium-sized particles stored in the ventral sac to create a thin thread of tube material. The fan worm rotates slowly, and as it lays down this material along the edge of its papery but flexible tube, gradually repairs and lengthens it.

Color the off-white tentacles of *Amphitrite*, as well as its red gills and all of the tan body segments. The drawing at the lower right illustrates the three ways a tentacle can transport a food particle to the region of the mouth, where the particle is sorted and either eaten or rejected.

The polychaete *Amphitrite* lives in semi-permanent burrows on mud flats or in rock crevices in the intertidal zone; it feeds on organic material deposited on the surface of the substratum. *Amphitrite* possesses a multitude of hollow, distensible, ciliated, prostomial *tentacles* that reach out over the substratum until a food particle is found. Tentacles move food particles to the mouth one of three ways, depending on the particle size. If the particle is very small, the tentacle forms a shallow ciliated food groove that carries the food to the mouth. For somewhat larger particles, the action of the ciliated food groove is augmented by peristaltic contraction along the length of the tentacle. Very large particles are wrapped securely by the tentacle which is then retracted, carrying the food to the mouth. Each tentacle is pulled separately through the folded upper *lip*, which sorts by size and rejects the particle or pushes it into the mouth.

Bright red *gills* filled with blood are located near the head region of the worm. Both the gills and the tentacles may be injured or removed by predators, but the worm is capable of regenerating these structures. The elastic tentacles of *Amphitrite* can reach out over a wide area: a worm 4 cm (1.5 in) long can cover a diameter of 20 cm (8 in) with a radiating maze of tentacles.

TENTACLE-FEEDING POLYCHAETES

SABELLA FAN WORM*
TUBE a
RADIOLE b
PINNULE c
FOOD GROOVE d
FOOD PARTICLE e
PALP f
MOUTH g
VENTRAL SAC h
COLLAR i
VENTRAL PLATE j
BODY SEGMENT k

AMPHITRITE*
PERISTOMIUM l
TENTACLE m
LIP n
GILL o

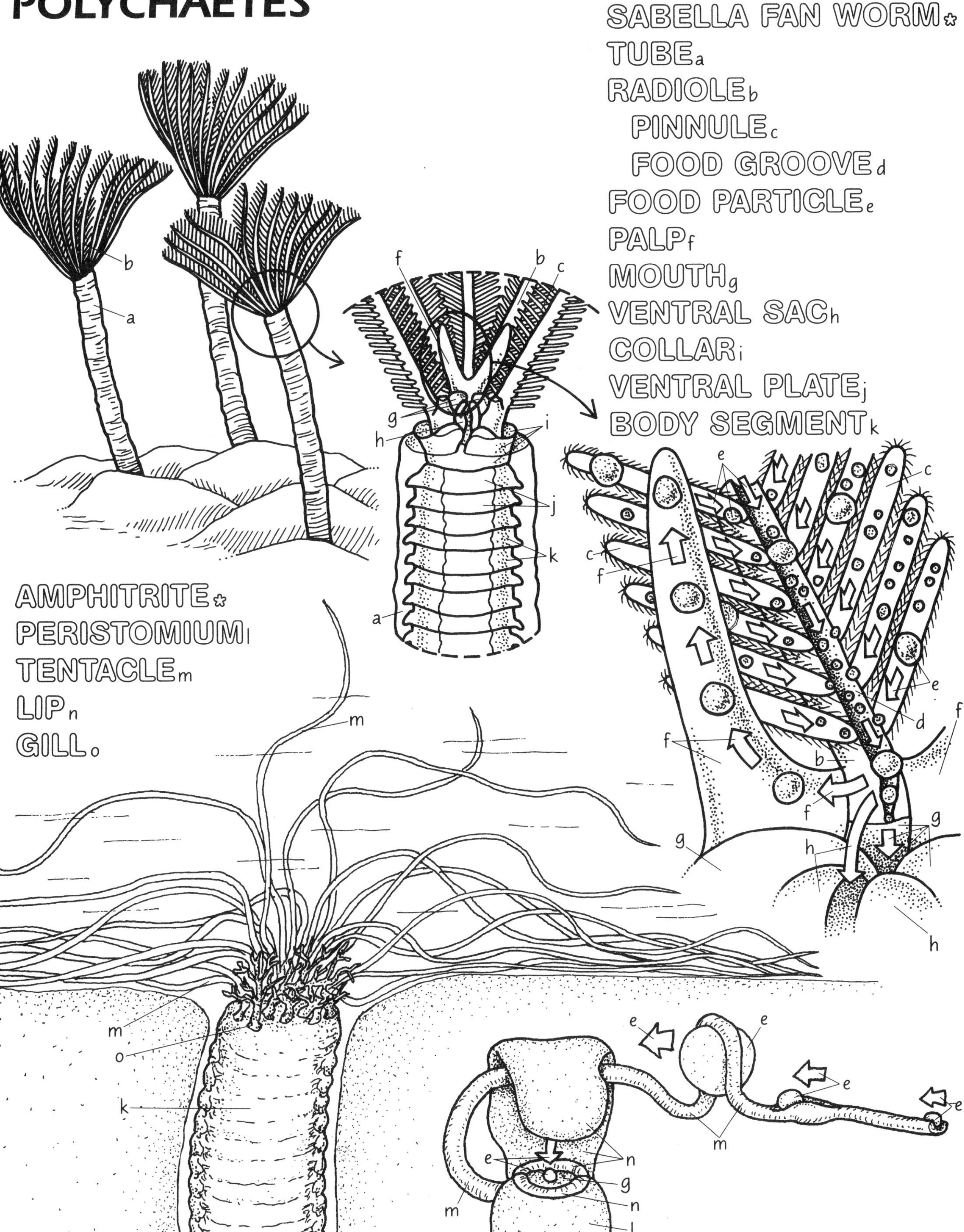

29 MOLLUSCAN DIVERSITY: INTERIOR OF A SHELL

A common introduction to marine biology is through the diversity and beauty of sea shells. Although these shells are interesting to collect and examine, the molluscs that once lived within are even more fascinating. This plate provides an understanding of how some molluscs are structured and how they function, using a clam (the cockle) as an example.

Start by coloring the shell interior at the upper right. In life, the shell is off-white with brown mottling.

On the empty right shell, or *valve,* shown here, is a peak where shell formation began. Called the beak, or *umbo,* of the valve, this peak can be used as a reference to indicate the dorsal side (back) of the clam. Below and to the right of the umbo (as pictured), is the *hinge ligament.* Made of protein, this compressible structure connects the valves and functions in the opening and closing of them. Below and to the side of the ligament are a number of projections called *hinge teeth,* which fit into corresponding recesses (sockets) in the other valve. This tooth and socket arrangement aids in the articulation of the valves by preventing one from riding over the other. This is important in burrowing, or when the clam is being attacked by a predator, since a tightly closed shell is its most effective defense.

Also visible in the empty valve are four oval *muscle scars.* These muscle scars are the sites of attachment of the *adductor muscles,* which pull the valves together and hold them shut, and of the pedal (foot) retractor muscles (not shown). The thin, curving line joining the two adductor muscle scars is called the *pallial line,* and it marks the point where the mantle attaches to the shell.

Color the end view and the cross section of the two valves on the left of the plate. In the cross-sectional view, the cut is made through a single adductor muscle. The visceral mass has been removed. Also, color the arrows that indicate the direction the valves move as they close upon contraction of the muscle. Use a light color for the mantle.

Looking at the two valves in the cross-sectional view, one can see that the adductor muscles and the ligament have opposing roles. When the adductors contract and the valves are brought together, the lower portion of the ligament is compressed and the upper portion is stretched. When the adductor muscles relax, the compressed part of the ligament expands, and the stretched upper part of the ligament contracts. This results in the valves gaping open so the clam can extend its foot and siphons.

Also visible in this cross-sectional view is the fleshy *mantle* which completely underlies the valve and is responsible for secretion and maintenance of the valve.

Color the clam at the top of the page and the internal view of the clam at the bottom of the page. Note that some names refer to related structures seen in the empty shell. Arrows that indicate the direction of the feeding current (below the gills) should be colored the same as the incurrent siphon. Those arrows above the gills should be colored the same as the excurrent siphon.

At the top of the plate, the cockle is shown in left side view with its *incurrent* and *excurrent siphons* extended from the posterior end, and its large *foot* extended from the anterior end. This is the normal position of a burrowed cockle while actively pumping water for feeding and respiration. The radial ridges on the outside of the shell add strength and help anchor the cockle in the sand.

In the illustration of the cockle at the bottom of the plate, the left valve with its underlying mantle has been removed to expose the internal organs. The large adductor muscles are clearly visible. It can also be seen how the extended siphons are continuous with the fleshy mantle that lines the inside of the shell. Note the large bilobed *gill,* the smaller *labial palp,* and the foot beneath the gills. The gills are covered with cilia, which beat in unison to create a current. The arrows indicate the direction of the water current. As water passes through the gills, small particles, such as phytoplankton and organic detritus, are trapped by specialized cilia, bound up in mucus, and transferred to ciliated food grooves along the gill margins. The food grooves carry the mucous-bound food particles in thin strands to the labial palps where the particles are sorted by size. Smaller particles are directed to the mouth, located beneath the labial palp. Rejected, larger particles accumulate below the gills, near the foot, and are periodically expelled.

INTERIOR OF A SHELL

COCKLE

VALVE a
UMBO b
HINGE LIGAMENT c
HINGE TEETH d
ADDUCT. MUSC. SCAR e
PEDAL RETRACT. SCAR f
PALLIAL LINE g

MANTLE ATTACHMENT g¹
ADDUCTOR MUSCLE e¹
MANTLE h
INCURRENT SIPHON i
EXCURRENT SIPHON j
FOOT f¹
GILLS k
LABIAL PALP l

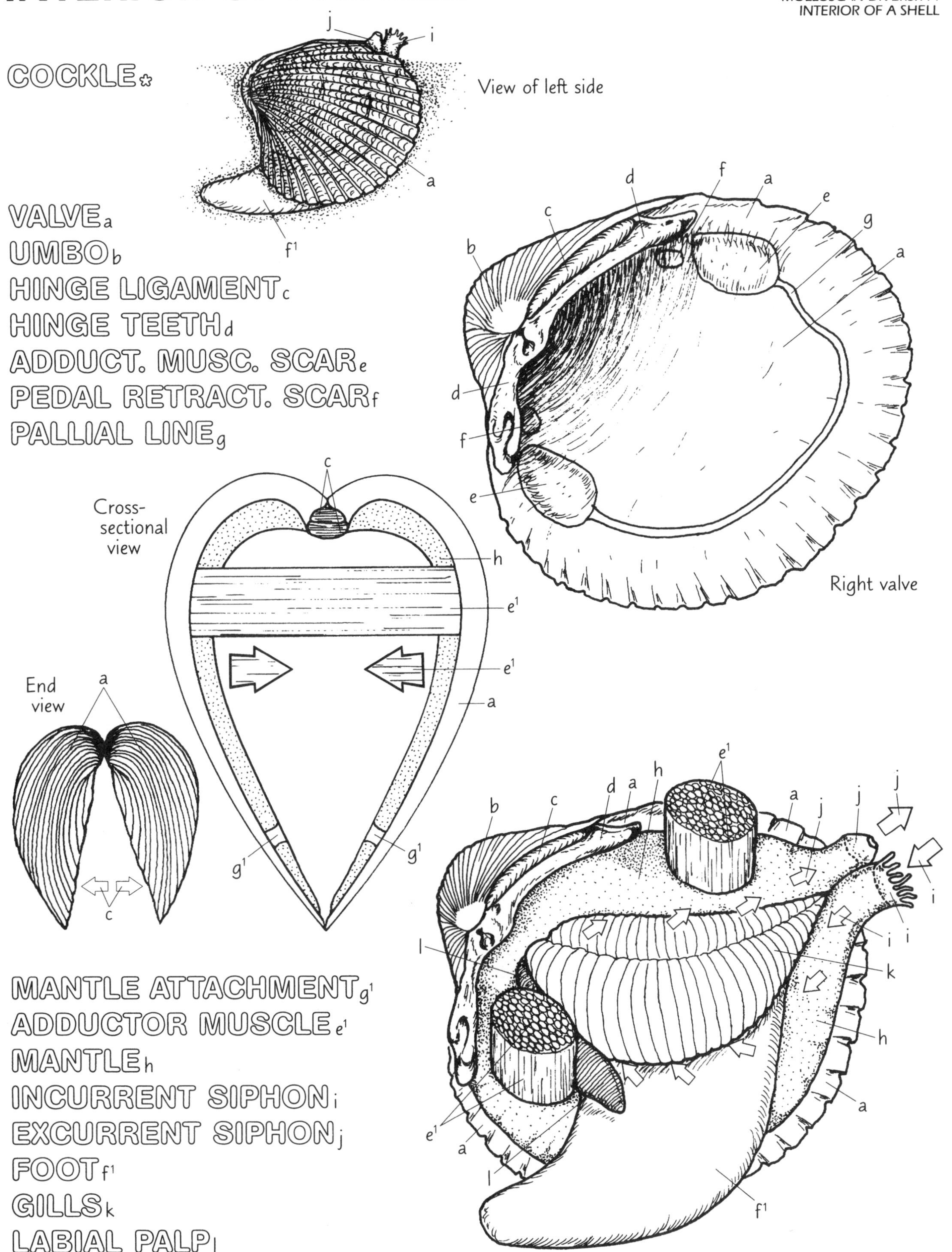

30 MOLLUSCAN DIVERSITY: THE BIVALVES

Variations in the structure and shape of bivalve molluscs reflect evolutionary adaptations to different environments. Most malacologists (scientists who study molluscs) concur that the bivalve group evolved from organisms originally adapted to living in soft sediments, such as sand or mud, and that some secondarily took up existence on top of the substratum. The discussion that follows includes the cockle, the softshell clam, and the bent-nosed clam (soft substratum dwellers, members of the infauna). The mussel and the scallop, that have adapted to living on top of the substratum (epifauna), are also presented.

Color the parts of each bivalve as the animal is discussed in the text. Use a color for the shells light enough that the texture of the surface will show. Use a light color for the scallop's mantle and then dot the "eyes" with a contrasting darker color. The tentacles on the scallop should be colored with the mantle color.

The cockle is found living very close to the surface, primarily in sandy substrata. Its siphons are very short; the *incurrent* and *excurrent siphons* face slightly different directions to ensure that the same water is not refiltered. Because the cockle lives so close to the surface, it is often exposed or dislodged by water movement. The cockle's large digging *foot* is extremely useful in reburrowing and escaping from predators.

The chalky-white softshell clam lives typically in very soft sandy mud, where it burrows very deeply. For filter feeding, it has elongated siphons that enable it to reach beyond the surface, into the water current. Water is carried across the gills where special cilia trap food particles and transport them to the mouth (Plate 29). As the clam grows larger, it burrows deeper, and its siphons lengthen accordingly. The softshell clam is rarely dislodged from its deep burrow; therefore its need for rapid burrowing ability is much less than that of the cockle, and its foot is much smaller.

The pale gray bent-nosed clam is not a filter feeder, but a deposit feeder. It uses its long incurrent siphon to probe along the surface of the sediment for deposited organic matter. The clam is named "bent-nosed" for the noticeable curve in the posterior region of its *shell*. The clam lies in the sediment on its left side with its bent-nosed posterior curved upward, and its pale yellow incurrent siphon extended beyond the muddy surface. Food is sucked up through the incurrent siphon, carried into the mantle cavity, and collected on the gills. Unlike its filter-feeding cousins who can remain relatively stationary as the water brings their food to them, the deposit-feeding bent-nosed clam soon depletes its food supply in an area and must move through the sediment to a new position. To facilitate this movement, the bent-nosed clam is very thin and possesses a broad, thin, and maneuverable digging foot.

The dark blue edible mussel lives on the substratum attached to pier pilings or rocks in the intertidal and shallow subtidal zones. It often occurs in large aggregations called mussel clumps. Mussels attach to the substratum by means of protein *byssal threads* which are secreted as a liquid from a gland near the foot. The dark brown threads harden upon contact with the water, holding the mussel in position. The mussel lacks large protruding siphons, and instead needs only a small ruffled area at its posterior end to direct the inflow and outflow of water for filter feeding.

The scallop has developed adaptations appropriate to its motile existence. The scallop is a filter feeder that lives on the sediment surface, completely unattached to any substratum. Because it is unattached, the scallop needs neither siphons for filter feeding, nor a foot for digging. It has rows of "eyes" along the edge of the *mantle* that can detect shadows and movement and aid in avoiding predators. The eyes may be blue, red, gold, or other colors, depending on the species. Along the mantle edge are mantle tentacles, which contain tactile receptors and chemoreceptors. These aid the scallop in perceiving its environment as it swims by clapping its valves together (Plate 103).

THE BIVALVES

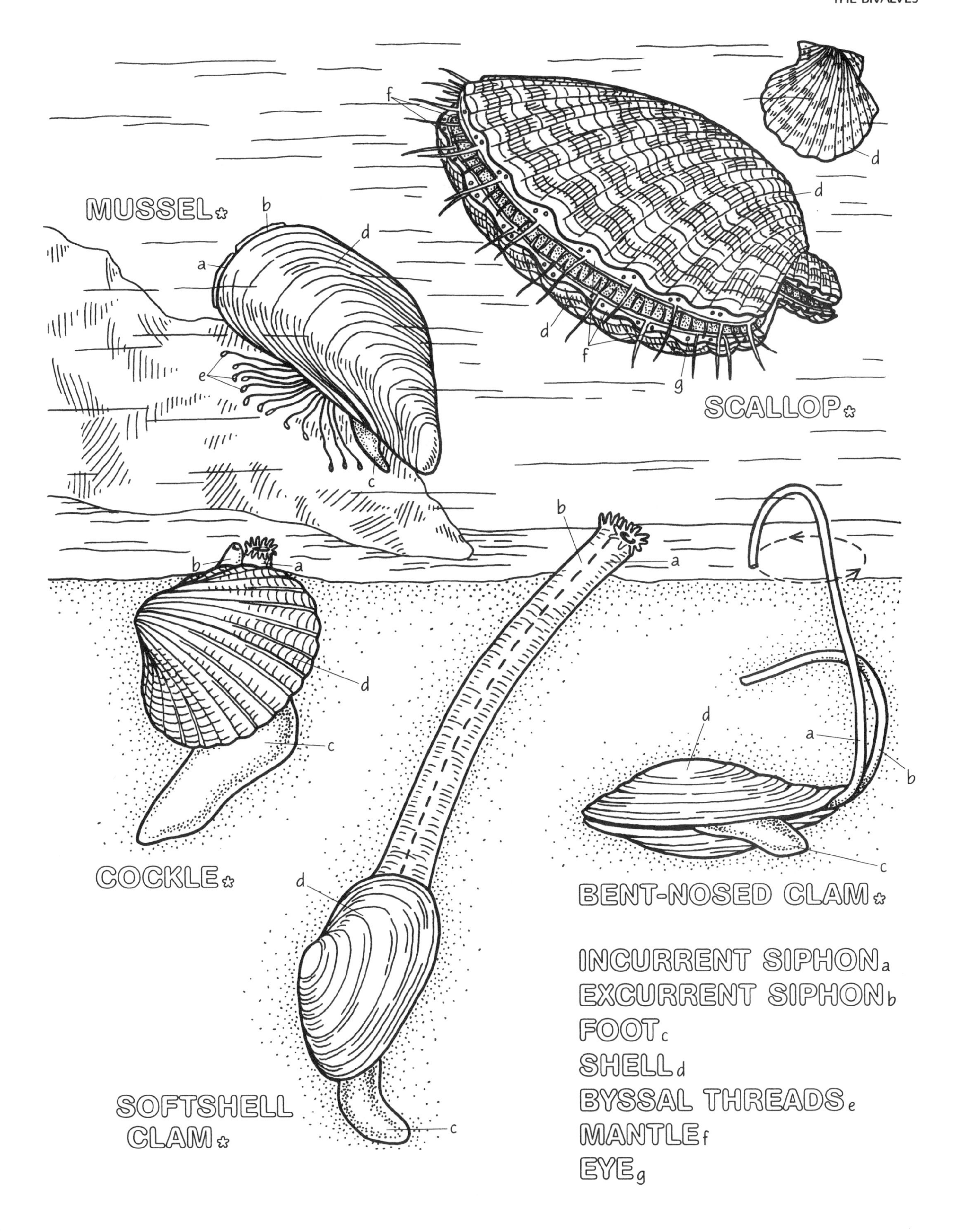

31
MOLLUSCAN DIVERSITY: SHELLED GASTROPODS

The gastropods (class Gastropoda) are the largest class of molluscs: 15,000 fossil forms have been identified, and living species number well over 35,000.

Color the lower empty shell of the tulip snail and use a light color for the body whorl. The tulip snail shell is light purple in life.

Most gastropod shells are built as a series of spirals called whorls. The tip, or *apex* of the shell, is the smallest whorl laid down by the snail in the early part of its life. As the snail grows, it lays down the intermediate whorls which form the *spire* of the shell. The final large spiral is the *body whorl*, terminating at the *aperture* or opening. The aperture is elongated into an anterior notch or *siphonal canal*, which holds the incurrent respiratory siphon in the living snail.

Color the illustration of the living tulip snail. Color over the eyes with a dark color.

The tulip snail possesses a tough, proteinaceous oval-shaped structure called the *operculum*, which is carried on its broad *foot* and is used to shut the snail snugly into its shell. When disturbed, the tulip snail retracts into its shell by first pulling in its *head*, then its foot, with the dark brown operculum brought in last to seal off the shell.

Projecting anteriorly is the elongated *siphon* used to carry water to the gills for respiration. Special chemosensory organs are located near the gills. The *tentacles* are chemosensory and touch-sensitive; the *eyes* are light-sensitive and can detect movement.

Tulip snails are predatory, feeding on other molluscs, particularly on bivalves. The tulips reach a length of 10 cm (4 in) and are commonly found in the Caribbean, the Gulf of Mexico, and along the southern coast of the United States.

Color the two illustrations of the abalone.

The abalone is a large herbivore common to the Pacific coast of North America. The abalone lives in shallow rocky areas of considerable wave action, and the flat shell shape reduces exposure to this water movement. The broad, flat body whorl terminates in an aperture that is as large as the whorl itself. The foot of the abalone completely fills the aperture. Because of the size and large surface area of the foot, the abalone can grip the substratum with amazing tenacity and remain securely fastened against strong waves and most predators.

The abalone has a pair of sensory tentacles and a large *mouth*. Around the foot of the abalone is a *mantle* from which the *epipodial tentacles* protrude. If the epipodial tentacles are touched, the mantle retracts and a strong muscular contraction of the foot brings the shell down tightly against the rock surface.

The abalone shell, measuring up to 37 cm (14 in) in some species, has several openings through which the respiratory water flows. Water enters the anterior, *incurrent openings*, passes through the gills, and exits through the rear *excurrent openings*. The excurrent openings also carry out the waste products of digestion and excretion and serve as the exit for sex cells when the abalone spawns (Plate 79). Shell color varies with species. Red, green, black, and pink abalone are named for their shell color.

Color the cowry and the moon snail as they are discussed in the text.

Some of the most beautifully patterned shells are those of the cowries, found in tropical and subtropical oceans. The cowry shell has a glossy, polished appearance, which is maintained by the cowry's ability to completely cover its shell with its mantle. The two large mantle lobes can be drawn up the sides of the shell to meet at the dorsal midline, or can be completely withdrawn into the shell. The mantle is often brightly colored and patterned and sometimes studded with small fleshly projections called papillae.

The cowry moves on its foot, probing with its tentacles, taking in the respiratory current through its short siphon. The cowry feeds on small bottom-dwelling invertebrates and dead animals.

The cowry shell grows up and over itself, so that in the adult animal shell only the body whorl is visible. Cowry species range in size from 6 to 150 mm (0.2–6 in).

The moon snail lives on the mud and sand flats, where a very large foot aids its movement through the soft substrata. Locomotion is further aided by the *propodium*, an extension of the foot that plows forward through the mud. The propodium has a flap that extends to cover the head of the snail, providing protection and leaving only the siphon and the tentacles exposed as the snail travels through the sand and mud. Note the spire and body whorl in the light brown moon snail shell.

SHELLED GASTROPODS

SHELL CHARACTERISTICS*
SPIRE a
APEX b
BODY WHORL c
APERTURE d
SIPHONAL CANAL d[1]
SOFT PARTS*
OPERCULUM e
FOOT f
SIPHON g
TENTACLE h
EYE i
MOUTH j
HEAD k
MANTLE l
EPIPODIAL TENTACLE m
INCURRENT OPENINGS n
EXCURRENT OPENINGS o
PROPODIUM p

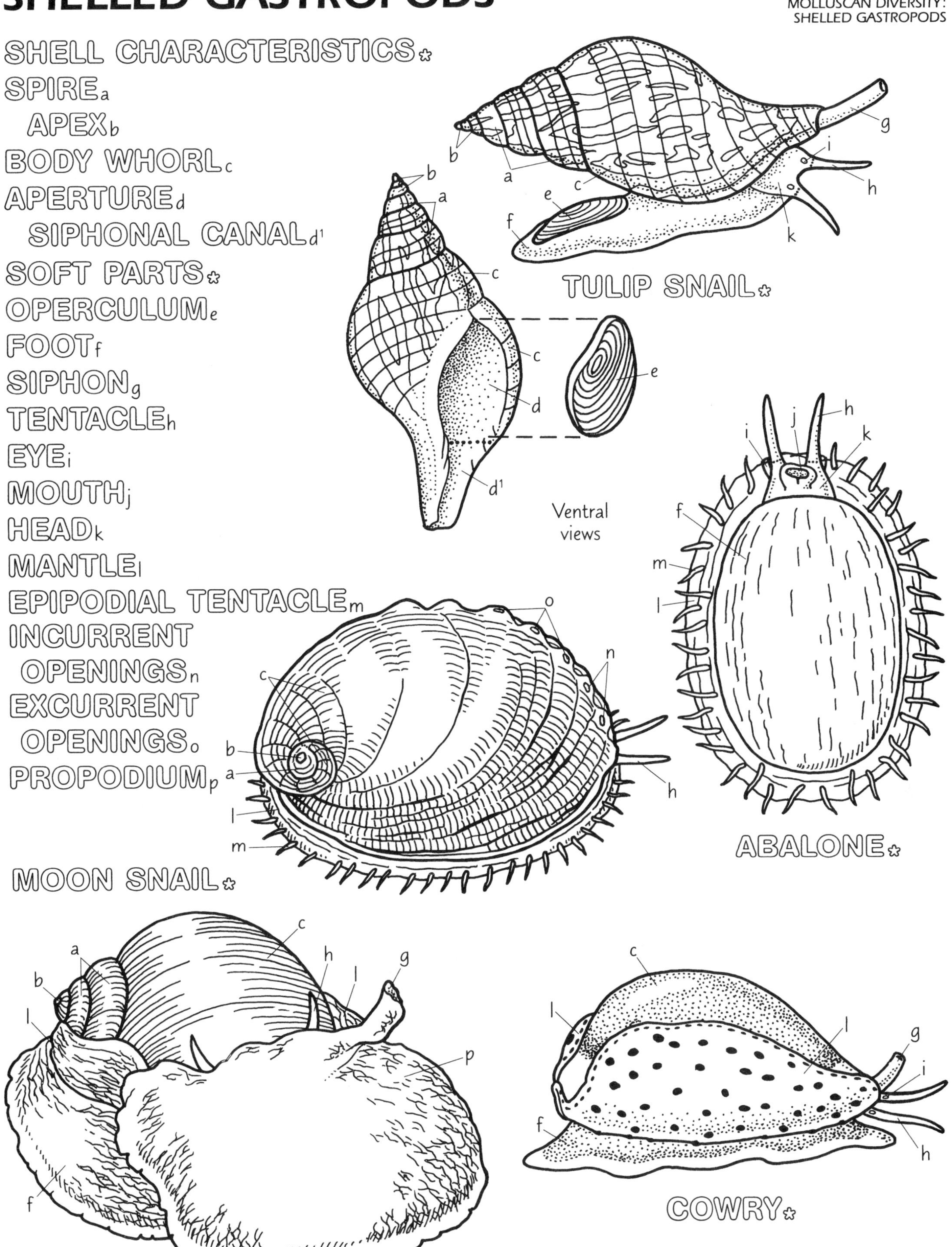

32 MOLLUSCAN DIVERSITY: SHELL-LESS GASTROPODS

The shell-less gastropods (sea slugs) are a group of marine organisms that have shed the passive defenses of the shell and operculum and instead have developed complex chemical and biological defenses to ward off their predators. They have adapted vividly colored warning patterns and are sometimes referred to as "butterflies of the sea." Illustrated and discussed here are two nudibranch sea slugs and the closely related sea hare.

Begin by coloring the dorid nudibranch as described below. If you wish, color the encrusting sponge, on which the dorid is feeding, purple or red.

Dorid nudibranchs dwell in rocky intertidal zones throughout the world, and are common along the Pacific coast of North America. The spotted *mantle* of the dorid covers the entire dorsal surface and hangs down over the *foot*. The mantle surface of many dorids is brilliantly colored or patterned and is thought to warn predators of the dorid's unpleasant taste. In a few cases, the color matches the background on which the animal feeds. This dorid nudibranch is gray or light brown with dark brown or black spots. The average size is 7.5 cm (3 in) and the large, broad foot and flattened body are adapted for feeding on its sole prey—encrusting sponges.

A circlet of off-white *gill plumes* protrudes from the dorsal surface and is capable of complete retraction into a special pocket. A pair of dorsal chemosensory tentacles, the *rhinophores*, can also be retracted into special pocket structures.

Now, color the aeolid nudibranch as described below. The hydroid colony is light pink on the stalk with red above and transparent tentacles.

The dorsal ornamentation of the dorid pales in comparison with that of the aeolid nudibranch. The aeolid has clusters of elongated dorsal structures called *cerata* which are often brightly colored with vivid contrasts. These are thought to draw attention away from the unprotected rhinophores and *oral tentacles*. The cerata of the species shown are brown with orange tips. If damaged or removed by predators, the cerata are quickly regenerated. The cerata function as a respiratory surface, and inside each is a glandular digestive lobe. The cerata may also contain specialized cnidosacs in which are located undischarged nematocysts taken from their cnidarian prey (Plate 100). Defensive mechanisms present in some species include poison glands, prickly bundles of calcareous spicules (sharp-pointed structures of calcium carbonate taken from their prey), or noxious mucous secretions.

Like the dorid, the aeolid has a pair of rhinophores; it also possesses pairs of elongated oral tentacles and *propodial tentacles* located at the front of the foot. These additional sensory structures aid the aeolid in finding and attacking its cnidarian prey.

The aeolid feeds primarily on hydroid colonies (Plate 23). Its elongated body and long, narrow foot (5 cm, 2 in) are adapted to clinging on the erect hydroid.

Aeolid nudibranchs are seasonally common on the Pacific coast of the United States, and occur both in the rocky intertidal zone and around floats and piers in quiet harbors and bays worldwide.

Now color the sea hare. Note that the parapodium is an extension of the foot and receives the same color. You may wish to color the algae green or red.

The sea hare, a close relative of the nudibranch, is named for its large, rabbit-ear-like oral tentacles and its voracious, herbivorous appetite. It is among the largest of the sea slugs; some species reach 40 cm (16 in) in length and may weigh up to 2 kg (about 5 lb). Its color varies from maroon to purple and is often mottled with lighter colors.

The sea hare possesses a pair of rhinophores; its gill is covered by the mantle. The foot has two broad flaps called *parapodia*, which can be folded over its back or flap to create a respiratory current over the gill.

Defensive adaptations of the sea hare include its secretion of a distasteful milky substance, and the glandular ejection of a vivid purple dye. In their mantles, sea hares store noxious organic compounds, garnered from algae, that further deter predators. They are also found worldwide in warm and temperate water habitats.

SHELL-LESS GASTROPODS

MANTLE$_a$
FOOT$_b$
RHINOPHORE$_c$

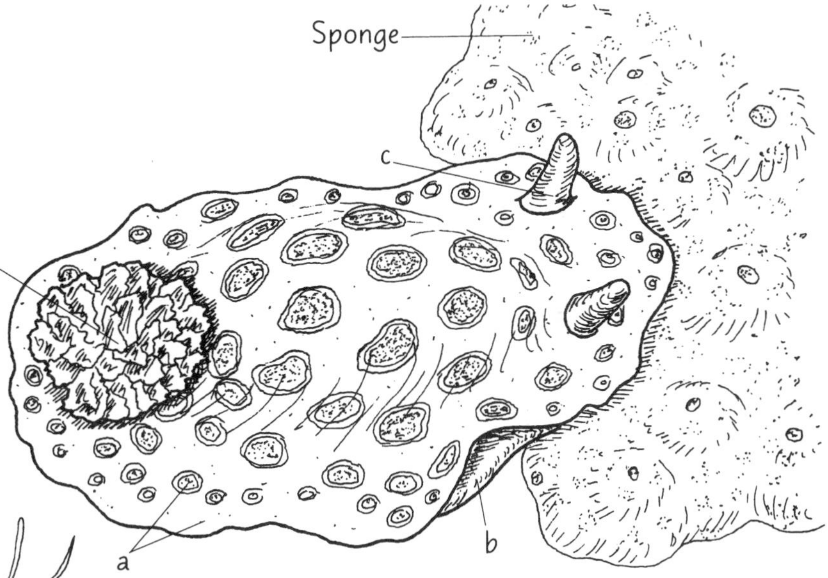
Sponge
c
d
a
b

DORID NUDIBRANCH*
GILL PLUME$_d$

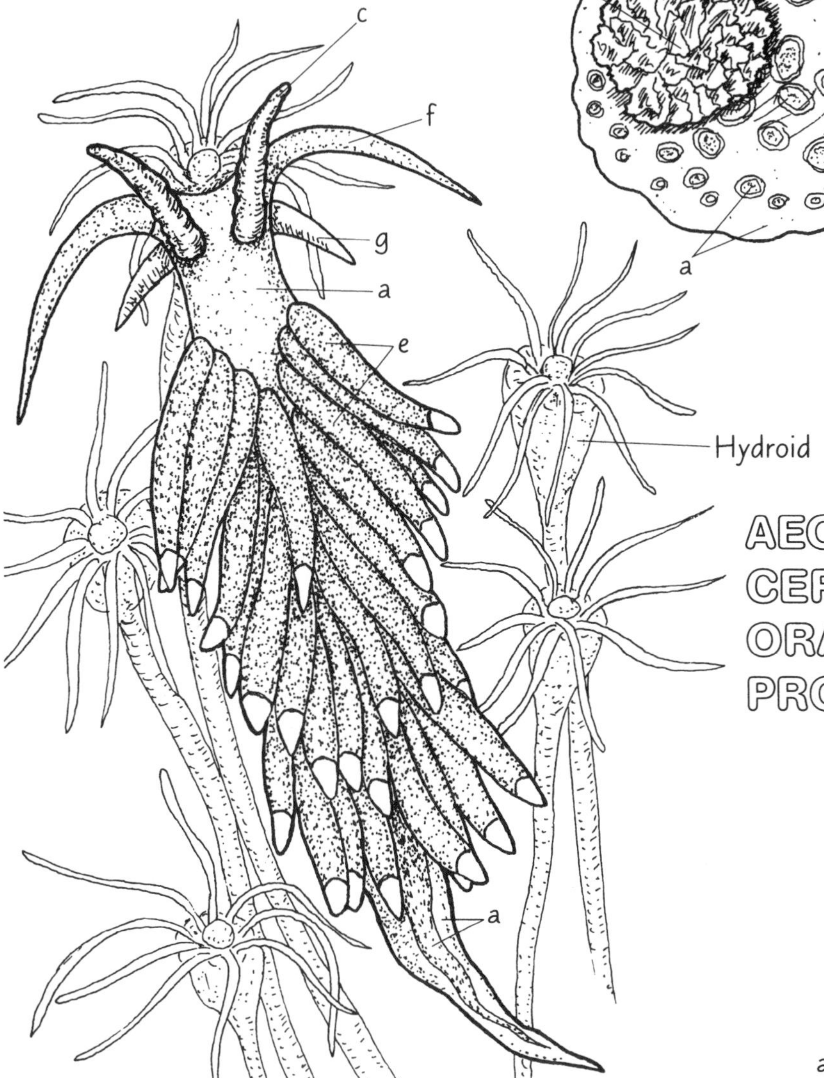
c
f
g
a
e
Hydroid
a

AEOLID NUDIBRANCH*
CERATA$_e$
ORAL TENTACLE$_f$
PROPODIAL TENTACLE$_g$

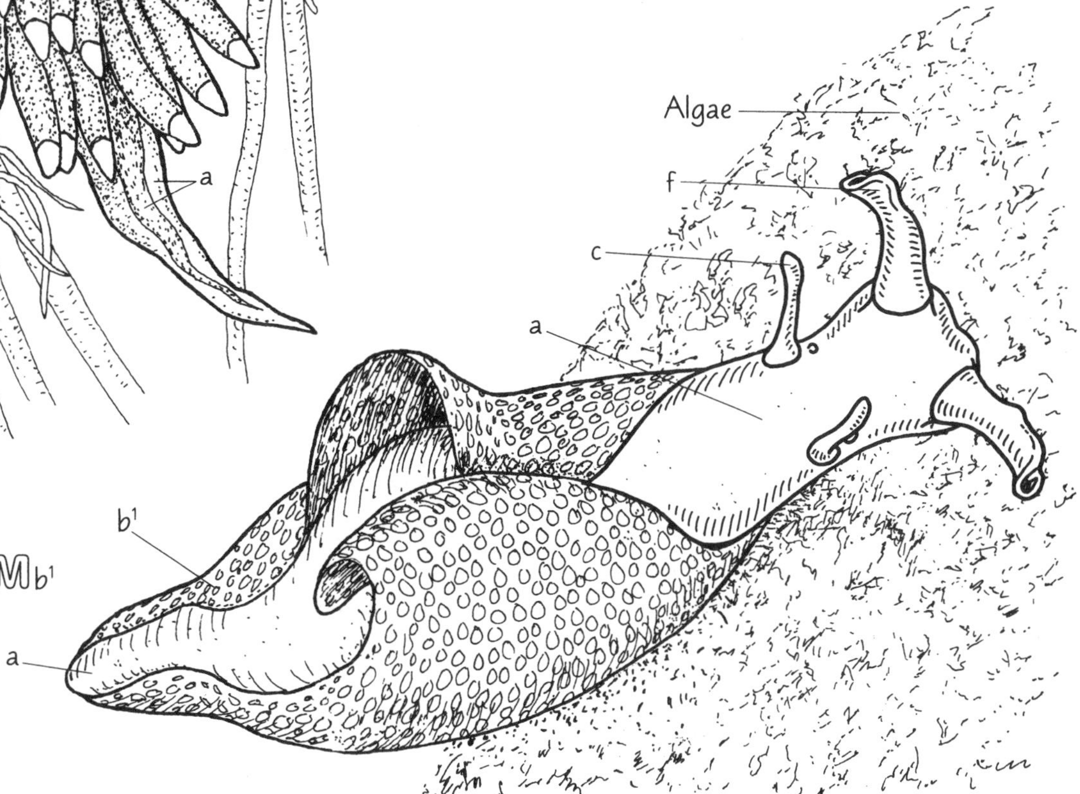
Algae
f
c
a
b1
a

SEA HARE*
PARAPODIUM$_{b^1}$

33
MOLLUSCAN DIVERSITY: NAUTILUS

The class Cephalopoda contains the squids, octopuses, nautiluses, and related forms. The cephalopods have a highly developed head with large, well-organized eyes.

Color the nautilus in its shell and the cut-away cross section.

The chambered nautilus, because of its *shell*, is considered to be the most primitive living cephalopod. The group dates back 450 million years, and through time 3500 species have existed. Today there are only six species remaining. The shell of the nautilus is thin, double layered, and pearly white inside, and dull white with reddish zebra-striping on the outside. The shell is partitioned into chambers by the *septa,* made of shell material. As the nautilus grows, it lays down a new, larger chamber at a rate of one chamber every few weeks, for a total of 38 chambers. When the animal matures, the septa thicken. The nautilus occupies only the last *body chamber* in the shell. The shell reaches from 15 to 25 cm (6–10 in) in diameter.

Color the illustration of the nautilus removed from its shell. The body of the nautilus receives a shade of the same color as the body chamber to indicate the animal's position in its shell. In life the body is translucent white and red.

The nautilus has a large number of *tentacles* arranged in two circles. The tentacles of nautilus differ from those of other cephalopods in that they are housed in sheaths and do not have suction cups, but instead hold on with special adhesive cells. The outer 38 tentacles are used to catch and manipulate prey. Males have an inner circle of 24 tentacles and females 48–52 (not shown). The tentacles can all be pulled into the shell and covered by a leathery *hood* that is formed by two specially folded tentacles. The tentacles have chemosensory cells that allow the nautilus to "sniff out" prey. Once a prey item has been detected, the tentacles extend up to twice the diameter of the shell to make the catch. Like the other cephalopods (Plate 34), the nautilus has strong, beak-shaped crushing jaws. They eat algae, fish, crabs, shrimps, and other invertebrates.

The *eyes* of the nautilus, less well developed than the eyes of other cephalopods, lack a lens and operate somewhat like a pinhole camera. The *funnel*, located below the tentacles, is an evolutionary development of the ancient molluscan foot. It consists of two lobes of tissue whose lower edges curl around each other. The saclike *body* contains the viscera and the gill chamber which opens to the outside via the funnel. The structure protruding from the rear dorsal area is the *siphuncle*, a continuous cord of special secretory tissue which penetrates the chamber walls, or septa, through *siphuncle openings*.

Color the bottom three drawings depicting the ways in which the nautilus moves. In order to show how the siphuncle and the chambers are used in buoyancy control, the third illustration is only of the nautilus shell.

Nautilus species live in the deep water of the tropical Pacific and until recently, relatively little was known about these animals. The Waikiki Aquarium in Honolulu has successfully kept a nautilus in captivity, and maintained its eggs through to hatching. Two hatchlings were reared for over a year, providing the first growth data for a nautilus. By attaching sonic tags to the shells of wild-caught animals, the Waikiki Aquarium found that the nautilus *rests* by day at depths up to 485 meters (1590 ft), possibly to avoid predators which include sharks, eels, and its fellow cephalopod, the octopus. It either rests with its tentacles retracted or by holding on to the bottom. When free of the bottom, the nautilus swims by jet *propulsion*, like all cephalopod molluscs. Water is brought into the mantle cavity through an opening behind each eye. Water is forced out of the mantle cavity through the funnel by contraction of the funnel musculature and retraction of the body into the shell, propelling the nautilus backwards.

At night, the nautilus rises to shallower waters by increasing its *buoyancy*. To achieve this, the siphuncle tissue secretes gas into the shell chambers and resorbs water. This lessens the weight of the nautilus and it floats upwards. Once in the shallow waters, the nautilus swims to reefs or rocky areas to feed. When increasing light signals the approach of day, the gas in the shell chambers is resorbed and replaced by water, and the nautilus sinks back into its deep water retreat.

NAUTILUS

SHELL$_a$
SEPTUM$_{a^1}$
CHAMBER$_b$
BODY CHAMBER$_{b^1}$
SIPHUNCLE OPENING$_c$
HEAD$_*$
EYE$_d$
TENTACLES$_e$
HOOD$_f$
FUNNEL$_g$
BODY$_{b^2}$
SIPHUNCLE$_{c^1}$

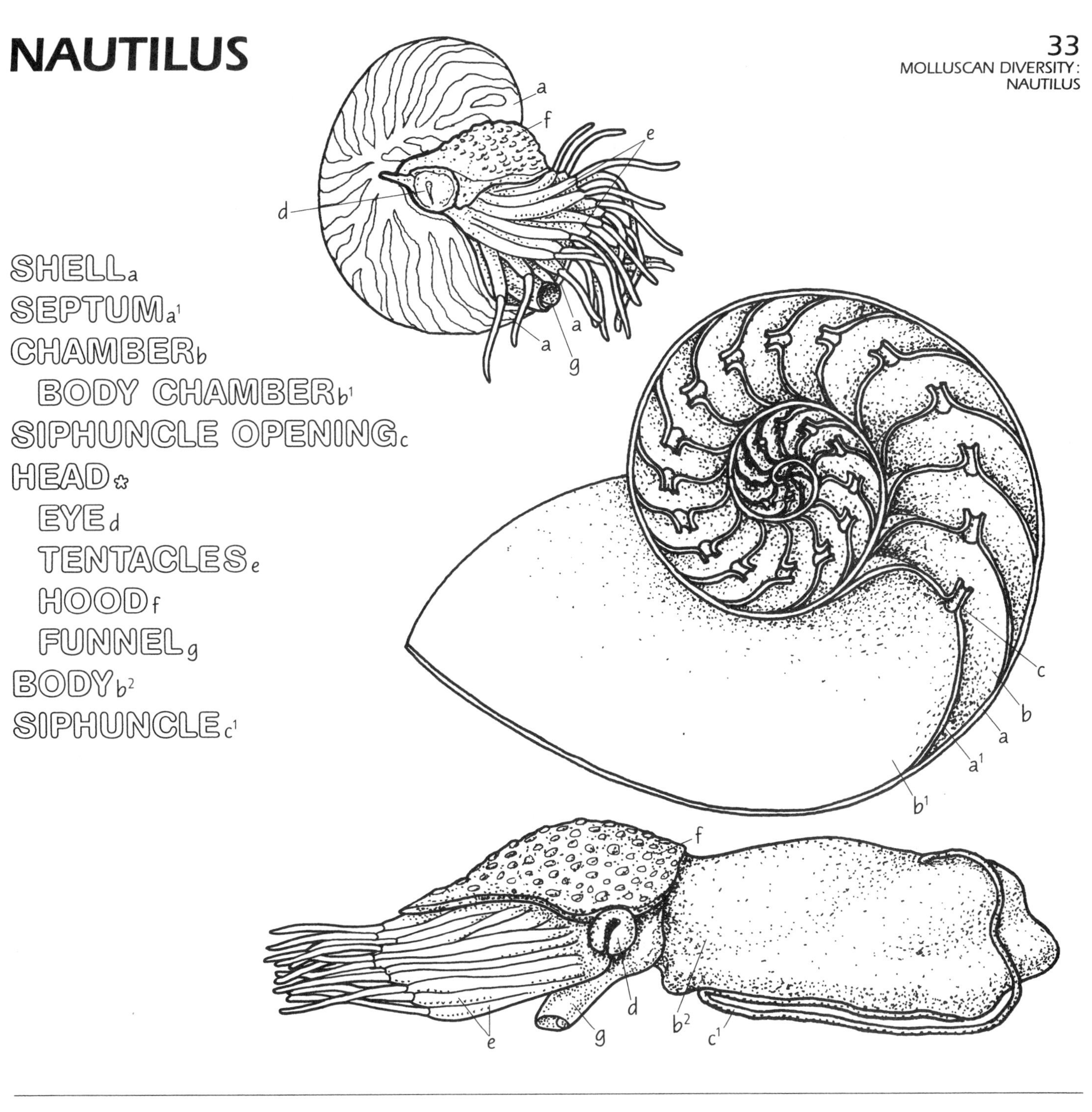

MOVEMENT$_*$
PROPULSION$_{g^1}$
BUOYANCY$_{c^2}$

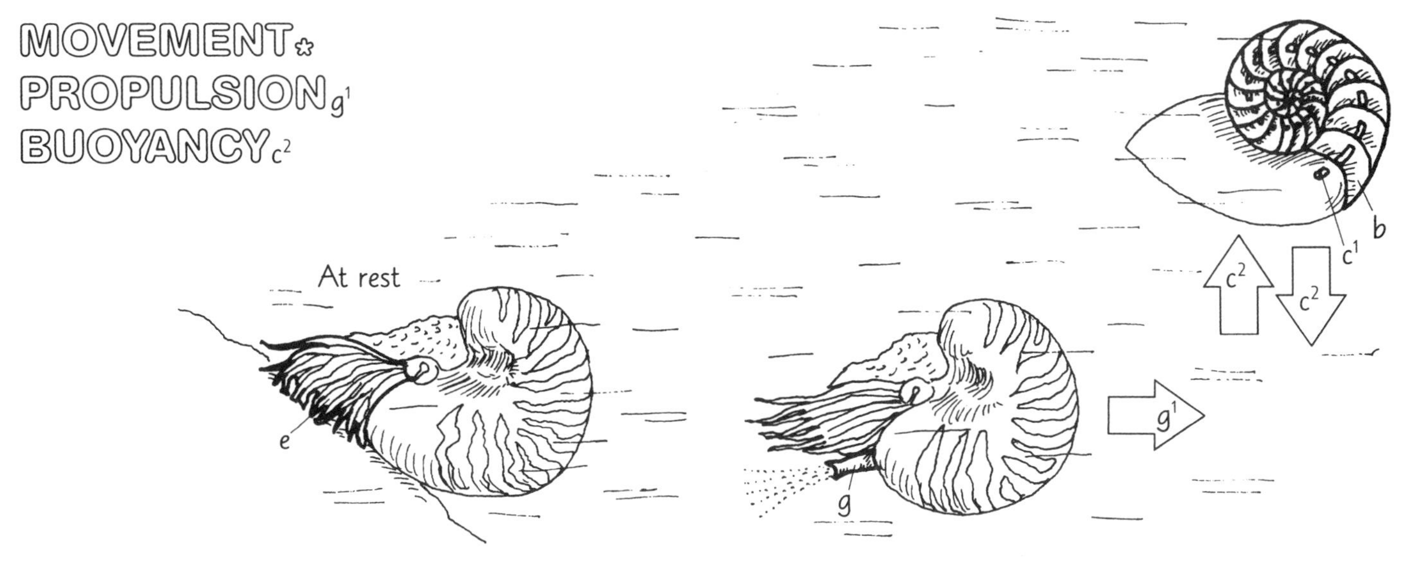

34
MOLLUSCAN DIVERSITY: SQUID AND OCTOPUS

The squid and the octopus are two highly developed members of the class Cephalopoda. While the chambered nautilus relies on the ancestral molluscan shell for protection and buoyancy, the squid has only a thin shell remnant within its mantle, and the octopus has lost the shell entirely.

Begin by coloring the large illustration of the squid.

The muscular *mantle* of the squid and the mantle cavity it houses are strengthened by an outer collagen sheath that maintains the mantle's shape and size. The squid swims by inflating its mantle cavity with water and forcing it out through the *funnel* in a jet-propulsion fashion. Normally the squid swims backwards, and its tapered body and broad, stabilizing *fins* make the squid a highly effective swimmer. Over short distances, squids are among the most rapidly moving of all marine organisms. Large squids can attain speeds of 24–32 kilometers (15–20 miles) per hour.

Squids may also swim forward by directing the moveable funnel posteriorly. When swimming forward, the squid extends its eight *arms* together like the streamlined prow of a ship to allow it to move smoothly through the water. Squids are also capable of hovering motionless in the water or sculling along slowly by undulating their fins. The combination of jet propulsion and fin undulation allows the squids to be highly maneuverable and very graceful in their movements. Many species of squid swim in schools, and their synchronized movement, and tight, rapid turns further attest to their superior swimming prowess.

Color the enlargement of the front view of the squid and the drawing of the beak. Also color the enlargement of the suckers, showing their shape and relative size.

Its swimming ability, coupled with its image-forming *eyes*, gives the squid great advantage as a predator. It can swim into a school of fish and quickly capture one with its long, sucker-tipped *tentacles*. The fish is dispatched with a bite behind the head from the *beak* of the squid. The beak is located in the center of the circle of arms, protruding from the mouth. The squid cuts its prey into small pieces with the beak and then pulls them into its mouth with its radula (not shown).

Located on the arms are stalked, adhesive discs, or *suckers* (see circled enlargement), which, in some species, are reinforced by horny rings or hooks. Contraction of the muscles attached to each sucker creates suction when the suckers come in contact with something solid. The tentacles, twice as long as the arms, have suckers only on their flattened ends.

Color both illustrations of the octopus. Note that the beak is not shown here.

The octopus does not normally swim about in the water. It will swim however, if threatened. It swims with its bag-like mantle held in the direction of movement, and its head and eight arms trailing behind. The funnel is directed rearward and the octopus moves in typical cephalopod fashion, propelled by inflating the mantle cavity with water and forcing it out the funnel in a propulsive jet.

The octopus lacks the streamlining that makes the squid such a successful swimmer; it prefers to remain in contact with a solid structure, pulling and pushing itself along using the suckers on its arms. It moves with a nimble quickness.

In most octopus species there are about 240 suckers on each arm, usually arranged in double rows. The suckers lack the stalk, the horny rims, and the hooks possessed by the squid. Octopus suckers vary in size from a few millimeters to 7 cm (2.75 in) in diameter. This range in sucker size provides the octopus with impressive dexterity, allowing it to manipulate even small objects with precision. A sucker 2 cm (0.75 in) in diameter requires a pull of 170 g (6 oz) to break its hold, so one can imagine the strength it would take to break the hold of two thousand suckers!

The octopus is generally a solitary dweller and seeks shelter or a permanent den in a cave or under rocks (Plate 104).

SQUID AND OCTOPUS

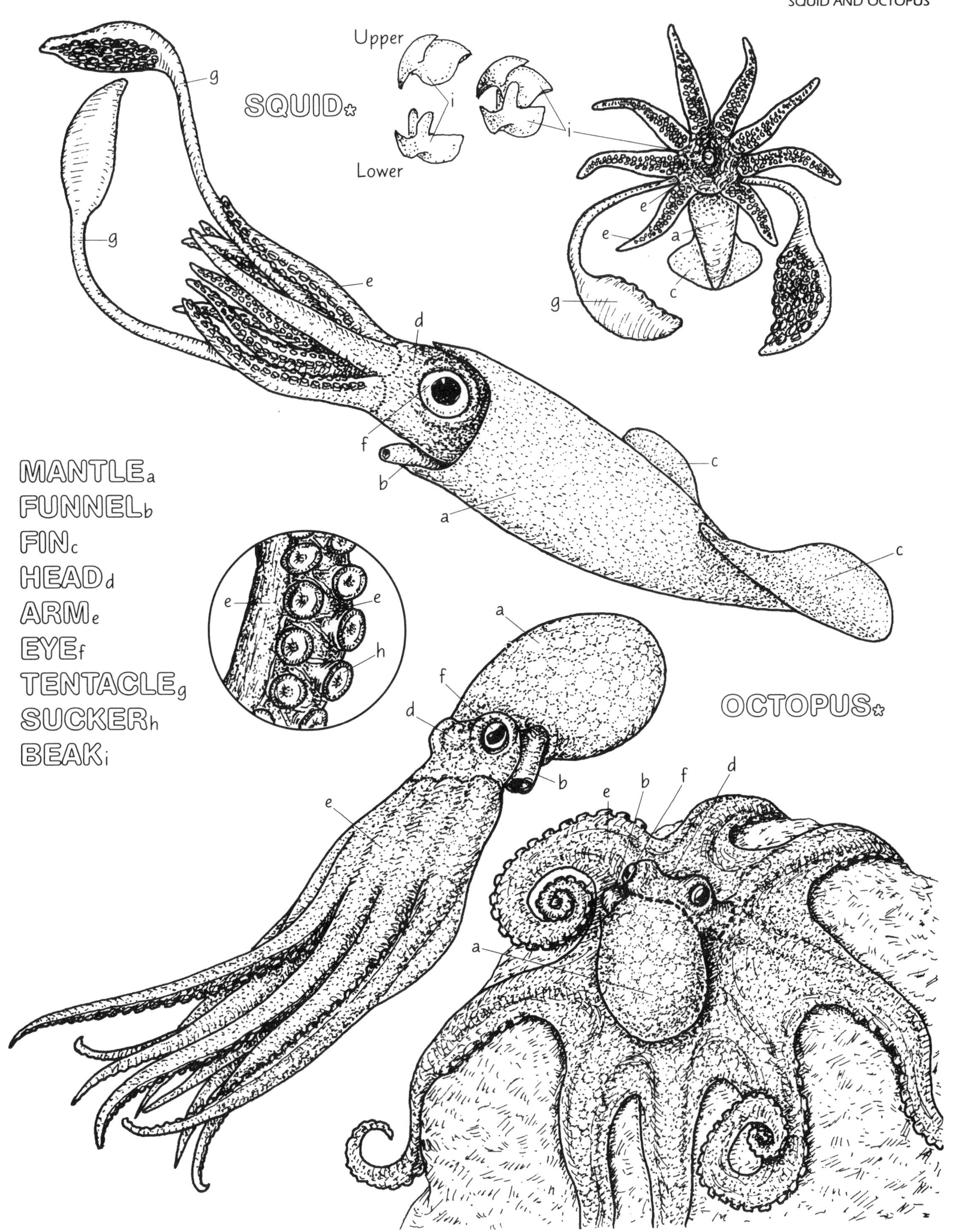

35 CRUSTACEAN DIVERSITY: SMALL CRUSTACEANS

Crustaceans belong to the phylum Arthropoda, which also includes the terrestrial insects (flies, bees, ants, etc.). Arthropoda is the most abundant animal group on earth, both in total number and in number of species. All arthropods share two characteristics: they are encased by an external skeleton, or exoskeleton, and their bodies and appendages are jointed, or segmented. Crustaceans also have a chemical complex of calcium carbonate within their exoskeletons, giving them a "crusty" texture, hence their name. Five very common small crustaceans are illustrated here.

Color the copepod in the upper right. The small arrows indicate the direction of water flow.

The copepod is among the smallest (0.5–10 mm, less than 0.4 in) and most abundant of crustaceans. It is usually a dominant member of marine zooplankton. The most prominent feature of the copepod is its elongated *antennae*, which extend at right angles from the anterior *prosome*. The copepod uses these structures as rudders when it is swimming slowly, and can flex them backward rapidly in a swift escape response. The large size of the antennae also provides a large surface area that assists in flotation. Normal swimming is achieved by the smaller pairs of antennae which are flexed backward in a rhythmic sculling motion that both propels the copepod and generates currents around the filter-feeding *mouth* parts. Another prominent structure is the single, light-sensitive *naupliar eye*.

The segmented copepod body bends only at the junction of the anterior prosome and the posterior *urosome*. The urosome terminates in the *telson* with its two extensions, called the *caudal rami* (singular: ramus).

Now color both the acorn and the stalked barnacles. Color over the fine hairs of the cirripeds on both barnacles with a light color.

Acorn barnacles (with a diameter of 5–50 mm, 0.2–2 in) attach themselves to solid substrata, such as rock, pier pilings, whales, and the bottom of ships. Accumulation of acorn barnacles (and other organisms) on ships is called "fouling." Fouling increases drag on the hull, decreasing speed and fuel efficiency by 20 percent or more.

From within the fortresslike gray *shell plates*, the barnacle extends three to six pairs of segmented, biramous (two branched) *cirripeds* which, depending on the species, is either held stationary like a plankton net, or swept rhythmically, catching whatever the water brings.

Stalked barnacles are characterized by fleshy, flexible, muscular *stalks*, which are capable of lengthening (moving the feeding apparatus out into the water column) and contracting (pulling the barnacle closer to the substratum). Stalked barnacles are found on floating organisms and debris, or attached to hard substrata. They range in size from 5–25 mm (0.2–1 in) in diameter and from 1–100 mm (0.04–4 in) in length.

Color the two diagrams of the isopod on the lower right.

Isopods vary in size from 1–275 mm (0.04–11 in). Illustrated here is the dark gray rock slater, a relatively large (50 mm, 2 in) isopod, related to the common terrestrial pill or sow bugs. Prominent in the dorsal view are the large *compound eyes* and antennae of the *head* as well as the serial body segments. The anterior-most seven segments, called the *pereon*, are shown in the ventral view together with the corresponding appendages called *pereopods* or walking legs, all of which are very similar in shape and give the isopods their name (iso: same, pods: legs). Also seen in the ventral view is the posterior body region called the *pleon* and its appendages known as *pleopods*, used in swimming and respiration. The final segment is the telson, which is flanked by the *uropods*. At the anterior end, as seen in the ventral view, are the mouth parts that are used to feed on scavenged material in the intertidal zone.

Now color the amphipod on the lower left. Note that only one of each paired appendage is shown.

This crustacean is from the gammarid amphipod (amphi: double, pods: legs) group, which includes the familiar sand-colored beach hoppers. Their structure is somewhat similar to the isopod just colored. Note that the front two pairs of pereopods are clawlike; these are called *gnathopods* and are used in feeding and mating. Behind these gnathopods are five pairs of walking legs, two pairs facing backward and three pairs facing forward. Behind the pereon is the pleon with three pairs of pleopods used for swimming. The pleon's most posterior segments constitute the urosome whose appendages (uropods) are used in jumping. First, the entire urosome is tucked under the body, then suddenly straightened, springing the amphipod into the air. This well-developed jumping ability of sandy-beach amphipods has earned them the common names "beach hopper" and "sand flea."

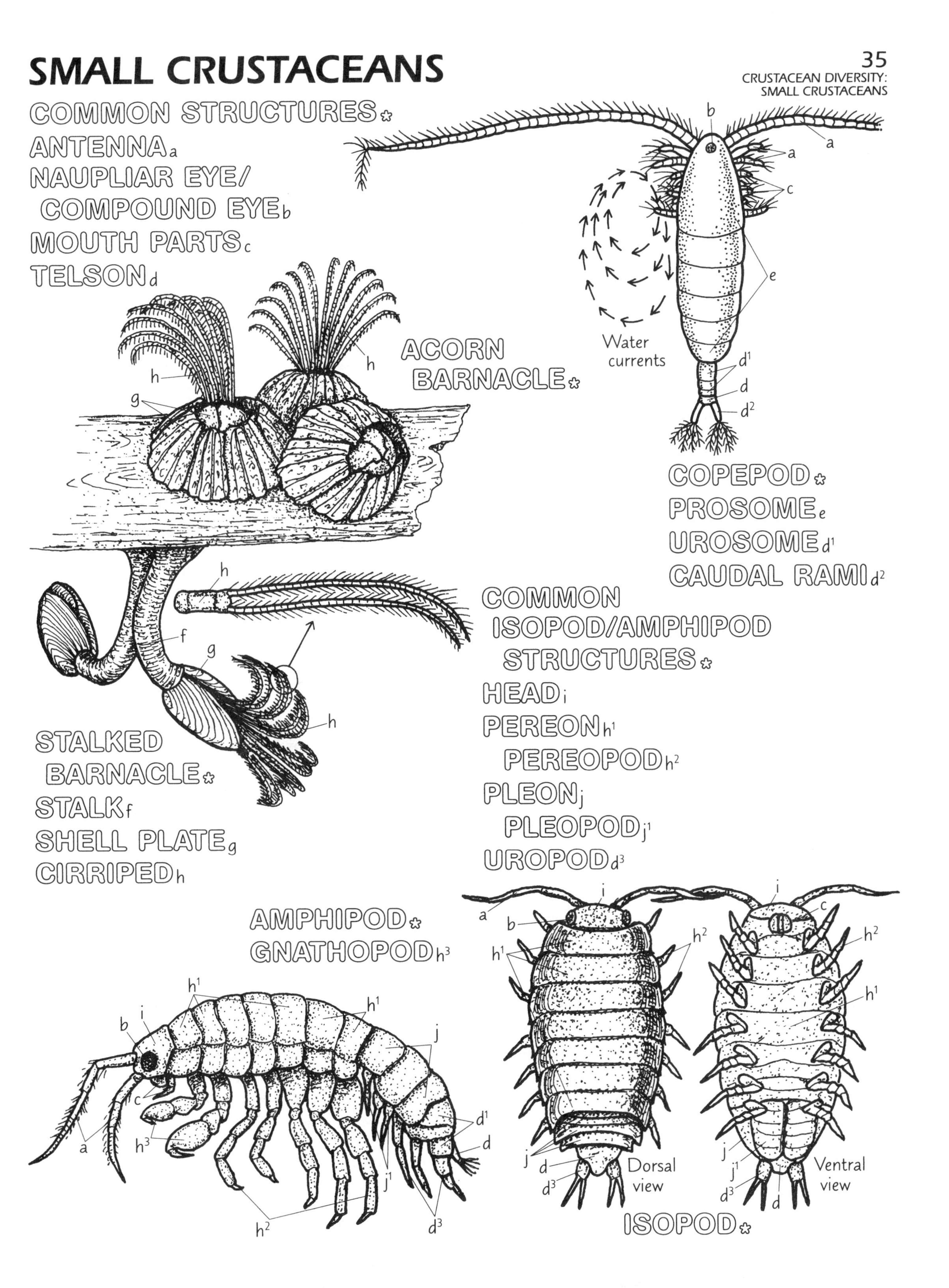

SMALL CRUSTACEANS
35
CRUSTACEAN DIVERSITY: SMALL CRUSTACEANS
COMMON STRUCTURES*
ANTENNA a
NAUPLIAR EYE/ COMPOUND EYE b
MOUTH PARTS c
TELSON d
ACORN BARNACLE*
COPEPOD*
PROSOME e
UROSOME d^1
CAUDAL RAMI d^2
Water currents
STALKED BARNACLE*
STALK f
SHELL PLATE g
CIRRIPED h
COMMON ISOPOD/AMPHIPOD STRUCTURES*
HEAD i
PEREON h^1
PEREOPOD h^2
PLEON j
PLEOPOD j^1
UROPOD d^3
AMPHIPOD*
GNATHOPOD h^3
Dorsal view
Ventral view
ISOPOD*

36
CRUSTACEAN DIVERSITY: DECAPODS

Most of the larger familiar crustaceans belong to the order Decapoda (deca: ten, poda: legs).

Color each decapod crustacean as it is discussed in the text. Begin by coloring the shrimp, for which only one of each paired appendage is shown, except for the uropods.

The long, jointed rear portion of the body of the shrimp is called the *abdomen*. Five of the abdominal segments have biramous appendages known as *swimmerets*, which beat rhythmically and are used in swimming. Located on the last abdominal segment are the *uropods*, a pair of biramous appendages that flank the *telson*. Together the telson and the uropods comprise the tail fan of the shrimp.

The large, smooth saddle-shaped structure at the anterior end of the shrimp is the *carapace*, which covers the head and thorax. Attached to the thorax are eight pairs of appendages. The anterior three pairs of thoracic appendages are called *maxillipeds*. These protrude anteriorly and are used in feeding. The rear five pairs of thoracic appendages are uniramous (one branch) and are the ten *walking legs* that give the decapods their name. In shrimp, the front two or three pairs of walking legs may have claws and are called *chelipeds* (claw feet). The head region has two pairs of biramous *antennae* and large compound *eyes* mounted on moveable stalks to achieve a maximum field of view. Also found on the head, but not visible in the illustration, are a pair of large chewing mandibles, and two pairs of smaller feeding appendages called maxillae.

Now color the lobster and note the difference in the shape of the right and left chelipeds. Only one of each paired appendage is shown here, with the exception of the uropods, chelipeds, antennae, and maxillipeds.

The large red and green American lobster is a bottom dweller, found in the colder waters off the Atlantic coast of North America. The lobster is heavy-bodied compared to the shrimp, with a large abdomen and huge chelipeds. The two chelipeds differ in shape, size, and function. The more massive cheliped (called the "crushing claw") is used to break through hard-shelled invertebrate prey, mainly molluscs. The more slender "cutting claw" is lined with sharp teeth and is used for shearing the prey into small pieces.

The abdomen has only rudimentary swimmerets (used by the female lobster to brood her fertilized eggs). However, a rapid flexure of the abdominal segments, coupled with spreading of the tail fan (uropods and telson), quickly propels the lobster backward. A series of these contractions will rapidly remove the lobster from immediate danger.

Color the hermit crab which is shown here in its shell, and also illustrated in a "see-through" shell. Do not color the shell in either illustration.

Hermit crabs are found in shallow water throughout the world's oceans. They generally inhabit the abandoned shells of gastropods, rarely leaving the security of this "borrowed" home. The abdomen of the hermit is large and its exoskeleton lacks the calcification of most crustaceans; thus the abdomen is soft and vulnerable when not protected by a shell. The abdomen is asymmetrically shaped to fit into the tapering spiral of the interior of a gastropod shell. The uropods are small, enabling them to cling to the interior of the shell at its apex. There are no swimmerets located on the right side of the coiled abdomen, and only females have swimmerets on the left side, to which the fertilized eggs attach. The crab's two posterior pairs of walking legs are small; the hooks on the ends hold the hermit in its shell.

Protruding from the shell are the well-armored chelipeds, two pairs of walking legs, the stalked compound eyes, and two pairs of antennae. Carrying its shell, the hermit travels over the substratum, scavenging for food. If danger approaches, quick contraction of its abdominal muscles pulls the hermit into the safety of the shell.

Now color both drawings of the sand crab. Note the enlarged antennae and the antennal hairs which are used for filter feeding.

The mole crab, or sand crab, is frequently found on wet, sandy beaches of both the Atlantic and Pacific oceans. Sand crabs follow the tide up and down the beach and must frequently reburrow in their wave-churned habitat. They take advantage of the abundance of suspended particles carried by the constantly moving water. Their sandy-gray, egg-shaped carapace is easily pulled into the sand by the rapid digging motions of their uropods and walking legs. The abdomen flexes underneath the carapace, and they burrow, facing into the waves, extending their long antennae into the moving water. The stout antennal hairs trap suspended food particles, which are then removed by wiping the antennae across the mouth.

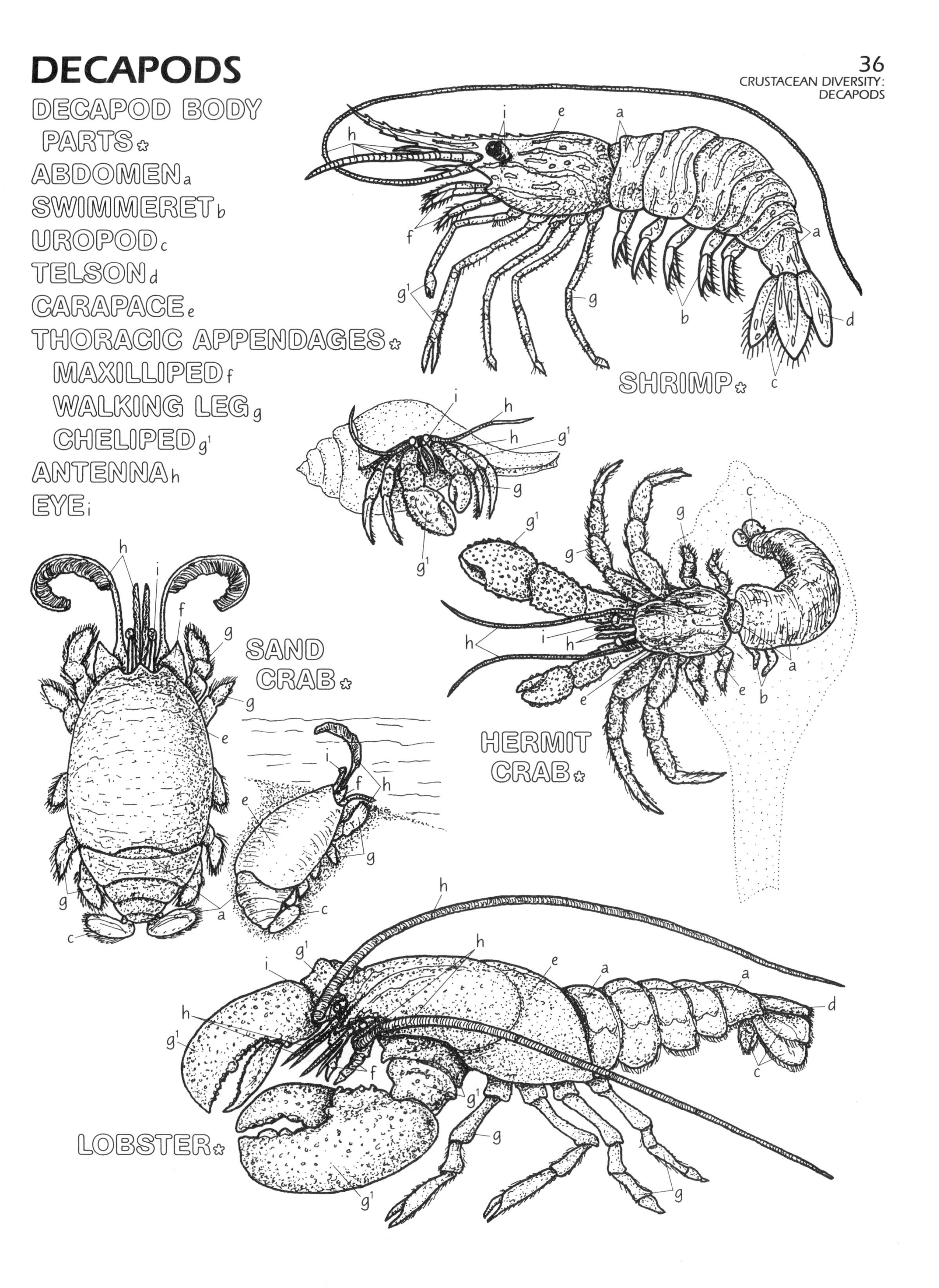
DECAPODS
36
CRUSTACEAN DIVERSITY: DECAPODS
DECAPOD BODY PARTS*
ABDOMEN a
SWIMMERET b
UROPOD c
TELSON d
CARAPACE e
THORACIC APPENDAGES*
MAXILLIPED f
WALKING LEG g
CHELIPED g1
ANTENNA h
EYE i
SHRIMP*
SAND CRAB*
HERMIT CRAB*
LOBSTER*

37 CRUSTACEAN DIVERSITY: CRABS

Color each crab separately as it is discussed in the text, paying particular attention to their differences. The small drawings show the crabs in their various environments, and need not be colored.

The crab body form differs from the decapods seen on the previous plate. The abdomen, so prominent in the lobster and the shrimp, is vulnerable to attack from the rear and hard to defend. In crabs, the abdomen is greatly reduced and flexed underneath the thorax, removing it from harm's way. Flexing the reduced abdomen under the thorax also shifts the crab's center of gravity directly over its walking legs, allowing it much more nimble movement, both forward and backward and sideways. The crab body type has achieved great success and there are approximately 4500 species of true crabs (Brachyura) found in a wide variety of marine habitats.

The cancer crab has four pairs of *walking legs*, a pair of chelipeds (Plate 36) terminating in a moveable finger called the *dactyl*, and a palmlike *manus*. The *abdomen* is visible where it folds behind the *carapace*, and tucks against the underside of the crab. The *antennae*, mouth parts, and the stalked compound *eyes* are also prominent features of the crab.

Cancer crabs inhabit intertidal and near-shore habitats in the colder and temperate waters of the earth's oceans. Some cancer crabs, such as the Pacific coast market crab, live on sandy bottoms and have relatively thin exoskeletons, allowing fast movement across the substratum. However, most (such as the one pictured here) have a very heavy, thick exoskeleton for protection. Although the weight of the exoskeleton slows the crab's movement, the crab's large chelipeds are quite strong and useful in crushing prey such as snails and bivalves, whose capture does not require speed. The claws of this brick-red male crab increase in size disproportionately to the body when sexual maturity is attained.

The shore crab inhabits the rocky intertidal zone, where it is a very agile climber and rapid crawler. The walking legs of the green colored shore crab are relatively short, and the last segment is pointed for clinging tightly to the wave-swept substratum. Shore crabs are among the most conspicuous of the rocky intertidal crabs, as they live higher in the intertidal zone and may be active during daylight hours and low tides in warm and temperate waters worldwide. Many shore crabs are herbivores with chelipeds specially modified for cropping the short algal turf growing on the rocks. The tips of the dactyl and manus form a spoon shape to enable the crab to bring algae and detritus to its mouth.

The blue crab of the Atlantic and Gulf of Mexico belongs to the group of swimming crabs. The two distal leg segments on the last pair of walking legs are flattened into paddles. The rear pair of legs, which are oriented vertically in non-swimming crabs, are pivoted to a near-horizontal position in the blue crab. These legs are flexible and can be rotated over the carapace in a sculling action that gives the crab both lift and propulsion. The blue crab swims sideways. The two sharp spines projecting laterally from the carapace cut through the water like the prow of a ship. Over short distances, blue crabs can swim at speeds up to one meter (3 ft) per second.

As a predator, the blue crab burrows into soft sediment, with only its eyes and antennae exposed, where it waits for prey. The blue crab swiftly attacks with its sharp-toothed chelipeds and can pluck a small fish right out of the water.

The name "box crab" refers to the squarish carapace that completely overhangs the walking legs when this tan colored crab is at rest. The hairy ridges on the box crab's chelipeds help it to burrow in the soft sandy bottom. The chelipeds are held tightly against the body, and the incoming respiratory current is filtered through the hairy ridges. This curious cockscomb ornamentation on the box crab's chelipeds has given rise to another common name for the crab: the rooster crab.

Box crabs specialize in eating gastropods. Most gastropods protect their vulnerable shell opening from predatory crabs by thickening the edge of the aperture. The box crab's chelipeds are asymmetrical; the right cheliped is equipped with a pronounced tooth on the dactyl that fits between two teeth on the opposing manus. This claw is inserted into the aperture of a gastropod, and the thickened edge of the shell is crushed between these teeth. A large box crab will then daintily snip along the gastropod's thinner shell with the scissoring dactyl of the opposite cheliped until the soft body of the prey is exposed.

CRABS

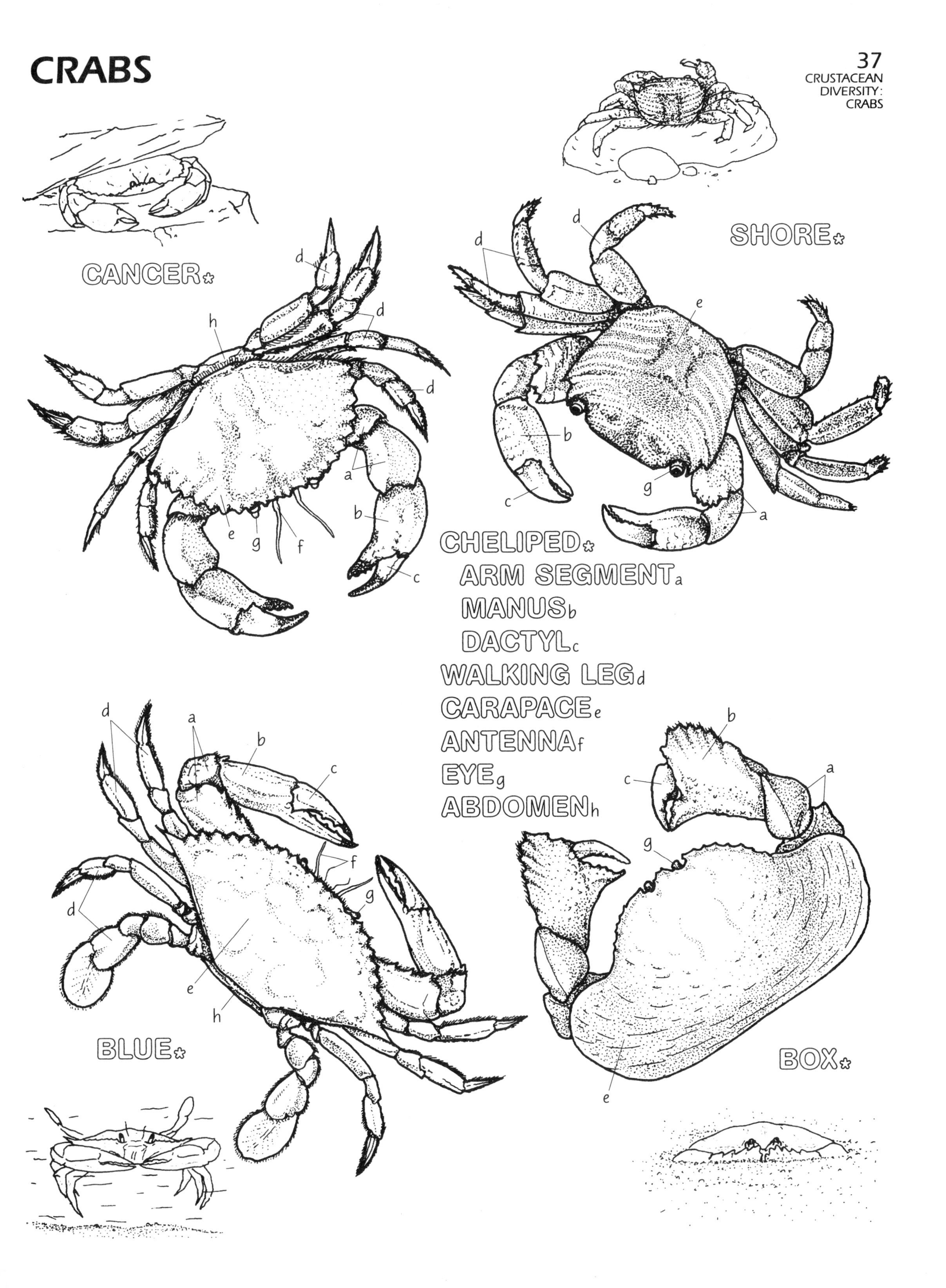
CANCER*
SHORE*
CHELIPED*
ARM SEGMENT a
MANUS b
DACTYL c
WALKING LEG d
CARAPACE e
ANTENNA f
EYE g
ABDOMEN h
BLUE*
BOX*
a
b
c
d
e
f
g
h

38 LOPHOPHORATE ANIMALS

The lophophore is a U-shaped or circular ring of hollow, ciliated tentacles used for suspension feeding. Three marine invertebrate phyla, the Phoronida, Bryozoa, and Brachiopoda, share this simple yet elegant feeding structure and are known collectively as the lophophorate phyla.

Color the phoronid worm. The tube is sandy gray, the body is light beige, and the lophophore is light green. The top view has part of the tentacles cut away to reveal the central trough.

On some sand flats in central California, patches of substrate in the low intertidal zone seem uncharacteristically firm under foot. Careful excavation will reveal the closely packed, flexible, chitinous tubes of phoronid worms. The phoronid bed may consist of many thousands of these slender, 7 to 10 cm (2.75–4 in) long worms. When the tide covers them, a green, double-spiraled *lophophore* protrudes from the opening of their vertically oriented tubes. Looked at from above, the lophophore reveals two ridges of ciliated *tentacles* separated by a central *trough*. The beating of lateral cilia on the tentacles creates a *feeding current*. Cilia on the face of the tentacles trap small food particles and carry them to ciliary tracks on the trough floor and, thus, to the centrally located mouth (not shown). The *anus* and the *kidneys* lie just outside the lophophore and make use of the *exiting current*. There are only twelve species of phoronids known. With the exception of one species that bores into rocks, all live in chitinous tubes oriented vertically in soft substrata or attached to rocks.

Color the brachiopod. Use light colors for the valves and color over the internal structures except the bright red lophophore.

Brachiopods are commonly known as lamp shells because they resemble the ancient Roman oil lamp in profile. There are many more species in the fossil record than currently living species. Because brachiopods live in a bivalved shell, they were classified as molluscs until the nineteenth century. The shell of a molluscan bivalve, such as a clam, consists of two side (lateral) valves (Plate 29). The 2 to 10 cm (0.75–4 in) brachiopod shell, however, consists of a *dorsal* (top) and a *ventral* (bottom) *valve*, clearly a different orientation. A proteinaceous stalk called a *pedicel* protrudes from the larger ventral valve and attaches the brachiopod to hard surfaces. When the shell gapes open a large horseshoe-shaped lophophore is quite prominent. In many species the arms of the lophophore end in tight spirals of tentacles. The direction of the feeding current and mode of food capture in the brachiopod are the same as described for the phoronid.

Color the bryozoans. The lophophore is transparent to pale tan and the zooecia are usually light brown or gray. The frontal membranes are usually dark, brown in some species, bright red in others. The operculum color varies. The front wall of the zooecium is absent to expose the internal anatomy.

Although they share the same feeding mode, bryozoans are very different in size and organization from the other two lophophorate phyla. The individual bryozoan lophophore has only a single row of ciliated tentacles arranged in a horseshoe or circle. The lophophore is just a few millimeters or less in diameter and protrudes from a calcium carbonate enclosure called a *zooecium*. In many bryozoans, the top of the zooecium has a flexible *frontal membrane* with *protractor muscles* attached to it and anchored on the floor and sides of the zooecium. When these protractor muscles contract, the frontal membrane is pulled downward, reducing the volume and increasing the pressure in the zooecium, forcing the lophophore to protrude. The opening of the zooecium in many bryozoans is covered by a hinged *operculum* that is pushed open with the protrusion of the lophophore. *Lophophore retractor muscles* attached to the lophophore's base pull the lophophore back in, and *operculum closer muscles* close the operculum behind it. Because the soft frontal membrane leaves the bryozoan vulnerable to attack by predators, an evolutionary trend to calcify the top of the zooecium has occurred over time, resulting in the emergence of more elaborate methods of lophophore protrusion.

Bryozoans are colonial animals. The small, boxlike zooecia are connected side by side, forming a basket-weave pattern characteristic of the Bryozoa. Colonies take on a variety of shapes. Many are simple sheetlike colonies that cover hard substrata such as rocks, pier floats, mollusc shells, and crab exoskeletons. Other sheetlike forms grow on the wide blades of kelp plants; here the connections between the individual zooecia are flexible to allow the colony to conform and bend with the plant. Other colonies grow in erect branching patterns which are often mistaken for hydroids or small red algae. However, close inspection of these suspect colonies will reveal the individual zooecia with their protruding lophophores.

LOPHOPHORATE ANIMALS

PHORONID ✿
- TUBE$_a$
- BODY$_b$
- LOPHOPHORE$_c$
  - TENTACLES$_{c^1}$
  - TROUGH$_d$
  - FEEDING CURRENT$_e$
- ANUS$_f$
- KIDNEY$_g$
- EXITING CURRENT$_h$

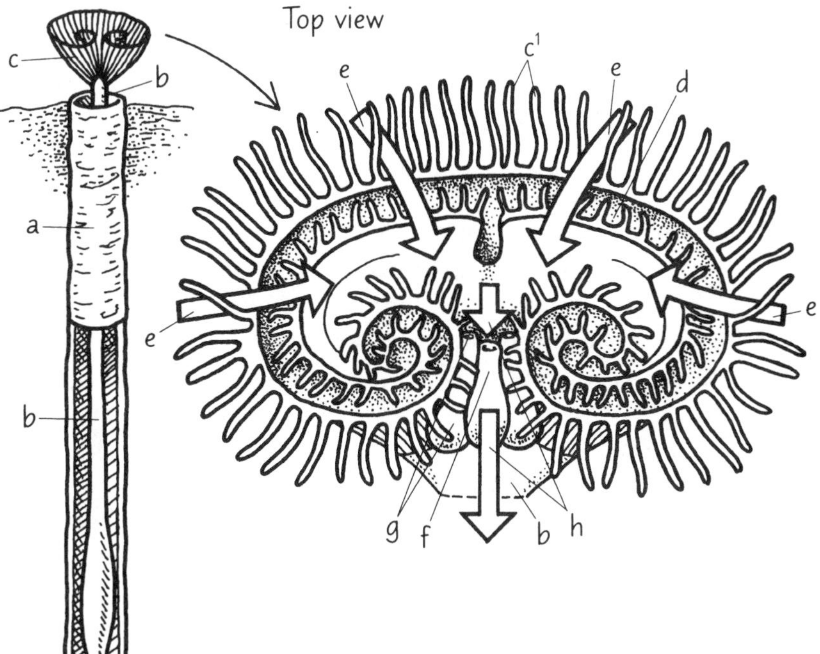

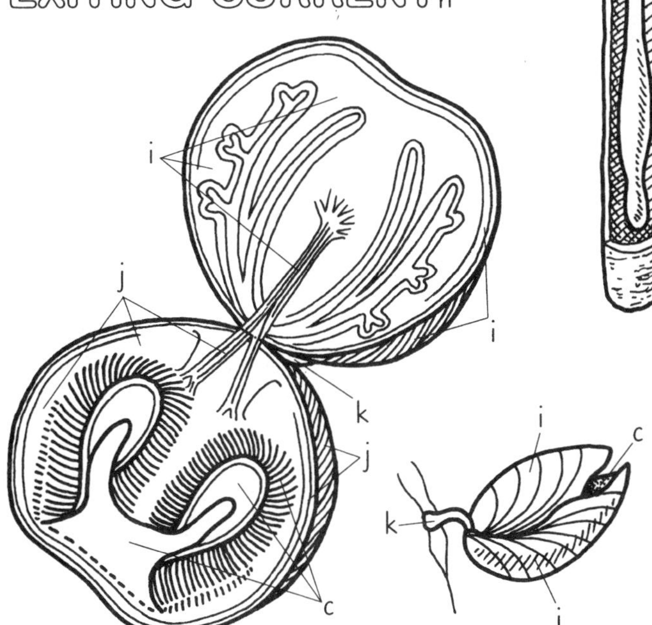

BRACHIOPOD ✿
- LOPHOPHORE$_c$
- SHELL DORSAL VALVE$_i$
- SHELL VENTRAL VALVE$_j$
- SHELL PEDICLE$_k$

BRYOZOAN ✿
- LOPHOPHORE$_c$
- ZOOECIUM$_l$
- FRONTAL MEMBRANE$_m$
- PROTRACTOR MUSCLES$_n$
- RETRACTOR MUSCLES$_o$
- OPERCULUM$_p$
  - CLOSER MUSCLES$_{p^1}$

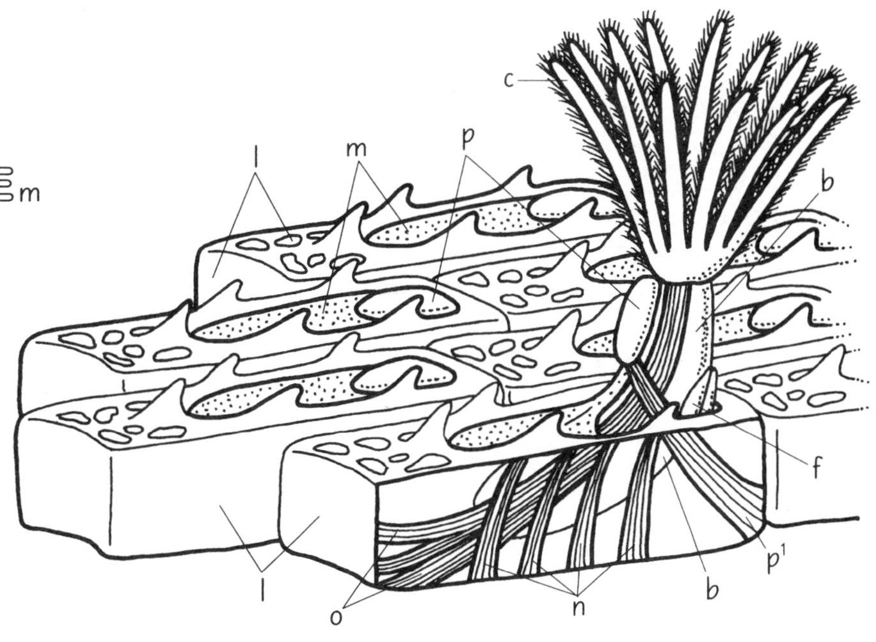

39
ECHINODERM DIVERSITY: SEA STAR STRUCTURE

The five-armed sea star is a widely recognized marine animal, belonging to the echinoderms ("spiny-skins," phylum Echinodermata).

Color the two views of the sea star. Note that the upper (aboral) surface is divided into a central disc and five rays. In life the spines are white, surrounded by a circle of blue, against a background that varies from tan to burnt orange to root beer brown.

The body plan of echinoderms shows a pattern of radial symmetry. That is, the basic and more or less equal units are arranged in a circle around a central disc, rather than astride a single midline, as in bilaterally symmetrical animals (like us). Many sea stars and other echinoderms have five of these similar units, giving them pentamerous (penta: five, mer: part) radial symmetry.

The *mouth* of a sea star is located on its oral surface, and the opposite side is called the aboral surface. The aboral side of the sea star has a highly textured, bumpy surface, sometimes with spines. These bumps and spines are parts of the endoskeleton, which is covered by a layer of epidermis. The endoskeleton consists of individual *ossicles,* or skeletal plates, which fit together closely in some species, more loosely in others. This tightness or looseness of fit determines the degree to which an echinoderm's body is stiff and tough or, on the other hand, soft and flexible.

The oral surface of the animal has furrows running from the mouth along the length of each *ray*. These furrows are called *ambulacral grooves* and are bordered by two or more rows of *tube feet*. The spiny endoskeleton is also visible on the sea star's oral surface.

Now color the labeled parts of the sectional cutaway illustration of the sea star. Note how the canals are connected in series from the madreporite to the ampullae of the tube feet. Also color the enlargements of the tube foot mechanism. Notice that the space within the rays is left open to highlight the water vascular system. In life, this area contains layers of muscle and elements of the reproductive and digestive systems.

The water vascular system of the sea star runs from the *madreporite* through the *stone canal* to the *ring canal*, from there through the five *radial canals*, and finally to *lateral canals* which lead to the individual tube feet. This unique system operates through a combination of muscles and hydraulic pressure (sea water acts as the hydraulic fluid).

Each tube foot consists of a hollow, muscular structure attached to a balloonlike fluid reservoir called an *ampulla*. The elastic ampullar surface is covered with a meshwork of muscle fibers. When these muscles contract, the ampulla is deflated and fluid is forced into the tube foot. This stretches the tube foot musculature and extends the tube foot outward beyond the ambulacral groove. Contraction of the tube foot muscles forces the fluid back into the ampulla and stretches the ampullar musculature. Many sea stars have muscle fibers in the bottom of the foot. When the bottom of the tube foot is pressed against a solid surface, contraction of these muscle fibers creates a vacuum, allowing the tube foot to operate as a suction cup. In addition to suction, the tube foot can stick to surfaces using unique cells called duocells. Duocells are so named because they secrete both a sticky adhesive chemical and a second chemical that breaks the adhesion.

The water vascular system, involving hundreds of tube feet, is the basis for the locomotor ability of the sea star. Each tube foot is under fine nervous control: not only can it be extended, but it can also be moved through a 360° arc by local contraction and relaxation of the musculature, playing against the hydraulic pressure from the ampulla. The sea star is capable of mostly quite slow, but very precise and well-coordinated movements on nearly any solid horizontal or vertical substratum. This system would not be effective, however, without some rigid structural support for the muscles. This is provided by the interconnecting ossicles of the endoskeleton.

The drawing at the top right shows a sea star prying open a mussel. Color this sea star, but not the mussel.

Using its tube feet, the sea star can capture live prey and manipulate it into its mouth. It can weaken the strong adductor muscles of a bivalve mollusc (like a clam or mussel) by exerting a strong and continuous pull with its suckered tube feet. The sea star need only force open the bivalve shells a few millimeters or so; it then extrudes its stomach into the opening and digests the bivalve externally.

Sea stars are usually carnivorous predators, and live on all types of ocean bottoms. Most sea stars have five rays and range from 10 to 25 cm (4–10 in) in diameter. Some species may be much larger and have more than five rays: the sunflower star of the Pacific coast of the United States has twenty-six or more rays and often reaches one meter (3 ft) in diameter (Plate 11).

SEA STAR STRUCTURE

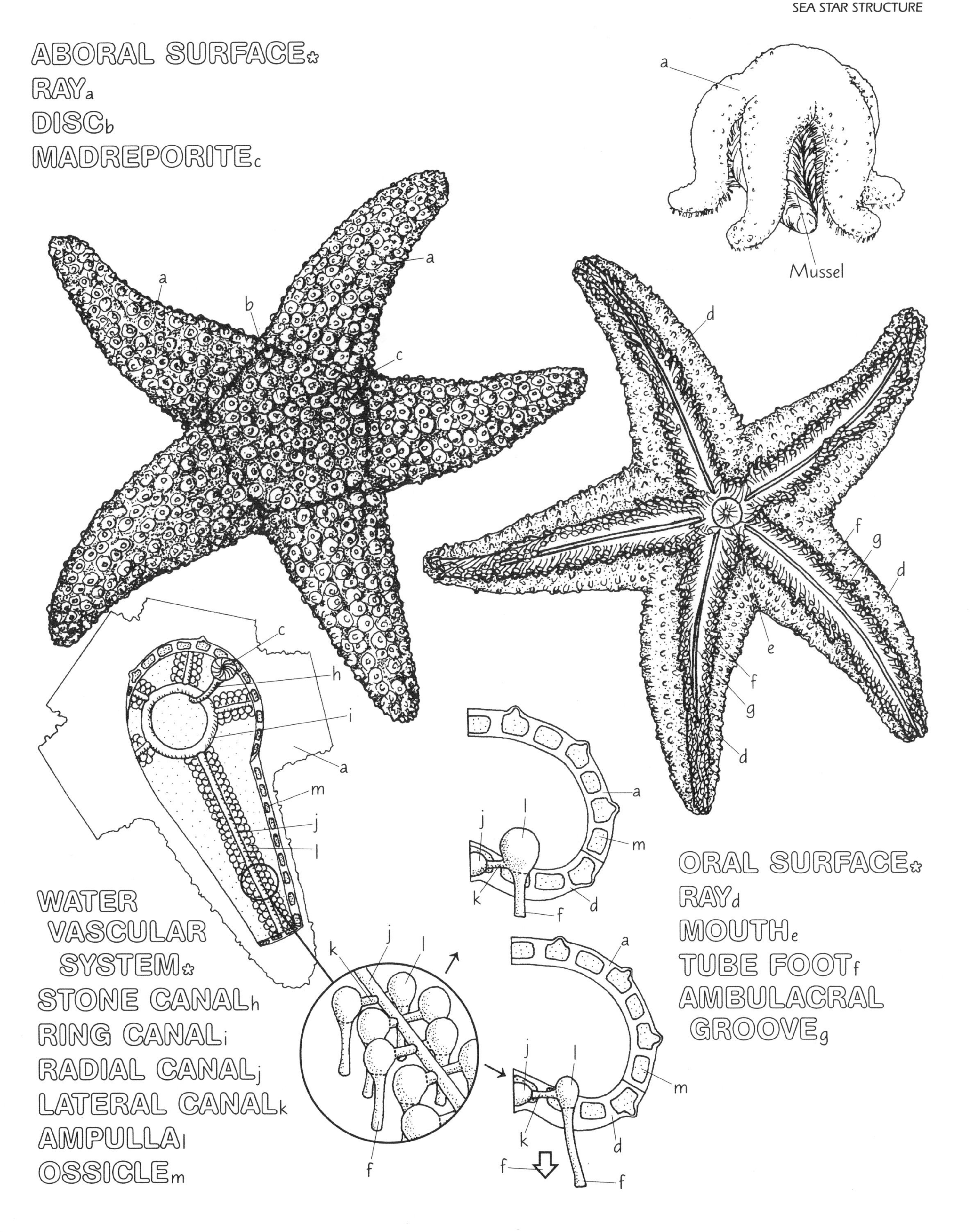

40
ECHINODERM DIVERSITY: OPHIUROIDS AND CRINOIDS

The ophiuroids and crinoids are two other classes, which, along with the sea star, are members of the phylum Echinodermata.

Begin by coloring the small sea star on the right side and the middle cutaway side view showing the location of its aboral surface, oral surface, and mouth. Color the cutaway illustrations of the ophiuroid (above) and the crinoid (below) showing the location of the structures in these two organisms.

Finally, color the illustration of the ophiuroid, the enlargement of the arm section, and the sequence at the upper right, showing ophiuroid movement and autotomy.

Like the sea star, the ophiuroid, or brittle star (also called serpent star) has a five-rayed pattern. The *mouth* is located on the *oral surface*. The opposite side is the *aboral surface*. Unlike the sea star, whose rays and disc merge together, the rays, or *arms*, of the brittle star are quite distinct from the disc. These arms are thin and usually spiny, each consisting of a row of large articulating structures, much like a vertebral column. These structures are skeletal *ossicles* which permit the arm an undulating, snakelike *movement*.

The muscles that link the articulating ossicles are capable of violent contraction, a defense mechanism the brittle star will use if it is captured or trapped. The contraction may cause the trapped arm to sever, in a phenomenon known as *autotomy* (self-cut), and allows the ophiuroid to escape, leaving part of the arm behind. The common name "brittle star" is derived from this ability to cast off an arm if handled. As with sea stars, severed arms are later regenerated.

Many ophiuroid species are omnivorous scavengers. They use their tube feet to catch suspended food material in the water, or to pick food off the bottom. The arms are used in locomotion: two arms are held forward, two out to the side, and one to the back. The *disc* is held up off the substratum and the laterally positioned arms move in a rowing motion that propels the animal along in short leaps. The *spines* on the arms help to keep the animal from slipping on the substratum. Of all the echinoderms, ophiuroids move the most rapidly over the bottom.

Ophiuroids comprise over 2000 species. Some smaller species that can withstand its rigors inhabit the intertidal shoreline among the rocks, but most species live in subtidal habitats, from shallow water to the deepest ocean bottom, or in coral reefs. Most ophiuroids are small, with a disc diameter of 1 to 3 cm (0.4–1.2 in) and arms from 5 to 6 cm (2–2.4 in) in length; a few species have long, tapering arms that reach a length of 15 to 25 cm (6–10 in).

Color the crinoid and the enlargement of one of its arms which shows the ambulacral groove and the tube feet. Crinoids come in a variety of colors, from drab tans to bright scarlets and blues.

The crinoids are the only group of echinoderms that have a mouth facing upward. Considered by many to be the oldest living class, crinoids include 5000 identified fossil forms but only 620 living species. Of these, some 80 species are called sea lilies, and remain permanently attached to a substratum by a long stalk. The sea lilies (not shown) live at depths of 100 meters (330 ft) or more.

The second group of crinoids are called feather stars. These animals can move about freely, and occur from shallow tropical waters into the deep sea. Feather stars can move along the bottom using clawlike *cirri* that sprout from the cuplike *calyx*. Also coming from the calyx are the long slender arms, from a minimum of five to as many as two hundred in some species. From each arm, rows of *pinnules* extend laterally, giving the appearance of a feather and the origin of these crinoids' common name, feather stars. An *ambulacral groove* extends along the length of each arm, and laterally onto the pinnules. The ambulacral grooves are flanked by tentaclelike *tube feet* with slender, mucous-secreting papillae along their length. Feather stars are nocturnal filter feeders, coming out from under hiding places at night to position themselves in a favorable current. Once in place, they position their arms upward into the moving water to passively filter feed by trapping plankton and other suspended material on the tube feet. Food is transferred to the ciliated ambulacral grooves, which then carry it bound in mucus to the mouth. When bothered, some species of feather star can swim by alternately beating up and down with groups of their long slender arms.

OPHIUROIDS AND CRINOIDS

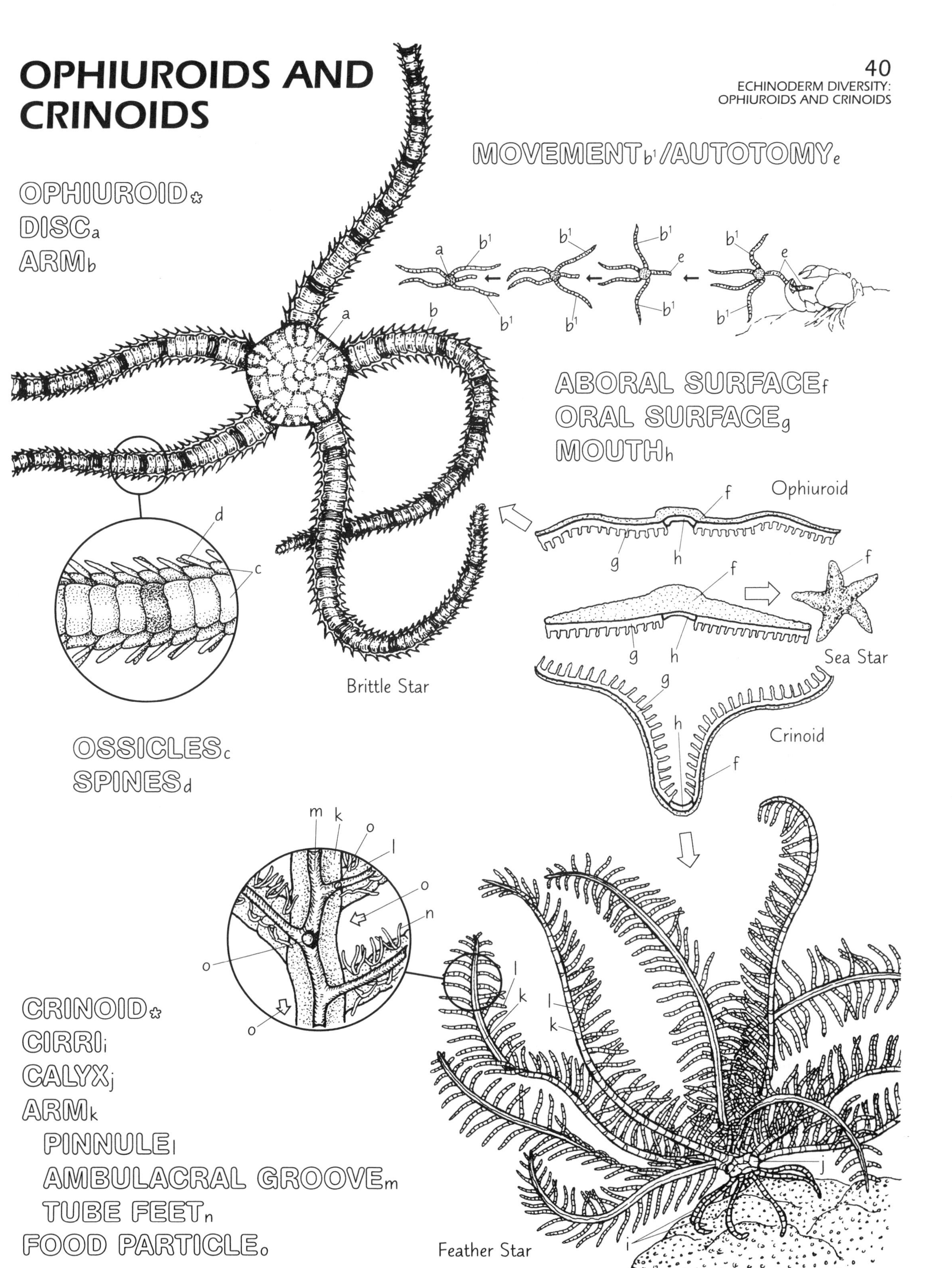

41 ECHINODERM DIVERSITY: ECHINOIDS AND SEA CUCUMBERS

Unlike other echinoderm classes, echinoids and sea cucumbers (holothuroids) lack arms and have bodies that are stretched along the oral/aboral axis.

Begin by coloring the sea urchin skeleton. Note that the alternating ambulacral and interambulacral sections of the skeleton are arranged like sections of an orange. Use light colors for these sections. They are purple in life. Then, color the enlargements of the individual ossicle and the tubercles, including the spines and other appendages.

Color the three drawings of the sand dollar. Note how flat the sand dollar is in the side view, and how the petals are positioned on the aboral surface.

The echinoids include the sea urchins and sand dollars. The individual *ossicles* (skeletal plates) of the echinoid are tightly sutured or joined, producing a rigid skeleton or "test."

The sea urchin is round or oval in shape, and the individual ossicles are organized in ten longitudinal rows, running from the oral to the aboral pole. Five alternating rows are called *ambulacral* plates, and the other five alternating rows are the *interambulacral* plates. The urchin's stalked *tube feet* are located on the ambulacral areas; its long, moveable *spines* and defensive *pedicellariae* are attached to both ambulacral and interambulacral areas.

Sea urchins live primarily on firm substrata. Their *mouths* and oral surfaces are kept toward the substratum; the *jaws* protruding from the mouth scrape and chew off algae and attached organisms. The jaws are mounted and articulated in an elaborate feeding structure known as Aristotle's lantern. Sea urchins can consume a considerable biomass, and where they occur in large numbers they have a devastating impact on the algae (Plates 107, 113).

Sea urchins move in two ways: by means of their long, stalked tube feet, with powerful suckers at the tips, and by means of their spines. Each tube foot emerges through a pair of holes *(pore pairs)* in the ossicle, as shown in the upper illustration, and moves multidirectionally. The spines come in two or more sizes and pivot freely on *tubercles*, or "bosses," on the ossicle. The urchin coordinates the tube feet and spines and walks and pulls itself along. Some urchins with very long spines walk quite briskly, using only the spines for locomotion.

The jaw-bearing pedicellariae protrude from the ossicles and are used for defense and cleaning. The *ampullae* and the other structures of the water vascular system are beneath the ossicles.

Sand dollars, purple gray in life, move solely on their small, short spines, which are visible as hairlike structures on the *oral surface* surrounding the mouth. The small tube feet (not shown) are found among the spines and assist in gathering food. These animals live on or in soft sediments and are primarily deposit feeders, although a few are filter feeders (Plate 105). The *petals* on their *aboral surface* relate to the five ambulacral areas of the sea urchin. These petals contain special respiratory tube feet.

Color the illustrations of the two types of sea cucumbers as well as the enlargement of the ossicle. Note the sea cucumber on the right lacks ambulacral/interambulacral areas.

Most sea cucumbers have thick, leathery bodies with unattached tiny ossicles scattered within. This gives the sea cucumber a wormlike flexibility with a potential for burrowing (in fact, many species are burrowers). The elongated body of the sea cucumber creates a "head" area, and "tail," and dorsal and ventral surfaces. Sea cucumbers have a prominent anus, and breathe through special respiratory "trees" that branch internally from the anus. Respiratory water is both brought in and expelled through the anus by muscular contraction.

The sea cucumber illustrated on the right is a surface-dwelling deposit feeder. It uses its *oral tube feet* to pick up food from the substratum. The tube feet, located on the well-developed "sole," are used for movement along the substratum. This light orange cucumber can grow up to 40 cm (16 in) in length.

The sea cucumber on the left is a filter feeder and lives among rocks. This red orange cucumber has five rows of tube feet for movement. The oral tube feet are highly branched and coated with mucus, which acts to trap suspended plankton and detritus when the tube feet are spread open. To feed, the cucumber moves each foot into its mouth in a rhythmic, systematic fashion, and removes the trapped food particles. Sea cucumbers are found on all types of substrata, from the intertidal zone to the deepest ocean depths (Plate 16).

ECHINOIDS AND SEA CUCUMBERS

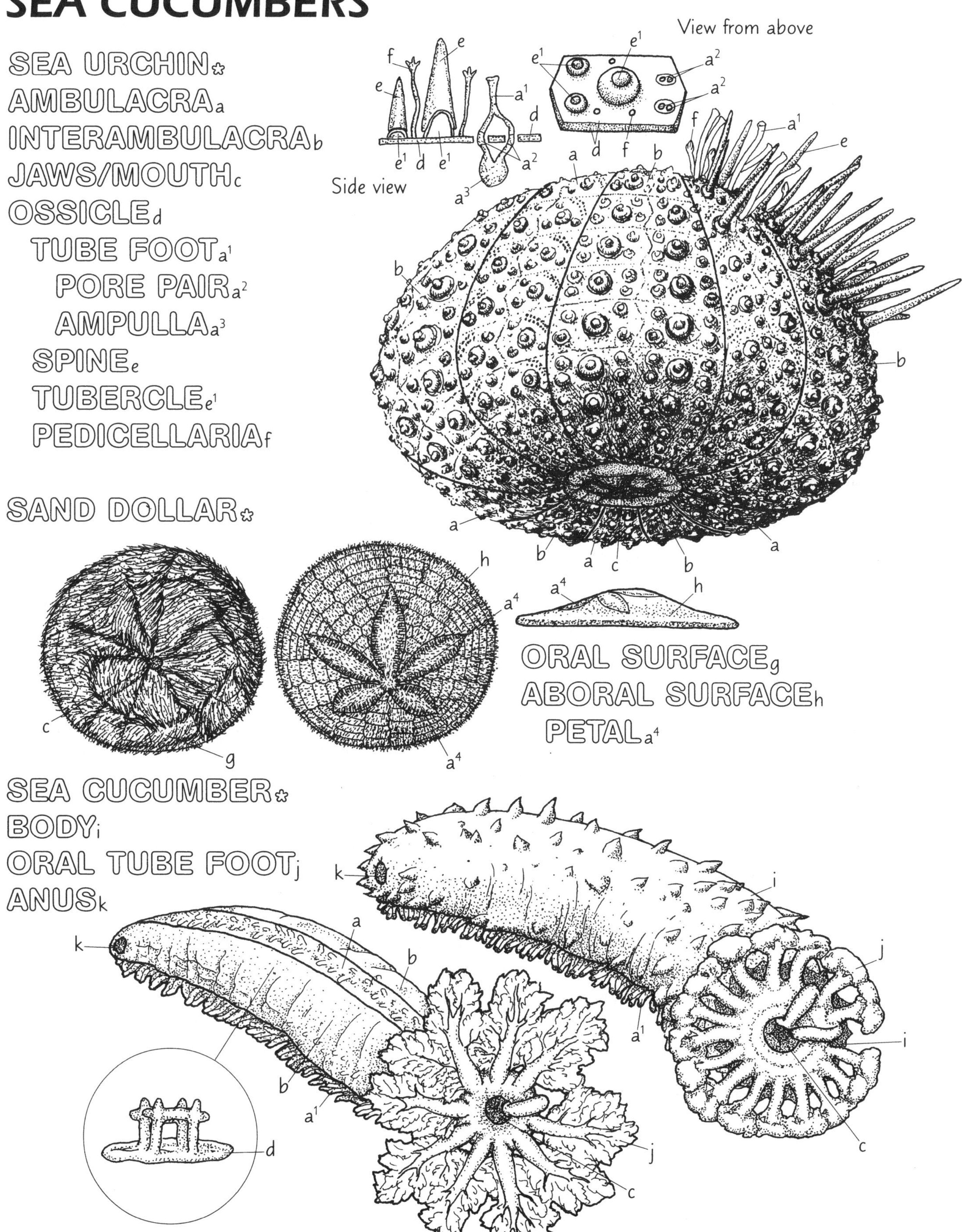

42
MARINE PROTOCHORDATES: SEA SQUIRT AND LARVACEAN

The protochordates (beginning chordates) are a group of marine animals with several characteristics that link them to the higher chordate animals (including fish and mammals). The most successful protochordates are the tunicates (urochordates), comprising 1300 species of sessile and planktonic forms. Within the urochordates, the largest number of animals belong to the group known as sea squirts.

Begin by coloring the solitary sea squirt attached to the pier piling. Use a light color for the tunic. Next, color the long-sectional side view and the cross-sectional view of the solitary sea squirt. The directional arrows indicate water flow, and the dotted arrows indicate movement of food particles.

Sea squirt species occur both as solitary individuals and in closely packed colonies. The first sea squirt illustrated here is a medium-size (7.5 cm, 3 in) solitary tunicate. The name "tunicate" is derived from the outer covering or *tunic* of the animal, which is thin and transparent in some species and quite thick and leathery in others. The tunic contains a tough structural hydrocarbon compound very similar to cellulose, called tunicin. Tunicin is secreted by the ectodermal tissue of the *body wall.*

Within the body wall are circular and longitudinal muscles, which, when contracted, cause water to be squeezed out the siphon in a thin stream (hence the name "sea squirt").

The sea squirt is a filter feeder with a simple anatomical organization. Water enters the large *buccal* (oral) *siphon* and flows into the *pharynx*. The pharynx is perforated by numerous ciliated *gill slits*. The cilia beat rhythmically and create a water current. Water flows through the gill slits into the *atrium*, then exits out the *atrial siphon*. The *endostyle* organ secretes a continuous supply of mucus, which is carried laterally by the ciliated surface of the pharynx to form a *mucous sheet*. Food particles are trapped in the mucous sheet, and the sheet is taken up by the *dorsal lamina* and shunted to the *stomach*, which opens at the bottom of the pharynx. The *anus* and *gonads* both empty into the atrium; their products are removed by the excurrent water flow out the atrial siphon.

Color the sea grapes at the bottom of the piling.

Some species of sea squirts bud asexually, at the base, to form groupings of individuals called social sea squirts. Illustrated here are the medium-size (4–5 cm, 1.5–2 in) light green sea grapes, commonly found on pier pilings and harbor floats on both the Atlantic and Pacific coasts of North America.

Color the compound sea squirts attached to the middle of the piling. Then, color the enlarged cross section of the compound sea squirt. The directional arrows indicate water flow.

A more complex internal structure is seen in the compound tunicates. These tunicates grow in asexual colonies that form thin (6-mm, 0.25-in thick) sheets over the surface of the rocks, boat bottoms, and almost any solid surface in relatively clean, quiet waters. Concentric groupings of seven or more individual white tunicates produce the flower pattern seen on these bright orange or purple tissue sheets. Each individual has its own buccal siphon and pharynx, but shares a common atrium and atrial siphon with its mates.

Now color the larvacean urochordate. The illustration shows this tiny (2–3 mm, less than 0.1 in) animal in its complex mucous house, which is actually transparent.

The larvacean tunicates retain the characteristics of a larva throughout adult life. A larvacean tunicate secretes a *house* of stiff mucus, completely surrounding its *body*; the transparent house may be simple or complex, depending on the species.

The larvacean filter feeds, creating a water current by beating its muscular *tail*. Water enters through a coarse-meshed *incurrent filter* that traps large suspended particles. The current then flows through an internal *fine mesh filter* and out the rear *exit valve*. When the incurrent filter becomes clogged with debris, the larvacean exits out an *escape hatch*. Once free, the animal secretes a new house; this may happen as often as every few hours.

SEA SQUIRT AND LARVACEAN

SEA SQUIRT*

TUNIC a
BODY WALL b
BUCCAL SIPHON c
ATRIUM d
ATRIAL SIPHON d¹
PHARYNX e
GILL SLIT d²
ENDOSTYLE f
MUCOUS SHEET f¹
DORSAL LAMINA g
STOMACH h
ANUS h¹
GONAD i
GONAD DUCT i¹

LARVACEAN*

BODY j
TAIL k
HOUSE l
INCURRENT FILTER m
FINE MESH FILTER n
EXIT VALVE o
ESCAPE HATCH p

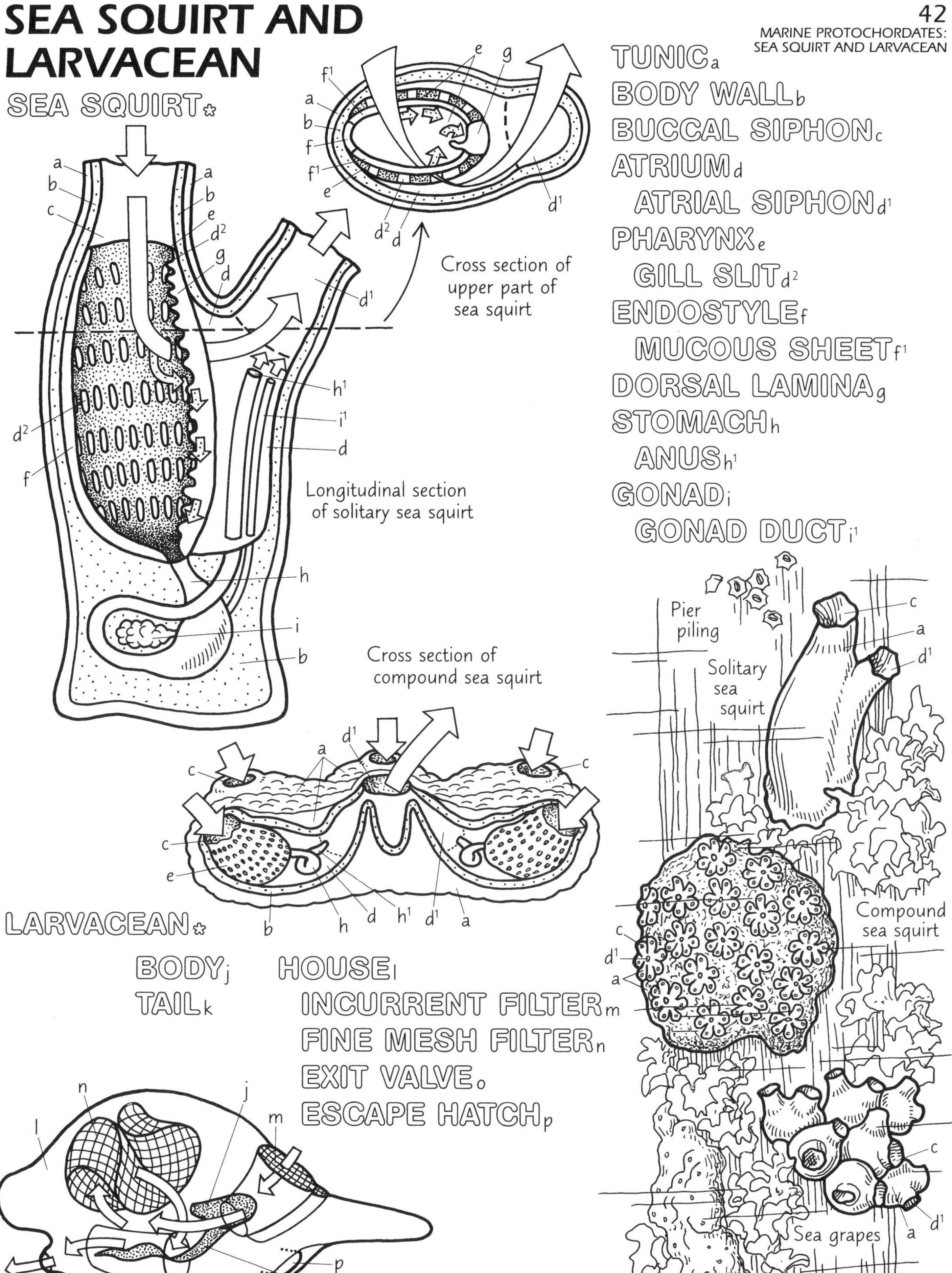

43
BONY FISH DIVERSITY: FISH MORPHOLOGY

To appreciate the diversity of marine bony fish, one needs to investigate the features that are common to most fish. The grouper, or sea bass, pictured here, is considered to be unspecialized, and has a basic fish morphology. In this plate, we will identify the major external body structures and introduce their function, and outline the internal support and muscle systems. In the next plate, we will explore the variety of body form and locomotion found in bony fish.

Begin by coloring the body of the sea bass in the two upper drawings. Color the external parts of the sea bass as they are mentioned.

The shape of the sea bass is fusiform—a streamlined shape offering the least resistance to movement through the water. The prominent fins are obvious in both the lateral and front view. The *caudal fin* usually provides the main thrust used in swimming. Also located in the midline are two unpaired fins, the *dorsal* and *anal fins*. These fins are used to stabilize the fish in the water and to lessen its tendency to pitch, especially while swimming slowly. They are also useful in preventing the fish from rolling over while turning at high speeds. The *pectoral fins* are located on the sides of the fish, behind the opening of the gill cavity, and *pelvic fins* are located ventrally, in front of the anal fin. These paired fins are used as stabilizers, but assist in turning and stopping as well. Fins are supported by fin rays of two types: bony pointed spines; and soft rays, which are jointed. The dorsal, anal, and pelvic fins have both spines and soft rays; the remaining fins consist of soft rays only.

In front of each pectoral fin is a large bony flap called the *operculum*, which covers the gill cavity that opens just behind it. Running the length of the fish, from the operculum to the base of the caudal fin, is the *lateral line*. The line consists of a series of very small canals that open to the surface and contain pressure-sensitive receptors. When the fish encounters movement in the water, such as the bow wave of an approaching fish, the water pressure pushes against the fish, entering the lateral line canals and triggering the pressure receptors. The lateral line, called the "sense of distant touch," is extremely sensitive and allows the fish to move in turbid water by "feeling" its way around obstacles, even when vision is greatly impaired. Other prominent external morphological features are the *eyes* (fish have fair vision), *nostrils*, and the bones of the *jaw*. Generally, fish have two nostrils on each side, which open to an olfactory pit and are used for scent, not for respiration.

The shape, size, and position of the jaws vary considerably in different fish, and are related to the type of feeding. The sea bass is a generalist carnivore, feeding on a wide range of prey and has a large *mouth* that opens terminally (at the front). Inside the stout jaws of the sea bass is the folded opening of the *esophagus*, which leads to the gut. Adjacent to and in front of the esophagus are the *gills*, whose arched gill bars support the gill rakers, sometimes used in feeding, and gill filaments, which provide the respiratory surface (Plate 49).

Now color the view showing the fish's skeleton. Note only a small region of the body musculature is illustrated. Color the fins the same colors as you used above. You may wish to color the pectoral and pelvic girdles with shades of the colors of their respective fins.

The bony fish's skeleton provides protection for the head and internal organs and support for the muscles. Prominent in the skeleton are the jaws, the fused bones of the *head*, the flat bones of the operculum, and the *supports* for the spinous and soft dorsal rays. The pectoral and pelvic fins are supported by girdles of bone, the *pectoral* and *pelvic girdles* respectively, that are anatomical homologues of our shoulders and hips. Also note the region just in front of the caudal fin known as the *caudal peduncle*, which is the pivoting point of the caudal fin. Notice how the vertebral column, or axial skeleton, is located towards the center of the fish instead of along the back as in ourselves and other terrestrial vertebrates. Because the fish lives in water, a buoyant medium, the skeleton does not have to support the body against the pull of gravity. Relieved of this chore, the axial skeleton is positioned more centrally to provide maximum support for the role of the trunk musculature in swimming (Plate 44). In the cross-sectional view of a single *vertebra*, the single *dorsal process* and articulated *ribs* show the sites of muscle attachment. The body musculature is organized into units called *myomeres*. The individual myomeres nest together like a series of stacked cones along the length of the fish.

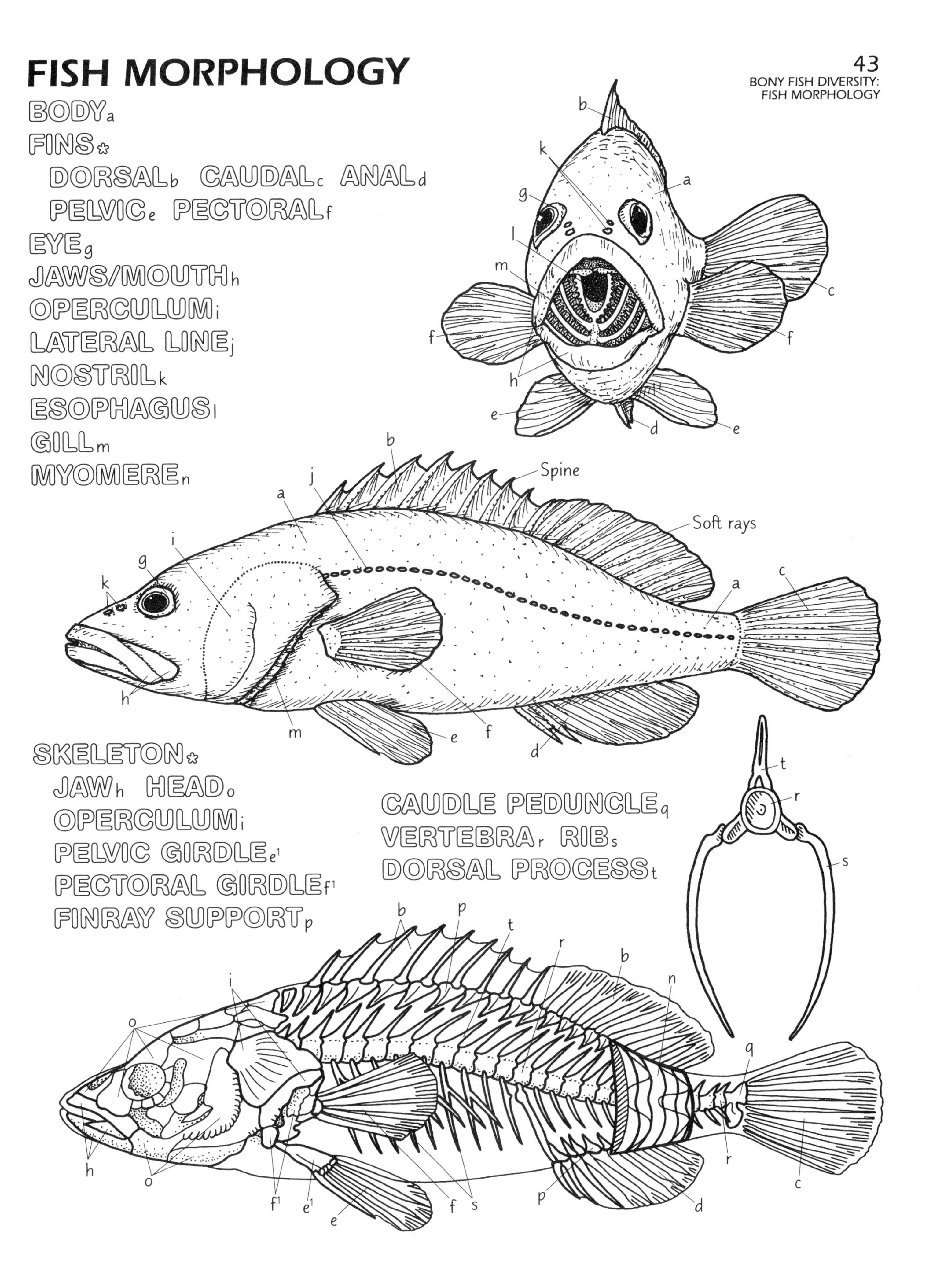

FISH MORPHOLOGY
43
BONY FISH DIVERSITY: FISH MORPHOLOGY
BODYa
FINS*
DORSALb CAUDALc ANALd
PELVICe PECTORALf
EYEg
JAWS/MOUTHh
OPERCULUMi
LATERAL LINEj
NOSTRILk
ESOPHAGUSl
GILLm
MYOMEREn
Spine
Soft rays
SKELETON*
JAWh HEADo
OPERCULUMi
PELVIC GIRDLEe1
PECTORAL GIRDLEf1
FINRAY SUPPORTp
CAUDLE PEDUNCLEq
VERTEBRAr RIBs
DORSAL PROCESSt

44
BONY FISH DIVERSITY: LOCOMOTION

Moving through water requires more energy than moving through air. The forces of friction and drag resistance are much greater on a body moving through water. Also, as a fish moves, it displaces water that flows around the body. A streamlined body shape is required to improve this flow. The fusiform or tear-drop shape of the sea bass shown here represents an effective morphological compromise to meet these demands.

Color the illustration of the swimming sea bass at the top left, noting the flexing of the body and the resulting thrust forward. Note that the myomeres are exposed on either side of the curve in the fish's body. Those on the inside of the curve are contracted while those on the outside are relaxed and are being stretched. You may wish to color the fins the same colors as used in Plate 43. Color the small illustrations of the moray eel and tuna moving across a grid to show the body undulations.

Most fish swim by undulating their body so that the *caudal fin* is whipped very rapidly from side to side in a sculling motion resulting in a powerful forward *thrust*. Beginning at the head, there is a sequential contraction of blocks of *myomeres* along the length of the fish. These waves of contraction alternate on opposite sides of the fish causing the axial skeleton to flex, the body to undulate, and the caudal fin to pivot at the *caudal peduncle* and scull through the water. The degree of undulation varies among species and depends on the stiffness of the vertebral column and the connection between the myomeres. The extremes can be seen in the body movement of elongate fish like *moray eels,* which have serpentine undulations, compared to a *tuna* whose body remains very rigid with only the caudal fin undulating rapidly.

Now color the different types of caudal fins.

The caudal fin usually provides the main thrust used in swimming. The size and shape of the fin is an indicator of the fish's ability to move through the water. The sea bass has a *rounded* caudal fin that is soft and flexible, but it also has considerable surface area. This fin gives effective acceleration and maneuvering, but is inefficient for prolonged, continuous swimming as it creates too much drag which tires the fish. A *forked* caudal fin produces less drag, and is efficient for more rapid swimming. Long-distance, continuous swimmers such as the tuna have *lunate* caudal fins, which are rigid for high propulsive efficiency and have a relatively small surface area to reduce drag. However, the rigid fin is ineffective for maneuvering.

Color the illustrations of the tuna, barracuda, and butterflyfish, noting the differences in the body shape and caudal fin.

Not all fish are engaged in rapid, *continuous swimming* like the pelagic tunas whose sustained swimming speeds range from 8–16 km/h, 5–10 mph (Plate 45). A fish's unique lifestyle places other demands on its body form and locomotion abilities. A fish such as the barracuda that swims leisurely in wait, then runs down its prey (Plate 109), relies on quick *acceleration* and short bursts of rapid swimming provided by its elongate muscular body and large caudal fin. Barracudas have been estimated to accelerate to 80 km/h (50 mph) in pursuit of prey.

The butterflyfish (Plate 47) needs great dexterity and *maneuverability* to capture its small prey. The body is compressed (flattened laterally) and disc-shaped in outline. The *dorsal* and *anal fins* are large and run along most of the body, providing excellent control in turning. They can be rapidly flexed for quick acceleration should escape be necessary. The *pectoral* and *pelvic fins* are also very large and maneuverable.

Finally, color the drawings illustrating different methods of propulsion. Color only the fins involved in propulsion.

Locomotion, in most fish, is a compromise among the demands for sustained swimming, quick bursts of acceleration, and maneuverability. Not all fish rely on the caudal fin for swimming propulsion. Electric fish swim by waves of undulations along the anal fin. Triggerfish use their dorsal and anal fins for locomotion. Sculpins and wrasses propel themselves using the sculling action of their pectoral fins. Seahorses and pipefish swim vertically, sculling through the water with their dorsal fins.

LOCOMOTION

SWIMMING*
- BODY a
- FINS*
  - DORSAL b
  - CAUDAL c
  - ANAL d
  - PELVIC e
  - PECTORAL f
- EYE g
- JAWS/MOUTH h
- CAUDAL PEDUNCLE i
- MYOMERE*
  - CONTRACTED j
  - RELAXED k
- THRUST l

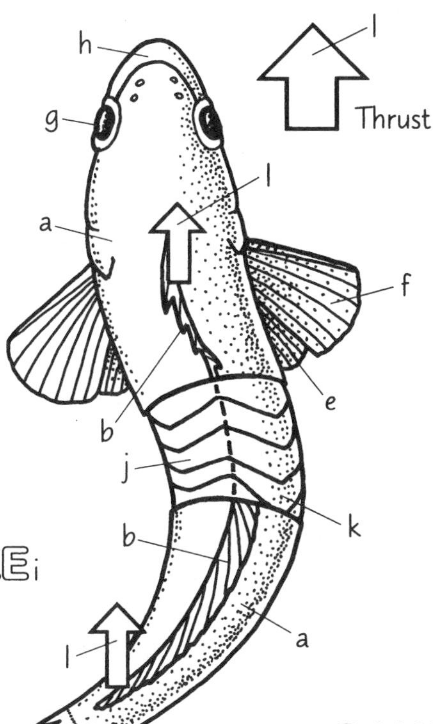

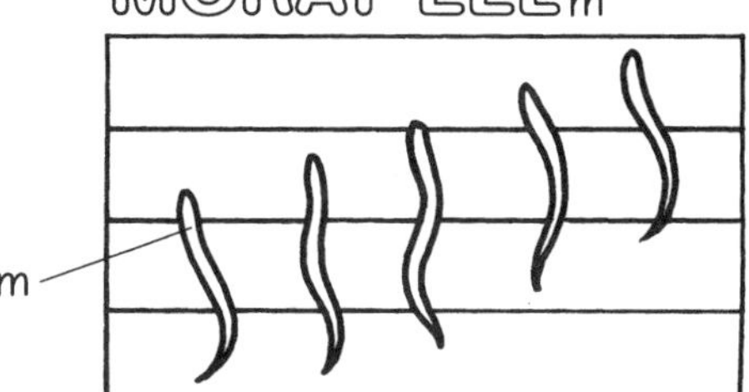

TUNA n

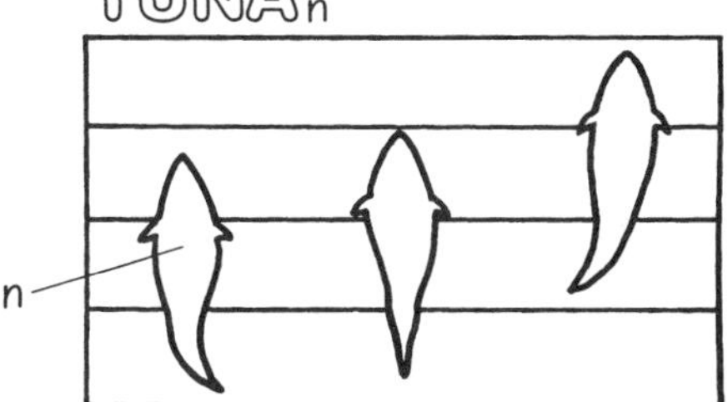

ROUNDED c^1

FORKED c^2

LUNATE c^3

PROPULSION*

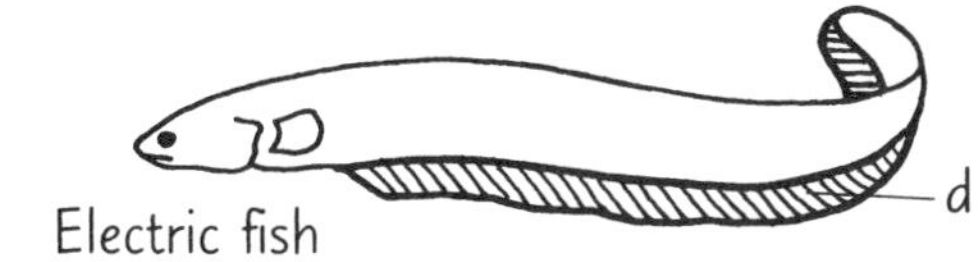

Electric fish

Triggerfish

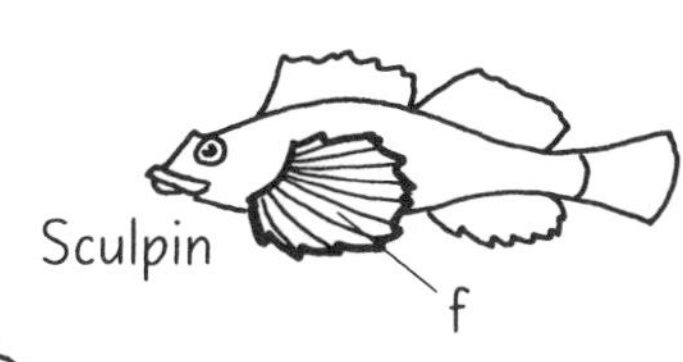

Sculpin

Seahorse

o

Tuna

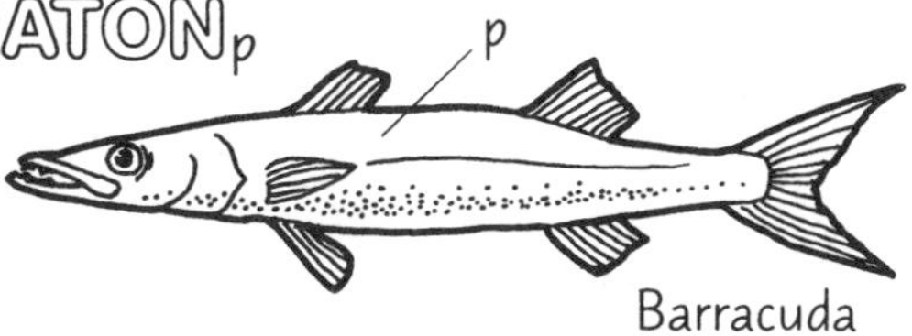

Barracuda

MANEUVERABILITY q

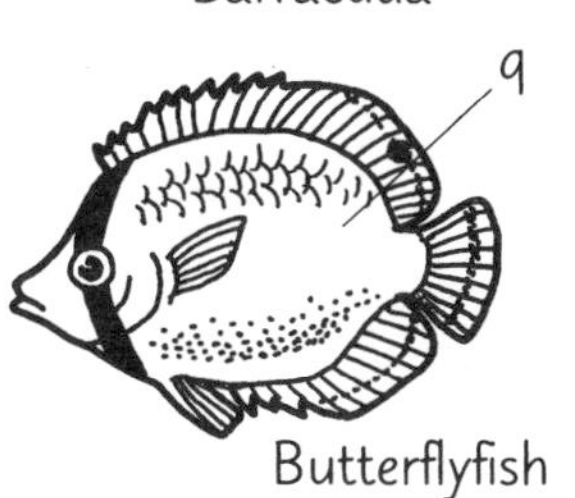

Butterflyfish

45
BONY FISH DIVERSITY: PELAGIC FISHES

In the marine environment, fish live either in the water column, or on the ocean bottom. The body structure of the fish is specially adapted to life in one or the other of these environments. Fish that live in the sunlit, open waters, constantly free of the ocean bottom, are called *pelagic fishes.*

All of the fish illustrated here have a similar coloration pattern: dark on the top half, light colored on the bottom. It is suggested that the body of each fish be colored a dark gray blue on the top half, and a silver or white on the bottom. Structures (b) through (h) may be colored the same colors that you used on the previous plates on fish morphology and locomotion. Color each fish separately as it is discussed in the text.

The flying fish has the capability of becoming airborne, gliding just above the water's surface. If pursued from below by a predator, the flying fish breaks the surface of the water at speeds up to 64 km/h (40 mph). The enlarged *pectoral fins* are stretched out at right angles to the body, and act as gliding "wings." The *pelvic fins* are similarly enlarged to increase the gliding surface. When the flying fish begins to slow down, it alights on the water, tail first, and the enlarged lower lobe of the *caudal fin* rapidly sculls the surface at a rate of 50 beats per second. The fish picks up speed and it again becomes airborne. Flying fish are found in the warm waters of the Atlantic and Pacific oceans. The largest species reaches a length of 46 cm (18 in) and lives off the southern California coast.

The northern herring is a very important small fish that feeds on zooplankton, especially copepods. The northern herring occurs on both sides of the Atlantic; a subspecies lives in the Pacific Ocean. The herring forms great schools (shoals) comprising billions of fish. These schools migrate to shallow breeding grounds, where, on the Atlantic grounds alone, two to three million tons of herring are caught yearly. Young herring are canned and sold as sardines; the older fish may reach a length of 30 cm (12 in), and are either canned or used for oil. Ownership of fishing rights to herring breeding grounds is highly contested, and several European countries have come close to war over this issue.

The swordfish is found in tropical and warm temperate seas worldwide. It is a continuous swimmer, following schools of mackeral, herring, and sardine. The swordfish swims through a school of fish, thrashing its upper *jaw*, or *bill*, and stuns the fish, which it then eats. The swordfish's bill may be as large as one-third of the fish's entire body length, which averages 1.8 to 3.6 meters (6–12 ft). Some swordfish as large as 6 meters (20 ft) have been caught.

The streamlined tapering of the swordfish makes it an excellent swimmer. The swordfish lacks pelvic fins, and the long, low-slung pectoral fins are kept folded against the body during rapid swimming. The *dorsal fin* is tall and remains permanently erect. The rigid caudal fin is lunate, for maximum swimming efficiency; the caudal area is strongly reinforced by a bony *keel.*

The sunfish is highly compressed and is a slow swimmer. This fish grows to be quite large (3–4 meters, 10–13 ft) and is the heaviest of the bony fish at 2000 kg (4400 lb). The sunfish is usually seen on the surface of the water "sunning" itself on its side, slowly flapping its small pectoral fins. Recent studies indicate, however, that the sunfish normally lives quite deep in the ocean's waters, and that those fish seen on the surface are actually quite abnormal.

The sunfish lacks pelvic fins, and the *anal* and dorsal fins are very large and set well back on the *body*, where they provide the swimming thrust. The caudal fin exists as a narrow band, but is not effective in swimming. The sunfish feeds on jellyfish and other small planktonic forms.

The albacore is a small tuna that averages 4.5 kg (10 lb) and is highly prized as a sport and commercial fish. Albacore schools are found in the Atlantic and Pacific, where they range into temperate waters to feed and spawn near the equator. Like the swordfish, the albacore is a continuous swimmer, with a rigid lunate caudal fin and reinforcing keel. In addition to the anal and dorsal fins, a series of smaller "finlets" are present and add to the hydrodynamic efficiency of the fish. The long pectoral fins are a unique feature of the albacore among the tuna family.

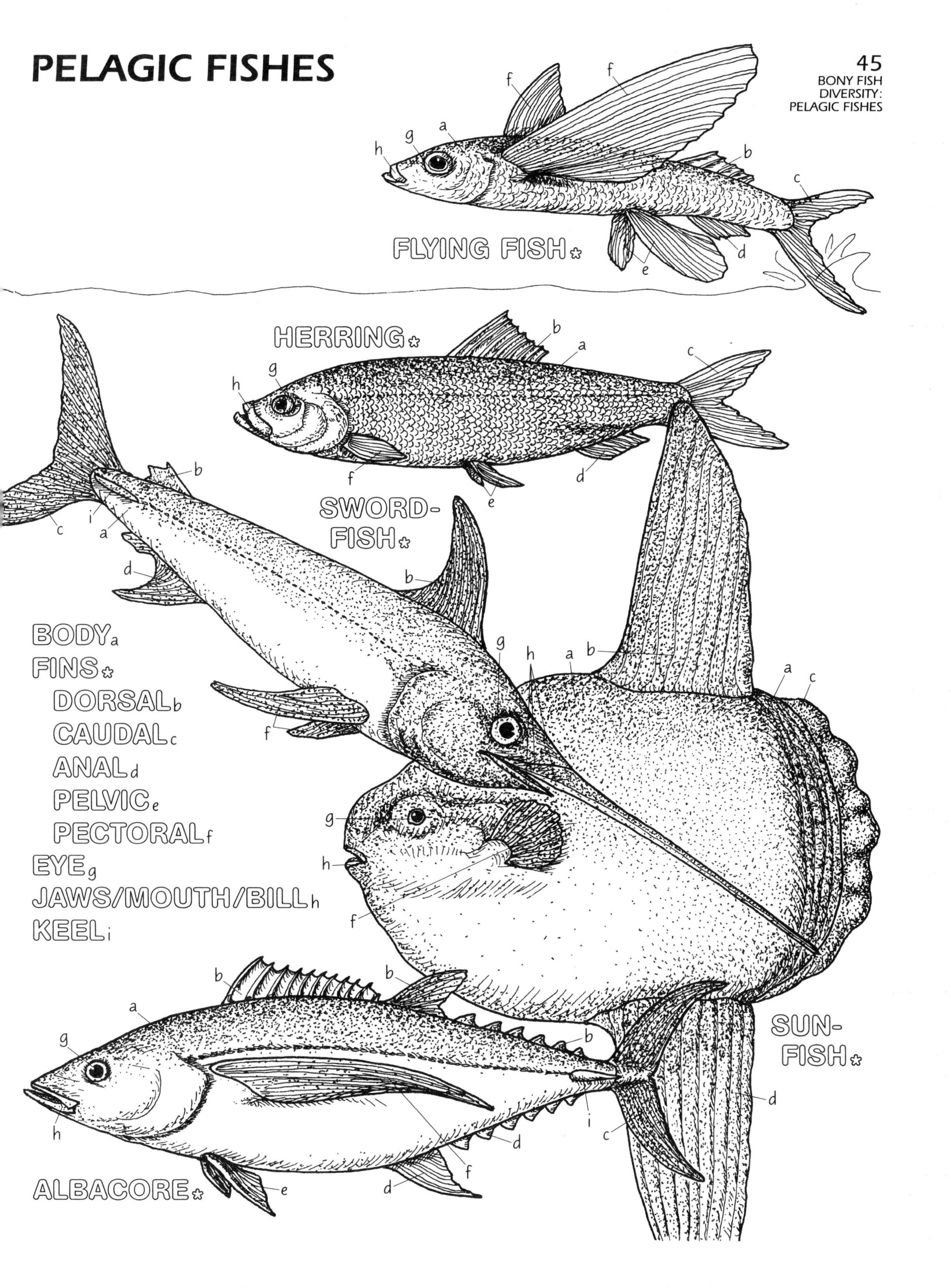
PELAGIC FISHES
45
BONY FISH
DIVERSITY:
PELAGIC FISHES
FLYING FISH
HERRING
SWORD-
FISH
SUN-
FISH
ALBACORE
BODY a
FINS
DORSAL b
CAUDAL c
ANAL d
PELVIC e
PECTORAL f
EYE g
JAWS/MOUTH/BILL h
KEEL i

46
BONY FISH DIVERSITY: BOTTOM DWELLERS

Fish that live on, or very close to, the bottom (benthic fishes) do not swim continuously and therefore do not require a streamlined body form. This plate considers the body modifications of the tidepool sculpin, the sea robin, the stargazer, and the highly modified starry flounder.

Color each fish as it is discussed in the text. The body color varies in these species, but generally, they are a dark gray or brown in the stippled areas, and light colors elsewhere. Structures (b) through (h) are, again, the same colors as on the three previous plates. The electric organs of the stargazer are located under the skin; they should receive a separate color, and then be colored over.

The tidepool sculpin is a small (8 cm, 3 in) fish, commonly found in Pacific coast tidepools of North America. The sculpin forages for small crustaceans and other invertebrates among the rocks and algae of tidepools. Its coloration varies from mottled reds to greens and grays, and this mottled coloration breaks up its profile against the heterogeneous background. The sculpin usually sits motionless on its enlarged *pectoral fins*, but it can dart about rapidly in the water for short distances. Tidepool sculpins are representative of a number of small, elongate, relatively unspecialized benthic fishes (such as blennies, clinids, and gobies), all of which frequent rocky areas.

The sea robin has a large head, encased in an armor of rough, bony plates. Sea robins walk along the bottom using the first three rays of their pectoral fins, balancing the *body* with the remainder of the pectoral fins and the *pelvic* fins. These three rays are articulated, and the fish uses them somewhat like fingers, pulling itself along and turning over rocks to uncover the crustaceans and molluscs on which it feeds. Sea robins are usually found on sand or sand-mud bottoms, at depths between 19 and 45 meters (62–148 ft). Species occur in most oceans except the very coldest; they are generally small, although some large species grow to 1 meter (3 ft) in length. In places where they occur abundantly, sea robins are sometimes fished commercially.

The stargazer, a medium-sized fish (up to 51 cm, 20 in), is a poor swimmer and usually lies buried in mud or sand. The Greeks called them "holy fish" because their small *eyes*, set on top of a rather square head, seemed to look toward the heavens. Some stargazer species have a fleshy lure attached to the floor of their large *mouths*. The lure attracts fish to within striking distance of the stargazers' *electric organs* whose discharge then stuns the prey. The electric organs, located behind the eyes, are thought to be derived from eye muscles and the optic nerve; they are capable of producing a stunning shock of 50 volts.

Another characteristic of stargazers is the presence of poisonous spines (not shown), located just behind the operculum and above the pectoral fins. These spines are grooved, with poison sacs at their base. Some species are believed to have a poison that can cause death in humans.

A far more benign, but equally specialized fish is the starry flounder which is a member of the family of left-eyed flatfish: both eyes are located on the left side of the fish. During its larval development, the right eye gradually migrates from its normal position to the other side of the body, while the mouth remains in the normal position. These fish are highly compressed and lie on one side on sandy bottoms, often with only their eyes and opercular opening uncovered. When the flounder swims, it does so in a sideways position. The starry flounder is very common in shallow, temperate waters of the Pacific, and is often found in nearly fresh water, especially when young. They grow to lengths of 90 cm (35 in) and weigh as much as 9 kg (20 lb). Starry flounders, unpigmented on their underside, get their name from the stellate scales that cover their body.

BOTTOM DWELLERS

BODY a
DORSAL FIN b
CAUDAL FIN c
ANAL FIN d
PELVIC FIN e
PECTORAL FIN f
EYE g
JAWS/MOUTH h
ELECTRIC ORGANS i

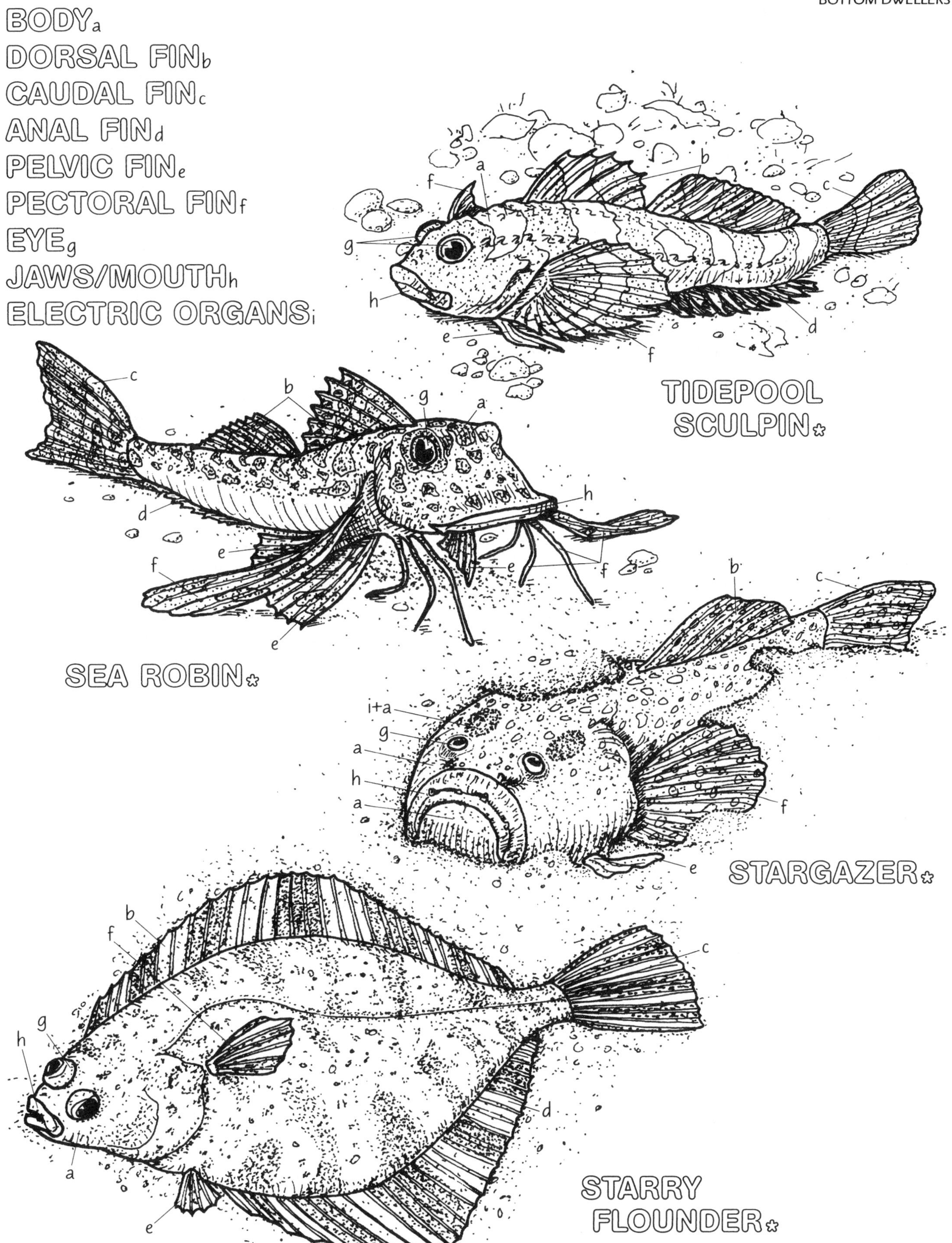

47
BONY FISH DIVERSITY: CORAL REEF FISHES

The pelagic zone and the homogeneous soft bottoms are monotonous environments compared to the coral reef, which is a mosaic of structure, form, and color. Few other marine habitats house such an assemblage of fish forms in such a small area. The five fishes introduced here represent the diversity of reef fishes, but do not come near to exhausting the variety of fish forms living there.

Color each fish as it is discussed. Structures (b) through (h) are again colored the same as on the previous plates. As you color, study the variations of each structure, comparing them to the other fish on the plate.

The coral grouper is a pink or coral color. Groupers are commonly seen on coral reefs, swimming leisurely just above or among the coral structures. The grouper can quickly accelerate from its slow forward motion and swim quite rapidly for short distances. Note its fusiform shape and broad, rounded *caudal fin*. Some species of grouper become quite large and fearless (230 kg, 500 lb; 2.1 meters 6.9 ft long). Large groupers will readily approach divers and allow themselves to be stroked and petted; they are easily trained to feed from a diver's hand.

The trumpetfish is an interesting fish sometimes seen swimming beside a grouper. This fish is long and slender (up to 76 cm, 30 in), and yellow or orange in color. It is often found hanging motionless, head down, among tall, swaying soft coral growths or sponges. This posture is a deception; the trumpetfish is a crafty predator with the ability to dart like an arrow and seize a small fish in its *jaws*.

The golden boxfish stands out against the coral reef with its bright yellow or gold color. The boxfish displays what is known as warning coloration, advertising that its skin is covered by a poisonous secretion. Predators learn to associate this color with the taste, and thus leave the boxfish alone (Plate 63). Boxfish, also called trunkfish and cowfish, have a second line of defense: their entire *body* surface is underlain by bony plates that encase them in an unpalatable armor. This bone structure gives the fish a box shape, and renders it rather inflexible and a very poor swimmer. Lacking pelvic fins, boxfish swim mainly by the sculling action of the *anal, dorsal*, and *pectoral fins*.

The stiffness of the boxfish contrasts with the graceful agility of the threadfin butterflyfish. The color of this fish is gray or white. The butterflyfish is short (13–20 cm, 5–8 in) and laterally compressed—a very maneuverable body shape. This fish can make extremely fast turns and quick stops; it can remain stationary in mid-water, back up, and move its head up and down. The pectoral and *pelvic fins* are important in these movements, and the fish uses its broad dorsal and anal fins to increase stability during fast turns. Butterflyfish swim around coral formations, poking in crevices and ledges for small invertebrate prey. When pursued, they dart in and out of these places with great facility, leaving faster, but less agile, predators behind.

Unlike the butterflyfish, the moray eel remains very close to its home. Eel body colors range from brown and green to gold and from solid colored to spotted. These fish have no pelvic or pectoral fins, and their body is long and snakelike. A dorsal fin runs the length of the body, and an anal fin extends from mid-body back, merging with the dorsal fin at the tail. Morays can swim in a slow, serpentine fashion, but are usually found entwined in the coral, under a ledge, or in a hole or cave. Their *mouths* are full of pointed teeth, but they will usually bite only when provoked or cornered. A few species are venomous. The prominent protruding *nostrils* are, as in most fishes, used only for smell. At night, morays leave their caves, crawling over the substratum, feeding chiefly on fish. Some of the larger species may grow to 2 or 3 meters (6–10 ft) in length.

CORAL REEF FISHES

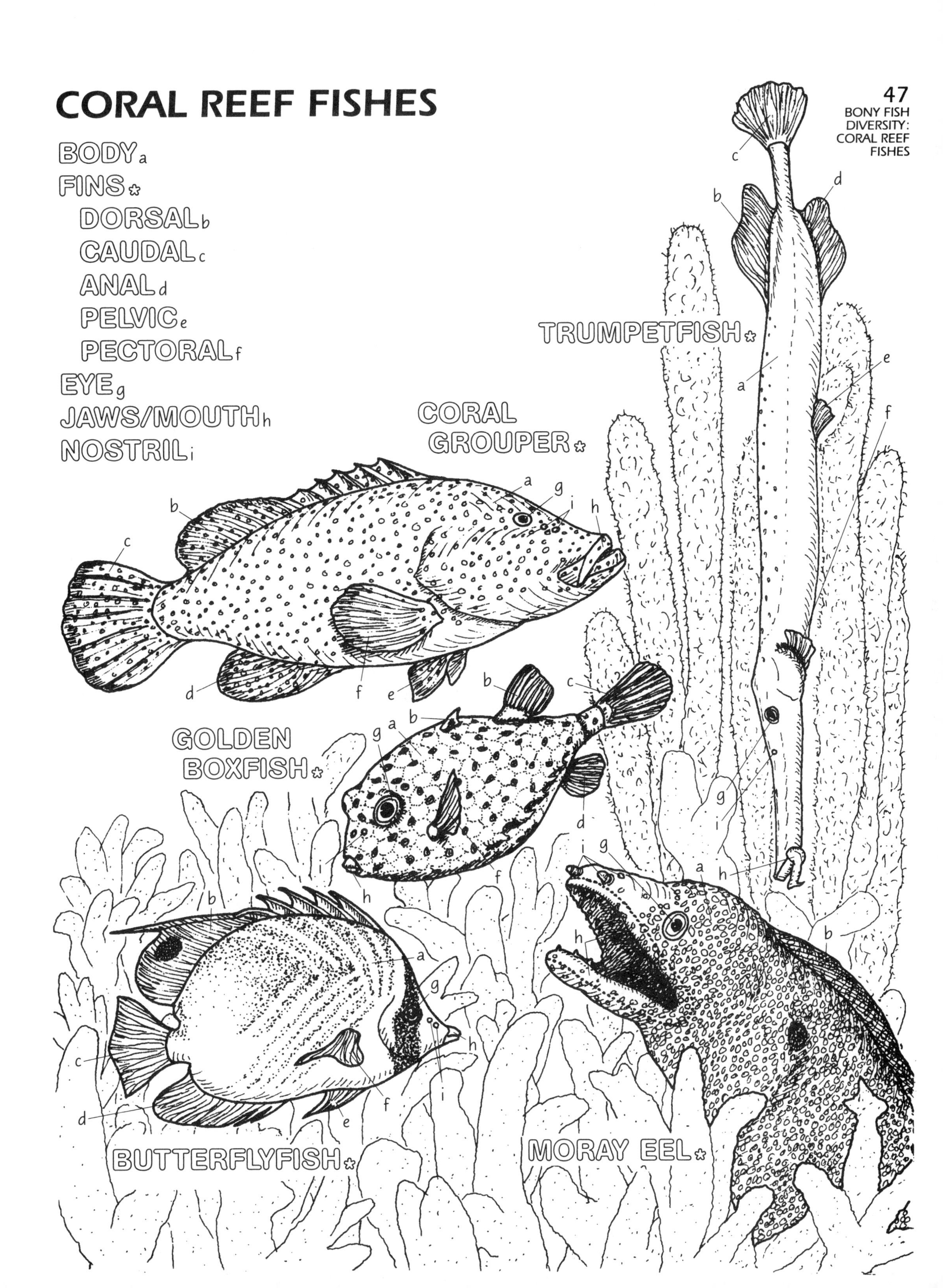

48
BONY FISH DIVERSITY: MIDWATER AND DEEP-SEA FISHES

The greater volume of the ocean lies below the sunlit photic zone (Plate 14), and remains in constant darkness. The water temperature falls with increasing depth to a relatively constant 4°C (39°F). Because plants cannot photosynthesize without light, there is little food in the depths of the ocean. Although such harsh conditions do not seem very favorable to life, a number of fishes have adapted to the midwater and deep-sea habitats.

The middle layer of the ocean, the mesopelagic zone, is where light fades into darkness. Here, many fishes create their own light, using it in intraspecific communication, to capture prey, and for other purposes. In the deepest ocean habitat, the bathypelagic fishes are mostly very small and have enormous mouths with long, sharp teeth that enable them to capture and swallow large meals. This plate presents six midwater/deep-sea fishes as an introduction to these fascinating creatures.

Color each fish as it is mentioned in the text. The three mesopelagic fishes in the top half of the plate can receive a light gray color; the three bathypelagic fishes can receive a dark gray. The light-emitting organs can be colored blue green, the color of light produced by them.

Because the upper layers of the water column hold an abundant food supply, many species undertake a nightly migration to the surface to feed. The lanternfishes are one such group of migratory species. Lanternfishes (4–10 cm, 1.5–4 in) are so named for the row of bioluminescent *light organs* (photophores) along their ventral surface. These light organs are used in intraspecific communication, allowing lanternfish to recognize potential mates, for example. Lanternfish have very large *eyes*, presumably an adaptation to night feeding in dimly lit waters.

The Pacific viperfish feeds on lanternfish and squid, and follows them upward in their nightly migration. Also equipped with light organs, the viperfish has an oversized *mouth* and fanglike teeth. In order to feed, the viperfish must tilt its head upward and extend its lower *jaw* to a point where its *gills* are exposed. The viperfish is small (22–30 cm, 9–12 in) and has an elongated *dorsal lure* that may be used to entice prey.

The hatchetfish is a mesopelagic fish that remains in waters several hundred meters below the surface. The upward-oriented eyes of the hatchetfish scan the faintly lit waters for zooplankton. The hatchetfish has binocular vision (somewhat unusual in fish), which allows it to gauge the position of prey hovering above. The light organs are used in an ingenious defensive behavior known as counter-lighting (Plate 70).

The black devil, also called anglerfish, has a large light organ mounted on a modified dorsal fin spine, which serves as a lighted lure and is held above the mouth. Prey is attracted to the light and quickly engulfed by the black devil's tooth-filled mouth. A gut completely lined with dark pigment is an interesting adaptation seen in the black devil and other deep-sea predatory fish. The black pigment is believed to mask the light organs of any swallowed prey so the luminescence does not reveal the presence of the predator.

Another deep-sea fish with an extremely large mouth is the pelican gulper eel, which can reach a length of 76 cm (30 in). The gulper eel has a distensible gut, and will often consume prey (primarily crustaceans) as large as itself.

The tripodfish lives in the deep ocean from the North Atlantic to the Caribbean Sea. It is considered a benthic fish, feeding on small zooplankton that hover just above the bottom. The "tripod" consists of the elongated *pelvic fins* and the ventral lobe of the *caudal fin*, and allows this fish to feed in the water column without having to expend energy in swimming. Tracks found in the vicinity of tripodfish suggest that these fish use their pelvic fins to walk along the bottom. Zooplankton bump into the threadlike extensions of the tripodfish's fins, alerting the fish to their presence. The tripodfish grows to about 25 cm (10 in) and has tiny eyes.

MIDWATER AND DEEP-SEA FISHES

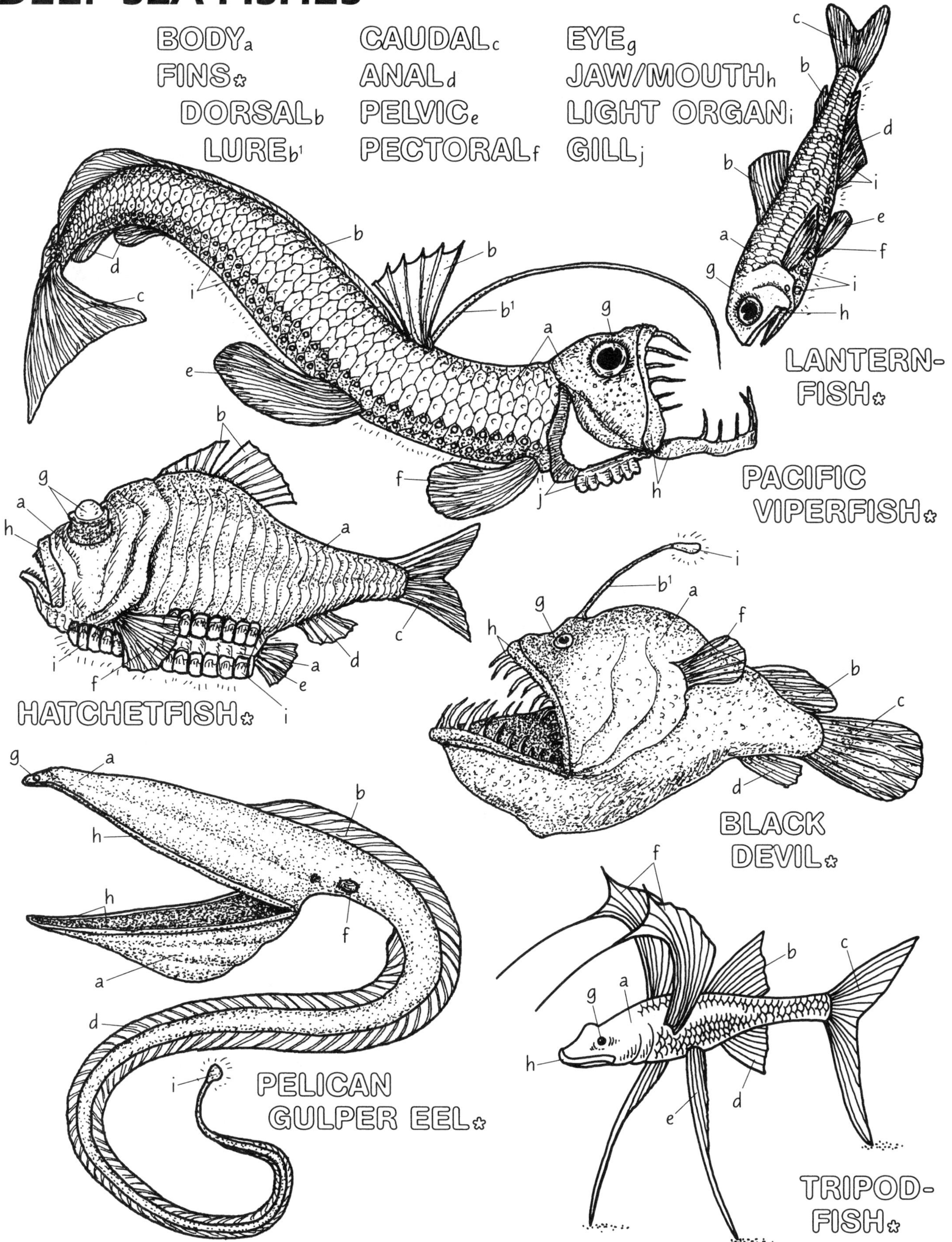

49
BONY FISH AND SHARK: COMPARISON OF STRUCTURE

The previous six plates dealt with the variety of forms found in the bony fishes. Before venturing forward in a study of the cartilaginous sharks and rays, a comparison of these two groups is in order. The major organs and systems, with special emphasis on the difference between the sharks and rays (elasmobranch fishes), and the bony fishes, are reviewed here.

This plate presents a highly diagrammatic comparison between the internal anatomy of a bony fish and that of a shark. Locate and color each organ or system in both illustrations as it is discussed in the text. Only those structures labeled and outlined with dark lines are to be colored.

Starting from the anterior, or mouth end of each fish, note the relatively small *brain*, which continues posteriorly as the *spinal cord*. The large *olfactory lobe* of the brain gives evidence of the importance of an acute sense of smell in both groups of fishes. The olfactory lobe terminates near the base of a blind sac, which opens at the *nostril*. Since the nostrils do not open to the throat—as they do in mammals for instance—most fishes must take in their respiratory water current through the mouth. Rays and other bottom-dwelling elasmobranch fishes take in water through an opening called the *spiracle*. Contraction of the throat musculature pumps water over the *tongue* and across the *gills*, which take up the sides of the throat. This oxygen-laden water passes over *gill filaments*, oxygenating the blood supply. *Gill rakers,* located on the inner face of the gill support (gill arch), prevent foreign matter from clogging the gills. Water is pumped out of the gill chamber past the operculum in bony fishes, or through the gill slits in sharks.

The relatively small *heart* is located near the base of the gills. It pumps blood through the gills and from there to the head and the rest of the body. *Kidneys* help regulate blood chemistry and deliver waste products to the exterior through the urogenital opening (not shown). The *gonads* (sex organs) also empty through this opening.

Another large organ linked to the circulatory system is the *liver*, whose main functions are to store surplus nutrients and to detoxify certain substances. In sharks and their relatives, the liver has an additional function: to contribute buoyancy to the body. This is because the liver stores oil which is considerably less dense than water. The presence of the oil in the shark liver is responsible for the latter being much larger in size than the liver of the bony fish. Most bony fish utilize a more efficient adaptation for buoyancy: the gas-filled *swim bladder*. Air is either gulped at the surface or secreted from the bloodstream into the swim bladder. Delicate regulation of the gas content in this organ allows a fish to maintain its position in the water column with a minimum expenditure of energy. By contrast, elasmobranchs function similarly to airplanes, requiring forward motion to keep from sinking.

The length and complexity of the gut has more to do with the diet of any particular species than with that species' relation to the bony fishes or shark group. The *spiral valve* is one gut structure found almost exclusively in sharks and their relatives. It serves to increase the surface area of the *intestine* for more efficient absorption of nutrients.

The characteristic difference between the two major groups, by which they are most commonly named, is the skeletal material. Elasmobranchs are cartilaginous fishes. Their skeletons are made of a relatively flexible material, but the skeletal structure is much less elaborately articulated than in bony fishes. The result is that the bodies of elasmobranchs are on the whole less maneuverable and adaptable than those of bony fishes. This is especially apparent in the structure and utilization of the fins.

Most fishes are covered by a protective layer of scales. In elasmobranchs, these are *placoid scales*, also called denticles. The word "denticle" indicates a relation to teeth, and sharks' teeth indeed originate from the skin layer, as do the scales. Placoid scales seen under a microscope have a sharp tooth-shaped projection, and collectively these scales give a sandpaper texture to the shark's skin. Bony fishes possess, as a group, several types of scales. The *ctenoid scales* shown here are thin and translucent. They lack the enamel and dentine layers of placoid scales, and instead have an outer surface marked with bony ridges that alternate with depressions. Ctenoid scales overlap to provide both protection and suppleness.

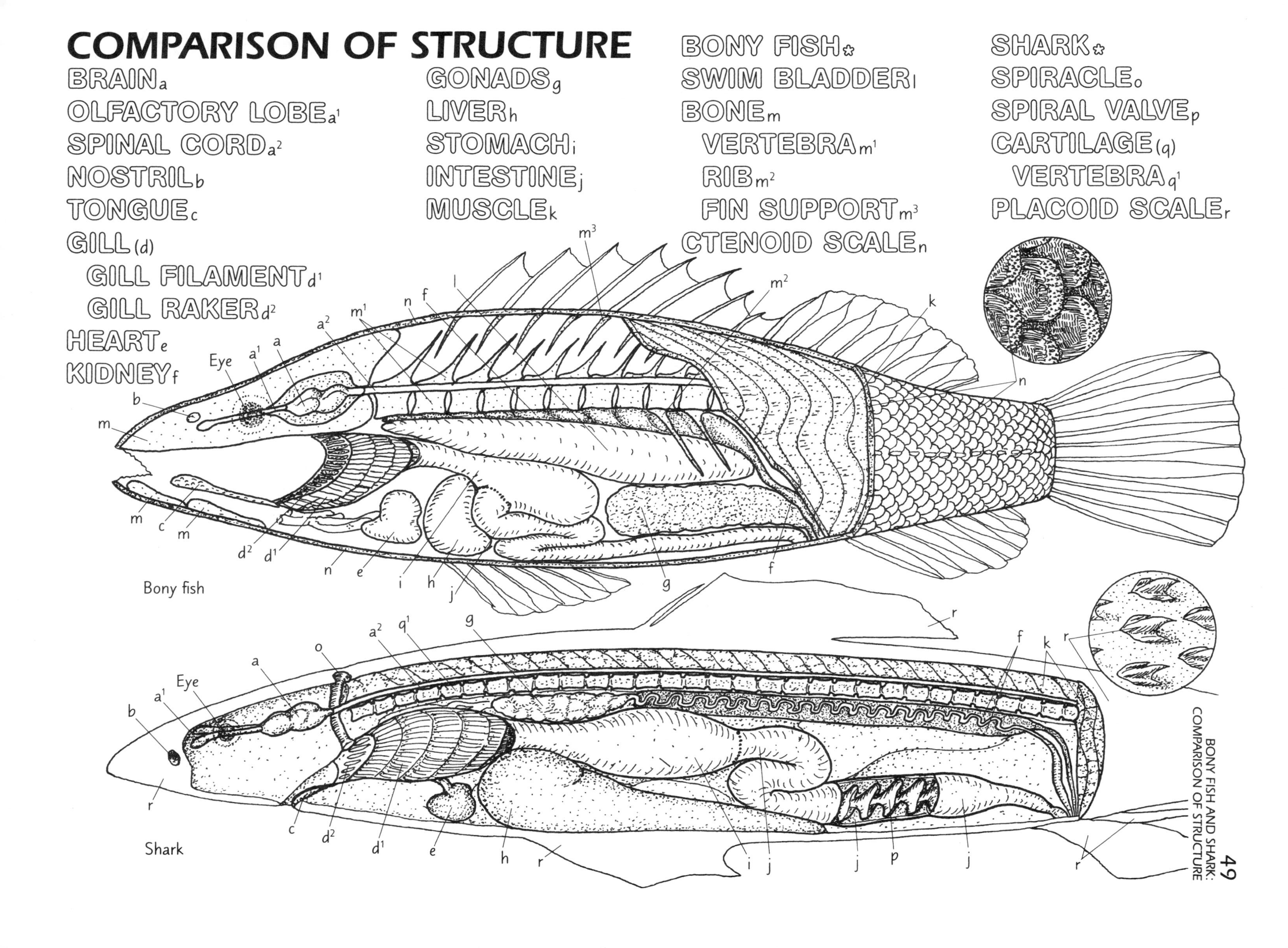
49
BONY FISH AND SHARK: COMPARISON OF STRUCTURE
COMPARISON OF STRUCTURE
BRAIN a
OLFACTORY LOBE a1
SPINAL CORD a2
NOSTRIL b
TONGUE c
GILL (d)
GILL FILAMENT d1
GILL RAKER d2
HEART e
KIDNEY f
GONADS g
LIVER h
STOMACH i
INTESTINE j
MUSCLE k
BONY FISH
SWIM BLADDER l
BONE m
VERTEBRA m1
RIB m2
FIN SUPPORT m3
CTENOID SCALE n
SHARK
SPIRACLE o
SPIRAL VALVE p
CARTILAGE (q)
VERTEBRA q1
PLACOID SCALE r
Eye
Bony fish
Eye
Shark

50
ELASMOBRANCH FISH DIVERSITY: FORM AND FUNCTION

The elasmobranch fishes (sharks, skates, and rays) are characterized by flexible, cartilaginous skeletons. They differ from bony fishes in two other major ways. They possess gill slits instead of an operculum, and they lack a swim bladder. These characteristics have a profound influence on the mode of existence of the elasmobranch groups. This plate introduces the three elasmobranch forms.

Color the dogfish views, including the upper right inset drawing which illustrates a single row of teeth in the lower jaw moving forward with use and being discarded.

The spiny dogfish shark (1 meter, 3 ft) possesses a streamlined dusky gray *body* with a complement of both paired, *pectoral* and *pelvic fins,* and unpaired, *dorsal* and *caudal fins* similar to the bony fishes (Plate 43). However, there are some differences. The dorsal fin of sharks is more rigid and is incapable of being folded down flush against the back. While the spiny dogfish possesses two dorsal fins, they do not straddle the midline and are not properly considered as "paired." The dogfish caudal fin is asymmetrical (heterocercal), with a larger upper lobe. This larger lobe, supported nearly to its tip by the vertebral column, gives the shark an upward as well as forward movement that counteracts its tendency to sink. Similarly, the larger pectoral fins are more rigid than those of bony fishes and are held horizontally with a slight upward cant that provides lift. The pelvic fins of the male each have an elongated *clasper* on the inside; this is used in fertilization of the female (Plate 85).

The absence of an operculum means that sharks cannot actively pump water over their gills, and this suggests that they have to swim constantly to maintain a flow of oxygen-bearing water over their gill surfaces. However, many bottom-dwelling sharks possess *spiracles* from the dorsal body surface to the gills through which the gills can be ventilated. Each spiracle is fitted with a non-return valve that opens and closes as the fish breathes; water is drawn in only and expelled out the *gill slits*.

The shark's *mouth* is usually underslung, or subterminal, with a pointed snout extending above it. When a shark strikes, it raises its snout and projects its wide-open *jaws* forward, allowing it to take a substantial bite. Shark *teeth* are usually pointed and sharp. As they break off or are worn down, they are continuously replaced from behind, as the arrows in the inset indicate. Unlike other vertebrates, the teeth of sharks are mounted in skin, not jaw bone. Shark teeth are actually modified placoid scales, or *denticles*, the same basic structure that covers the shark's body, giving it a texture like sandpaper. Teeth are lined up in five or six rows (fewer in some cases; more in others) and grow forward as the leading tooth wears or falls out.

Color the underside (ventral side) of the big skate.

The skate is a much flattened (depressed) elasmobranch that spends its daylight hours buried in the sand on the bottom, with only the eyes and large gill-ventilating spiracles uncovered. At night, the skate emerges and swims, using undulations of its long pectoral fins like wings. The skate feeds on bottom-dwelling crustaceans and molluscs that it captures in its ventrally located mouth and crushes with its flat, block-shaped teeth. The skate's body is moderately slender and often has rows of enlarged denticles along the back. The male's pelvic fins possess long claspers for mating. The anal and caudal fins are absent, and the dorsal fins are very small. Skates are found in all seas, and species range in size from one meter to the "barndoor" skate which is over three meters (10 ft) long. They are white ventrally and tan to dark dorsally.

Color the stingray, including the enlarged spine at lower right.

The southern stingray inhabits Atlantic and Caribbean waters. Stingrays are similar to skates in the shape of the pectoral fins, position of the mouth, spiracle, and gill slits, and in behavior, feeding, and diet. They spend the daylight hours buried in soft substrata and the nighttime foraging for bottom-dwelling crustaceans and molluscs. Rays differ from skates in having a long whiplike tail that lacks dorsal fins and possesses one or more spines at its base. The spines are modified denticles and, like the shark's teeth, are replaced in series when broken or pulled out. Buried in the sand, these rays are almost invisible, and the unwary wader who steps on one is often impaled on one of the spines by the lashing motion of the tail. Stingray "stings" are complicated by the presence of venom glands on each side of the spine; the venom flows along grooves in the spine and into the wound. The pain is often excruciating, and some stings have proven fatal to humans. Southern stingrays are medium brown dorsally and white ventrally.

FORM AND FUNCTION

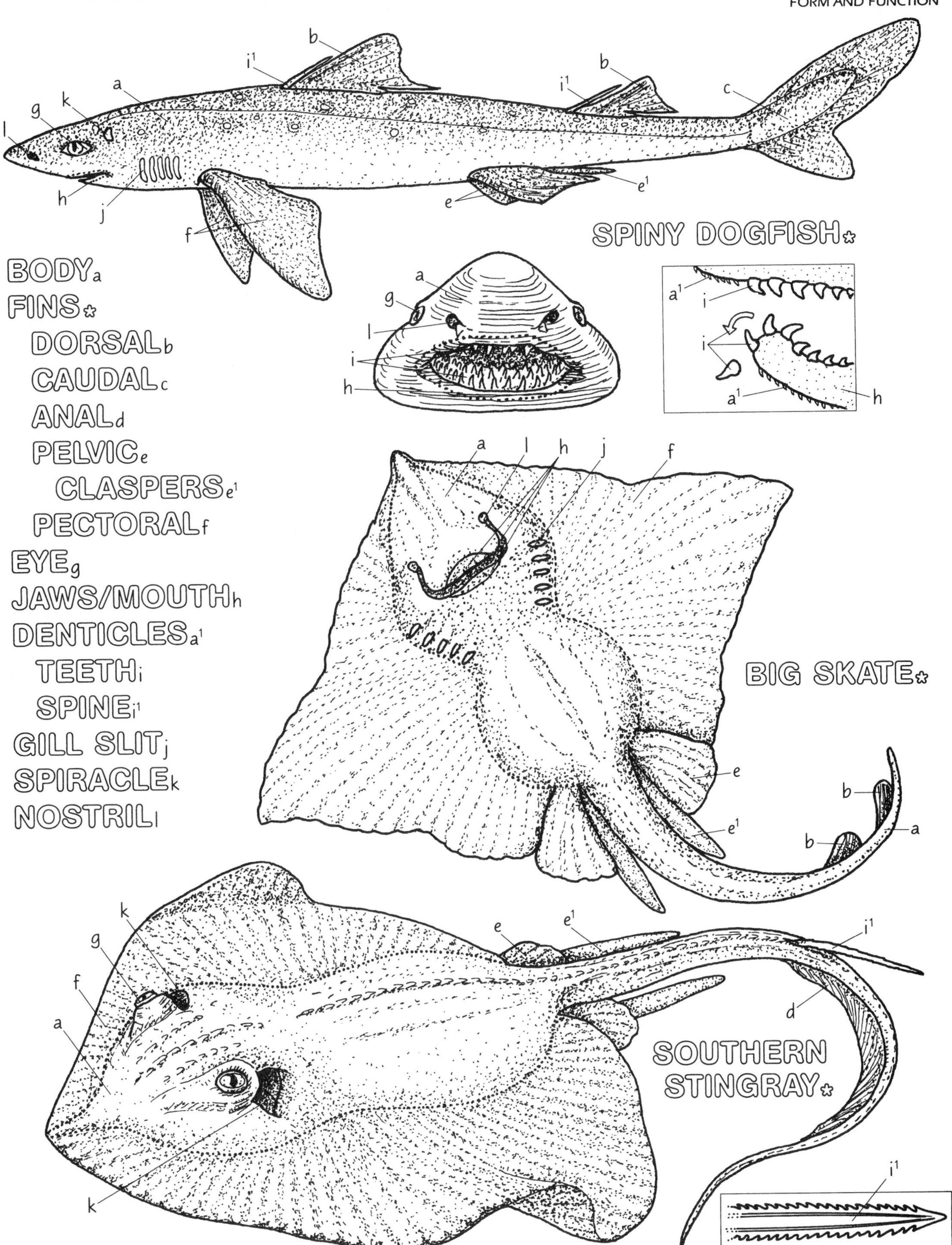

51 ELASMOBRANCH FISH DIVERSITY: SHARKS

There are about 250 species of sharks, and many of these are widely distributed in the world's oceans. This plate introduces some of the unusual or more notorious members of the group.

Begin with the basking shark and color each animal separately as it is discussed in the text. Most sharks are gray in color. The arrows indicate the feeding current in the basking shark. Note that the jaws/mouth of the hammerhead do not show in the larger view, and it and the thresher shark lack a keel. The lower half of the great white shark is not to be colored in either drawing so as to give the natural appearance of countershading.

The basking shark is the second largest fish in the sea, reaching lengths of 9 meters (30 ft) or longer. The filter-feeding basking shark is depicted with its cavernous *mouth* open, in feeding position as it swims slowly along, engulfing large zooplankters and small fish. Water passes out through the large *gill slits* while small prey are caught on the gill raker surfaces, then swallowed. Basking sharks are thought to be harmless to humans. Traveling in schools of up to 100 individuals, basking sharks are found in the temperate zones of the Atlantic and Pacific oceans on both sides of the equator. The basking shark is sometimes hunted for its oil-rich liver because of its high vitamin content.

The large (6 meters, 20 ft) hammerhead shark has all the typical shark characteristics except for the head, where flattened cephalic lobes create a rectangular shape suggesting a hammer. The *eyes* and *nostrils* are located at the tip of these lobes. As the hammerhead swims, it swings its head back and forth. This behavior is thought to increase the likelihood that the shark's sensory receptors will detect food by broadening its swath through the water. In addition, the flattened head may act to increase the shark's maneuverability. Hammerheads are found in tropical and subtropical waters of all oceans, and some attacks on humans have been recorded.

The great white shark has a cosmopolitan, temperate water distribution, ranging into cool waters like those found off central California. The great white accounts for the majority of verified shark attacks on humans, but the reasons for this are not agreed upon. Some experts believe great whites are territorial and see humans as intruders in their territory. Others feel divers and surfers are mistaken for seals or other marine mammals; still others feel that the movements of swimmers, mistaken for those of distressed fish, attract the shark. Another plausible theory is that the great white shark is mainly curious and, unlike most sharks that "feel" an unknown substance by bumping it with their snouts, the great white "feels" with its mouth. Even a gentle "feel" or taste with its mighty tooth-studded *jaws* can be fatal to a human. Observations of great white shark attacks on seals and sea lions suggest the shark lunges at its prey and takes a large bite. Backing away to avoid the dangerous thrashing flippers, the shark waits for the prey to bleed to death before devouring it. If the "prey" happens to be a human diver or surfer, the victim may have time to get out of the water before the shark returns. Perhaps this explains why many shark attack victims survive. The great white is recognizable by its large black eyes and its lunate *caudal fin*, which is not the typical shark heterocercal form. The great white is also stouter than most sharks. A record great white, over 6 meters in length, was taken off the coast of Cuba. Such a shark weighs well over 1600 kilograms (3500 lb).

The thresher shark is immediately recognized by the greatly elongated dorsal lobe of its caudal fin. The caudal fin may account for up to one-half the shark's body length, which may reach 6 meters; threshers weigh up to 450 kilograms (1000 lb). They have never been implicated in attacks on humans. Threshers are found in the subtropical and temperate waters of the Atlantic and Pacific oceans, often feeding in groups on schools of small fish. They swim circles around their prey, using the large caudal fins to stun and scare the fish into a group that can be fed upon more easily ("threshing behavior").

SHARKS

BODY a
FINS *
DORSAL b
CAUDAL c
ANAL d
PELVIC e
PECTORAL f
EYE g
JAWS/MOUTH h
GILL SLIT i
KEEL j
NOSTRIL k

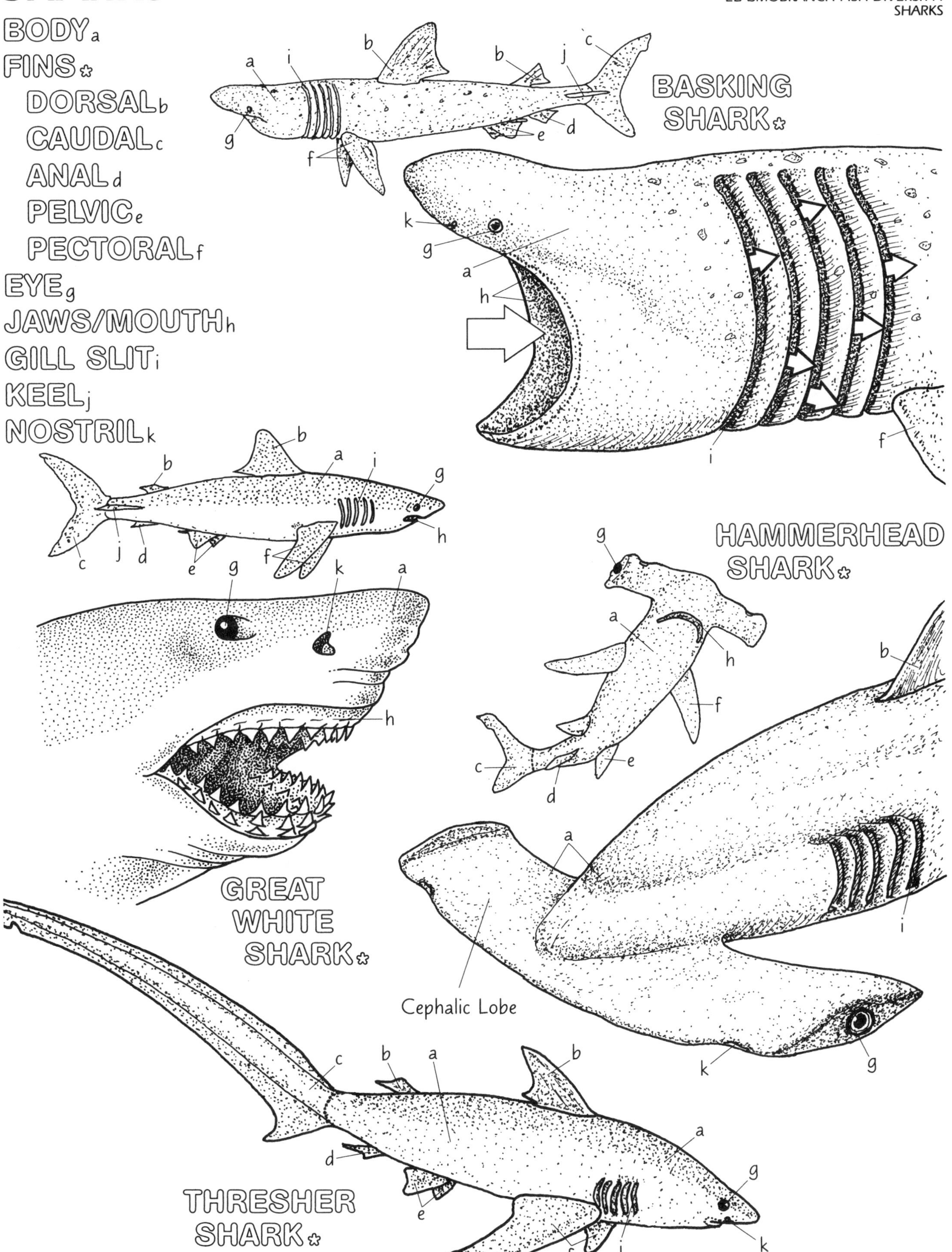

52 ELASMOBRANCH FISH DIVERSITY: RAYS AND RELATIVES

Color the manta ray at the top of the page.

The Atlantic manta ray, like all mantas, possesses special *cephalic fins* on either side of its *mouth*. These fins are the anterior portion of the *pectoral fins* and serve to form a funnel through which plankton and small fish are directed into the mouth of the filter-feeding manta. Because the cephalic fins are widely separated and held in front of the *body,* they were called "horns" by fisherman who referred to the manta as a devilfish. Mantas are graceful creatures which swim by slowly beating their huge, winglike pectoral fins. The Atlantic manta shown here may reach a width of over 6 meters (20 ft) and weigh in excess of 1360 kilograms (3000 lb). It is dark gray on the dorsal surface and creamy white underneath. Mantas are renowned for their habit of leaping high into the air and coming down like a shot; this may serve to loosen parasites or perhaps to stun fish.

Now color the spotted eagle ray.

The spotted eagle ray is found inshore in all tropical and warm temperate seas. This large, active ray (over 2 meters, 6 ft in length; 230 kg, 500 lb) is generally seen swimming near the bottom joined in formation with several other rays. Its long tail (more than three times its body length) is characterized by one to five *spines* near its base and is held straight out behind the body. The placement of the spines makes them a questionable defensive weapon compared to the posteriorly located spines of other stingrays. The eagle ray's body is thicker than other stingrays, and the head projects forward from the disc formed by the pectoral fins and body. The head is thick, box shaped, and harbors powerful jaws with pavementlike teeth used in crushing bottom-dwelling crustaceans and molluscan prey. The *eyes* and *spiracles* are mounted on the sides of the head rather than on top. Eagle rays are green with dark spots.

Large numbers of eagle rays often enter estuaries at high tide to feed on commercial shellfish. They use their pectoral fins like plungers ("plumber's friend") to pop shellfish out of their burrows; their passage can leave a mudflat looking like a minefield. Oystermen often construct fences and other deterrents to foil the marauding rays.

Next color the sawfish.

The gray sawfish looks distinctly un-raylike. However, it shares with the rays a bottom-dwelling existence and ventrally placed gill slits. Its form is dominated by the long, paddlelike extension of the head that bears pairs of *teeth* (actually modified denticles like shark teeth and stingray spines). The sawfish uses its "saw" to dig shellfish out of the bottom and to stun fish. It will swim through a school of fish slashing wildly and impaling or stunning fish, which it then eats. Sawfish also use their weapon defensively against natural predators and humans.

The body of the sawfish is decidedly more like a shark than a ray, and it swims by using its *caudal fin* in typical shark fashion. These animals can be quite large (over 6 meters, 20 ft, in length and 365 kg, 800 lb) and are found inshore in all tropical seas. They often enter brackish and even fresh water, as demonstrated by a large population of sawfish landlocked in Lake Nicaragua.

Now color the electric ray. The electric organ is located beneath the skin and is given a separate color for illustration purposes.

The giant electric ray is thick and flabby compared to other rays, and its almost circular body is smooth and without scales. The dark gray body is continuous with a relatively thick tail characterized by two *dorsal fins* and a distinct caudal fin. It is white underneath.

Two large kidney-shaped *electric organs* are located on the pectoral fins. These organs are modified muscle tissue directly innervated by the brain. They contain special hexagonal-shaped tissues that are, in essence, storage batteries wired in series. The electric organs account for one-sixth of the ray's total weight and can generate an initial shock in excess of 200 volts, enough to electrocute a large fish.

Ichthyologists (scientists who study fish) diving at night off southern California reefs have found that the electric ray emerges from its half-buried, daylight repose and actively hunts along the reefs. When it physically encounters other fish it shocks them with electricity and then uses its pectoral fins to move the stunned, inactive fish into its ventrally located mouth.

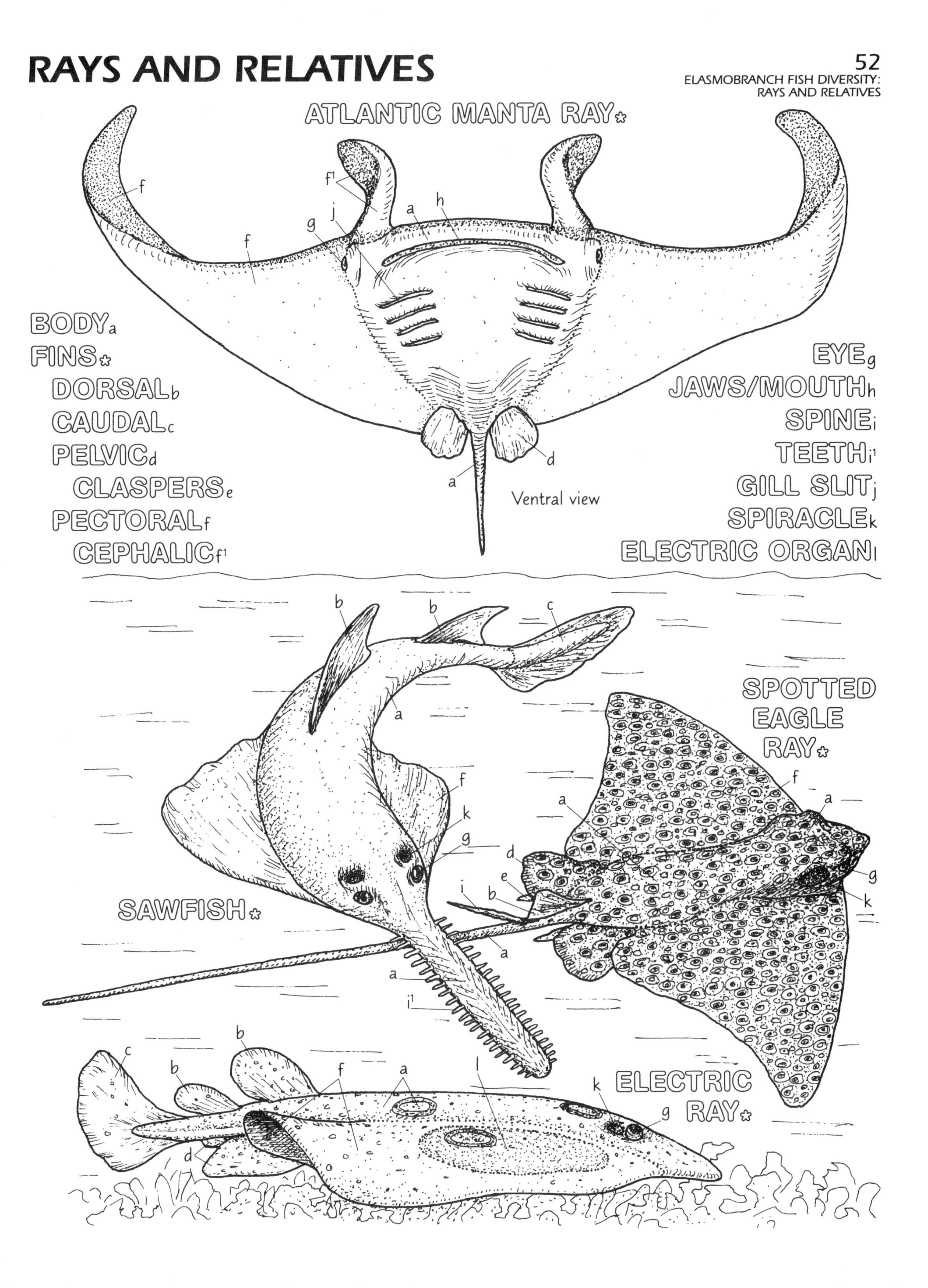
RAYS AND RELATIVES
52
ELASMOBRANCH FISH DIVERSITY:
RAYS AND RELATIVES
ATLANTIC MANTA RAY*
BODYa
FINS*
DORSALb
CAUDALc
PELVICd
CLASPERSe
PECTORALf
CEPHALICf1
EYEg
JAWS/MOUTHh
SPINEi
TEETHi1
GILL SLITj
SPIRACLEk
ELECTRIC ORGANl
Ventral view
SPOTTED
EAGLE
RAY*
SAWFISH*
ELECTRIC
RAY*

53
MARINE REPTILES: TURTLES AND SEA SNAKES

In terms of both number of individuals and numbers of species, fishes are by far the most successful group of vertebrates in the sea. Reptiles, on the other hand, are represented by only a few species. Most of these are still tied to the land because they must lay their eggs on shore; three of the four reptiles discussed in this and the next plate share this requirement. Here two reptiles are introduced: the green sea turtle and the yellow-bellied sea snake.

Begin by coloring the adult green sea turtles at the top of the page. Color the four illustrations of egg laying, hatching, and juveniles. Note that the broad tracks receive the same color as the forelimbs that created them.

The green sea turtle, an endangered species of the Caribbean Sea and the Atlantic and Pacific oceans, is one of several species of turtle that spend their lives at sea. Others include such exotically-named species as the hawksbill, leatherback, and loggerhead. Turtles have roamed the seas since before the dinosaurs and were very successful until humans entered the picture. Many sea turtles feed on gelatinous marine zooplankton, jellyfish and such. The turtles mistake floating plastic bags and deflated helium balloons that have drifted out to sea for their prey. The inedible plastic and latex blocks the turtles' intestinal tracts and they starve to death. Many other sea turtles perish in the nets of shrimp fishermen, although a turtle exclusion device (TED) is available and effectively allows turtles to escape from the net.

Sea turtles nest on broad sandy beaches. Once every three years, an adult female green sea turtle undertakes a journey back to the beach where she was hatched to lay her own *eggs*. For some turtles, these migrations may be several hundred kilometers long. Males and females mate in the surf just offshore from the rookery. The male grasps the female with his large *forelimbs* and transfers his sperm to her, as shown in the upper drawing. After a few days, the female makes a nocturnal trip onto the beach. She pulls herself up the beach with her forelimbs, all the way to the dry sand of the upper beach. She digs a broad pit with her forelimbs and then delicately excavates a bottle-shaped *burrow* with her agile *hind limbs* (center illustration, far left). The female lays approximately 100 leathery-skinned eggs in the burrow and carefully covers them with sand. She buries the pit entirely and throws sand all about to disguise the location of the nest. Her job completed, the female returns to the sea. Her broad *tracks* left behind indicate the difficulty this turtle has in moving on land (center illustration, second from left). The forelimbs are modified into highly effective swimming flippers, but they cannot lift her bulk off the sand. Similarly, her *carapace* is much reduced and streamlined for swimming; it does not serve as a fortresslike retreat, unlike those of many freshwater and terrestrial turtles.

Before leaving the breeding grounds, the female may return to the beach to lay eggs as many as five times at 15-day intervals. The eggs incubate in the warm sand for about 60 days, and the young hatch all at once and begin to dig to the surface (center illustration, second from right). They emerge at night and instinctively find their way to the ocean (center illustration, far right). The green turtles are most vulnerable to predation during their time in the burrow and during their scramble to the sea.

The young turtles remain at sea and do not reappear in sea grass beds until at least one year later. When four to six years old, the females will return to the exact stretch of beach where they hatched and contribute to the next generation.

Color the light portion of the yellow-bellied sea snake golden yellow and the dark pattern black. Note that the flattened tail receives a different color.

Most zoologists agree that snakes evolved from lizards and are the most modern of the reptile groups. Sea snakes are found in shallow tropical and subtropical waters. All are related to the cobra family. They have a potent venom that can cause severe injury to humans. The *yellow-bellied sea snake* is found in the Pacific off the coast from Panama to Mexico. It is commonly seen on the surface, often in aggregations of several hundred individuals.

Sea snakes are well adapted to a marine existence. Many give birth to the young alive at sea, and the newborn snakes can immediately swim on their own. The sea snake has a *flattened tail* used as a paddle in swimming. Sea snakes generally feed on fish, and can remain submerged for thirty minutes or longer between breaths. Most species are docile, although some attacks on divers have been reported. They are best appreciated from a distance.

TURTLES AND SEA SNAKES

GREEN SEA TURTLE*
HEAD a
FORELIMB b
TRACK b1
HIND LIMB c
CARAPACE d
TAIL e
EGG f
BURROW g
JUVENILE h

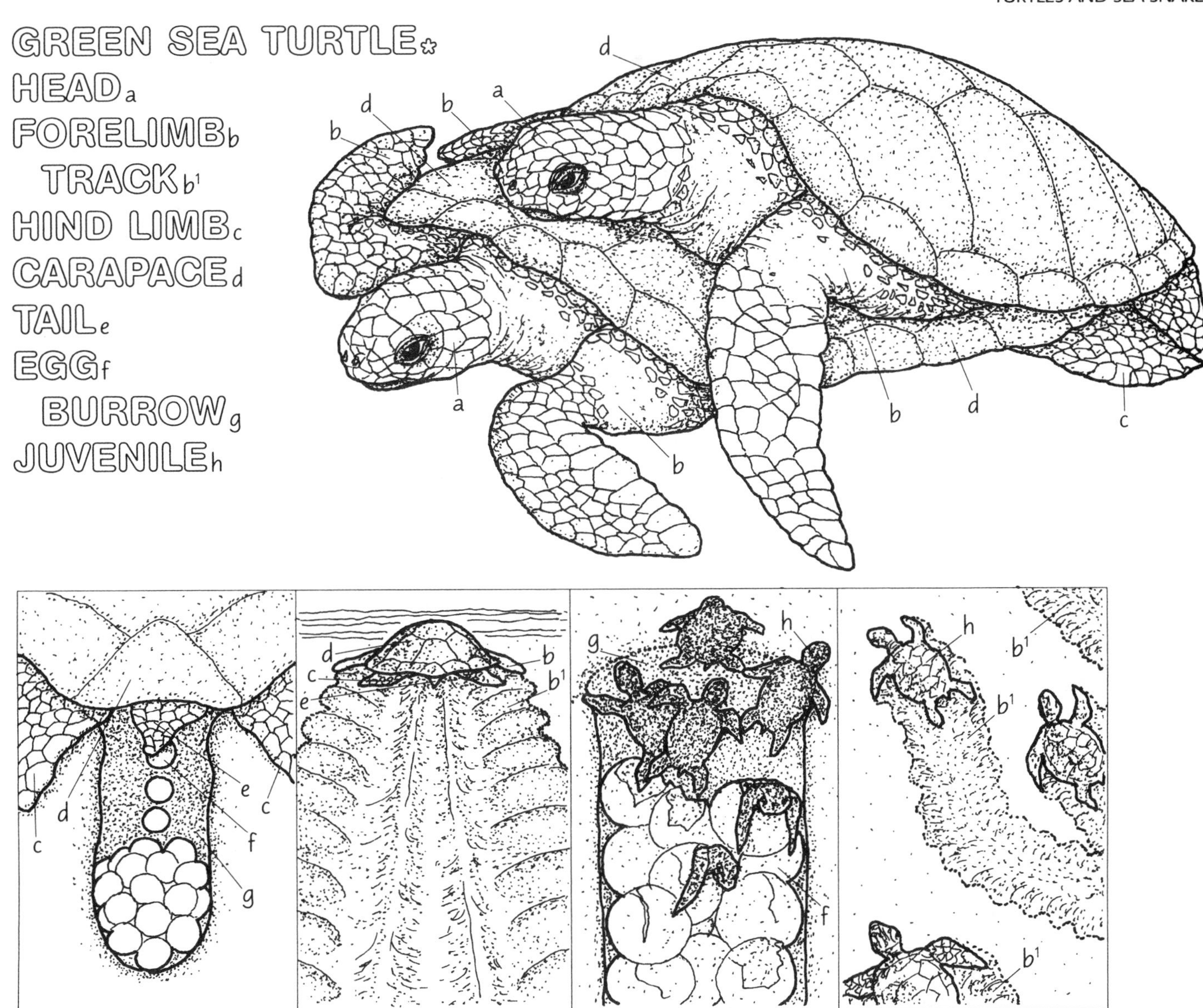

YELLOW-BELLIED SEA SNAKE i
FLATTENED TAIL j

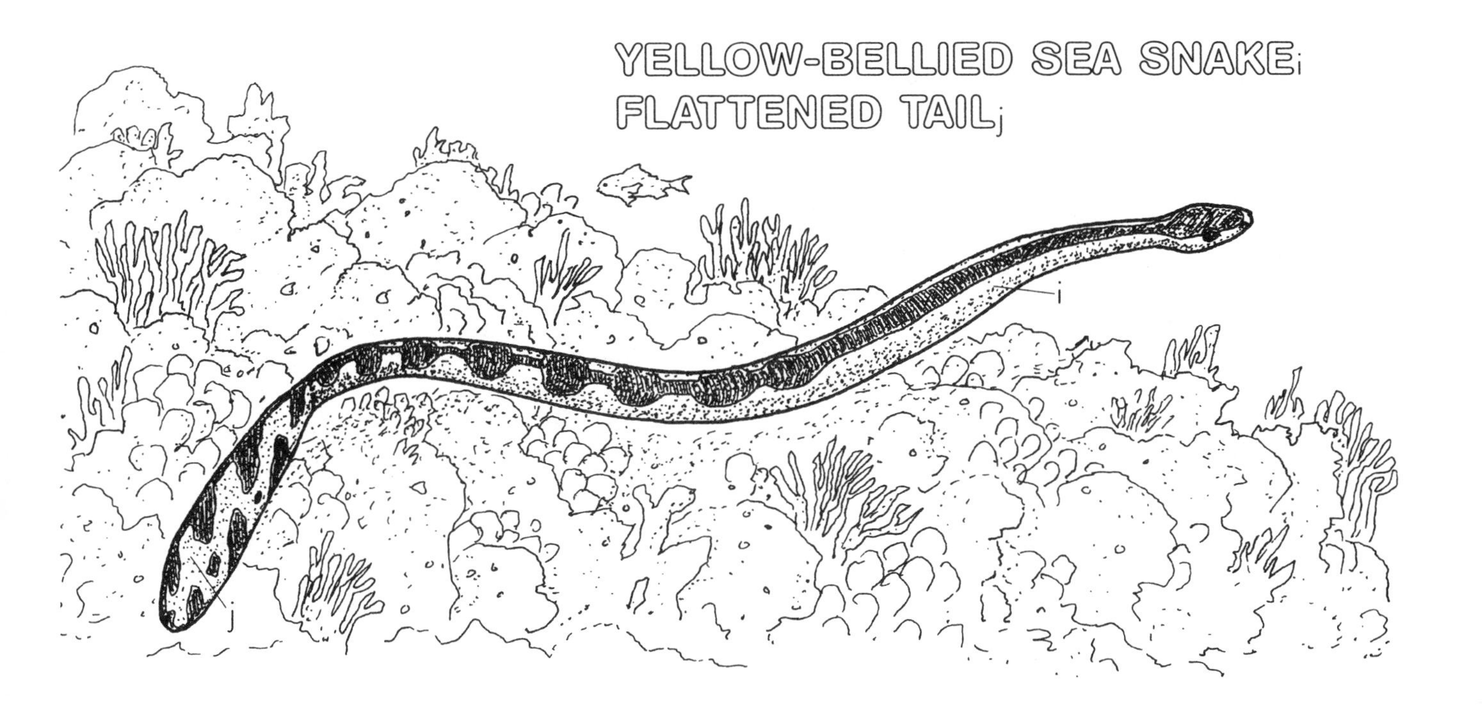

54
MARINE REPTILES: IGUANAS AND CROCODILES

In this plate, we examine the remaining reptilian groups, the lizards and crocodilians. Each has a single marine species: the marine iguana and the saltwater crocodile.

Color the iguanas on the rocks and in the water.

Marine iguanas are found exclusively on the Galapagos Islands of the southeastern Pacific Ocean. At one time, they were estimated to number between 200,000 and 300,000. As many as 4500 of these lizards were counted along a single mile of the lava rock intertidal zone. Unfortunately, in modern times their numbers have significantly declined because of the introduction of feral dogs and cats to the islands.

Marine iguanas are fearsome in appearance, but are actually quite docile and can be easily approached by humans. The larger males in some island populations can reach 60 cm (2 ft) in length. The *head* is blunt and a ridge of stout light-colored *spines* runs from the back of the head to the base of the *tail*. Typically dark colored, the *body* can be splotched with green and red in local populations. Marine iguanas are estimated to live up to 20 years.

Marine iguanas are herbivores, subsisting on intertidal and subtidal *algae*. They occasionally supplement their diet with grasshoppers and crustaceans. The juvenile and female iguanas feed primarily on intertidal algae while the larger, older males swim offshore sculling with their large tails, and dive to depths of over 14 meters (45 ft). Here they feed underwater, staying submerged as long as 60 minutes, but more typically 6–10 minutes per dive.

Two problems arise from marine feeding. First, the lizards swallow salt water while feeding and must purge their salt load. Marine iguanas have very efficient salt glands which are located above their *eyes* and drain into their *nostrils*. The iguanas frequently sneeze to expel the accumulated saline secretion and this accounts for their often white-encrusted heads. The second problem is heat loss. The cold-blooded iguanas can lose as much as 10°C (18°F) of body heat from their once-a-day feeding bout in the cold water. To recover, they sprawl out on the lava rocks under the tropical sun. Overheating is regulated by the dilation and constriction of blood vessels in their chests. At dusk, the marine iguanas pile together to conserve heat against the cool night air.

Now color the saltwater crocodile.

Unlike the recently-evolved lizards, crocodilians were well established when the dinosaurs came into their own. Crocodilians include crocodiles, alligators, caimans, and gharials. Although several species of crocodile and alligator are known to occur in estuaries and brackish water, only the saltwater crocodile of northern Australia and Southeast Asia can truly be considered "marine." Saltwater crocodiles have efficient salt excreting glands on their tongues, and are known to have crossed 1000 km (600 mile) stretches of salt water to reach remote islands such as the Seychelles in the middle of the Indian Ocean.

Dubbed the "Saltie" by Australians, the saltwater crocodile is the largest living reptile. Stories of males 7 to 8 meters (22–26 ft) long persist, although these have most likely been hunted out. In modern times, a male is considered large at 5 meters (16 ft), and the females are much smaller at 2.5 meters (8 ft) long. They are estimated to live between 70 and 90 years.

Saltwater crocodiles live in tidal rivers and estuaries. The juveniles are chased out by territorial adults during mating season and move along the coast looking for their own tidal river. Salties propel their huge bodies with broad sweeps of their powerful tails. However, they are very inactive animals and not capable of sustained exercise. They float with only the tops of their long snouts above water, exposing the nostrils and eyes. Their hunting strategy relies on surprise with a quick lunge and capture in one fluid motion. They eat a large variety of prey including fish, crabs, feral pigs, buffalos, and domestic cattle. One common name, man-eating crocodile, is probably an overstatement, although several people are killed or injured every year by Salties in Australia. The animals are capable of quick bursts of speed on land, but quickly tire and can be outrun by a fit human. This is not a recommended activity.

Color the mother crocodile and the nest with the incubating young.

Salties are doting mothers. After courtship and mating, females build *nests* of mud and vegetation well above water level to avoid flooding and lay clutches of 40–60 eggs. The females stay close by the nest during the three month nesting period while the warmth of the decaying vegetation incubates the developing embryos. The mothers listen for the chirps of the emerging *young*, and then excavate the nest and move the young to the safety of the water. About 30 cm (1 ft) long at hatching, the young Salties grow quickly, reaching 1.5 meters (5 ft) in about two years.

IGUANAS AND CROCODILES

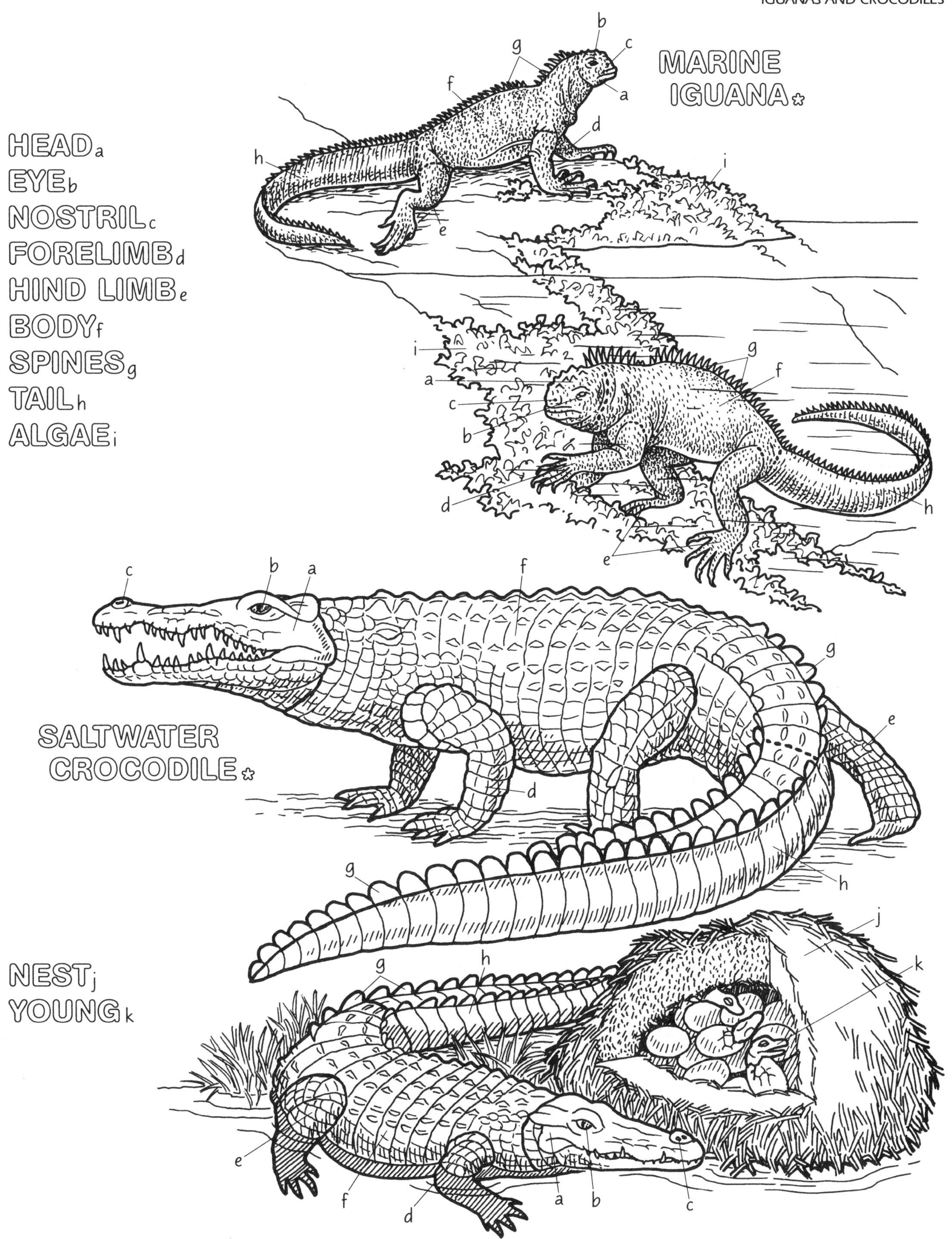

55
MARINE BIRDS: FORM AND FUNCTION

Like their fellow warm-blooded marine animals the whales, birds travel long distances to areas of high productivity using their superior power of locomotion. A bird is a flying machine and every part of its body is adapted to the necessities of flight. The structures must be as light and aerodynamic as possible while maintaining strength. Bones are hollow with strengthening internal struts to provide a rigid airframe. They can also be fused with another bone for strength. The lungs have a series of associated air sacs that ramify throughout the body to increase ventilation rate and capacity. The bird's metabolism is rapid and food is processed quickly. Therefore, the weight does not have to be borne any longer than necessary.

Color the regions of the sea gull's body. Note that the bones of the right wing are exposed in the drawing and should be colored with the wing color. Most sea gulls have a white body with wings of gray or black. Their bills and legs are usually yellow, and their heads are either black or white. Also color the types of feathers; note the asymmetry of the primary flight feather. Finally, color the cross-sectional view of the wing, showing the wing's function as an airfoil.

Wings and *feathers* are the key to successful flight. Wings are modified forelimbs and have all the same bones as our arms. The bones of the wrist and fingers are fused for strength. Feathers consist of keratin, the same material that makes up our hair and nails. It is a tough, versatile protein that gives the feather its light weight and strength. Birds have two basic types of feather: down and contour. Short, fluffy *down feathers* serve as insulation. They are next to the skin of the *body* and are overlain by *contour feathers* which provide shape and a smooth outer surface. The wing feathers, called flight feathers, and the *tail feathers* are special types of contour feathers.

The inner wing, which generates lift, is covered by *secondary flight feathers* or secondaries. In cross-section, the inner wing has the classic configuration of an airfoil, curved on the top and flat on the bottom. Air flowing over the upper surface must move faster to get around the curved shape and this increased speed reduces the air pressure on the upper surface. Air can flow unimpeded under the wing so there is no reduction in air pressure. The net result is upward lift.

The outer wing, which provides propulsion, is covered by *primary flight feathers* or primaries. The side branches or barbs of each feather are short on the leading edge, and long on the trailing edge. Birds move through the air using a strong downward and backward power stroke with the large surface area of the trailing edge of the primary feathers. This is followed by an upward and foreword recovery stroke during which the shorter barbs on the leading edge provide less resistance as they are pulled through the air. If you watched from the side as the bird flies, you would see that the wing tip traces counterclockwise circles, with the primary feathers functioning like the propellers of an aircraft. Some birds are able to generate enough lift to take off by rapidly flapping their wings in place. Others must get a running start along the land or water or jump from some height before sufficient lift is achieved to become airborne. The bird's tail functions as a stabilizer or rudder in flight, and a braking device when landing.

Wing shape and size determine how a bird will fly. A short, wide wing provides superior lift and maneuverability, but requires considerable energy to move it through the air. It does not allow sustained, long distance flight. A long, narrow wing tapered at the tip moves more easily, and allows sustained, fast flying but less maneuverability. Many birds take advantage of updrafts and use their very long, narrow wings to soar and glide with few wing beats.

The bird's *bill* consists of a horny sheath covering the bony components of the upper and lower jaws. There is considerable variation in bill shape, from long and pointed in a fish-catching heron to blunt and chisel-shaped in a mollusc-feeding oystercatcher.

Some diving birds use their hind limbs to swim underwater. Their short, stout *legs* are positioned towards the back of the body and their feet are webbed. These birds are excellent swimmers, but awkward on land. Herons, on the other hand, have very long legs for wading in ponds and estuaries and are quite agile on land. However, their legs are so long they can not be tucked up against their bodies, and are held well out behind them as they fly.

FORM AND FUNCTION

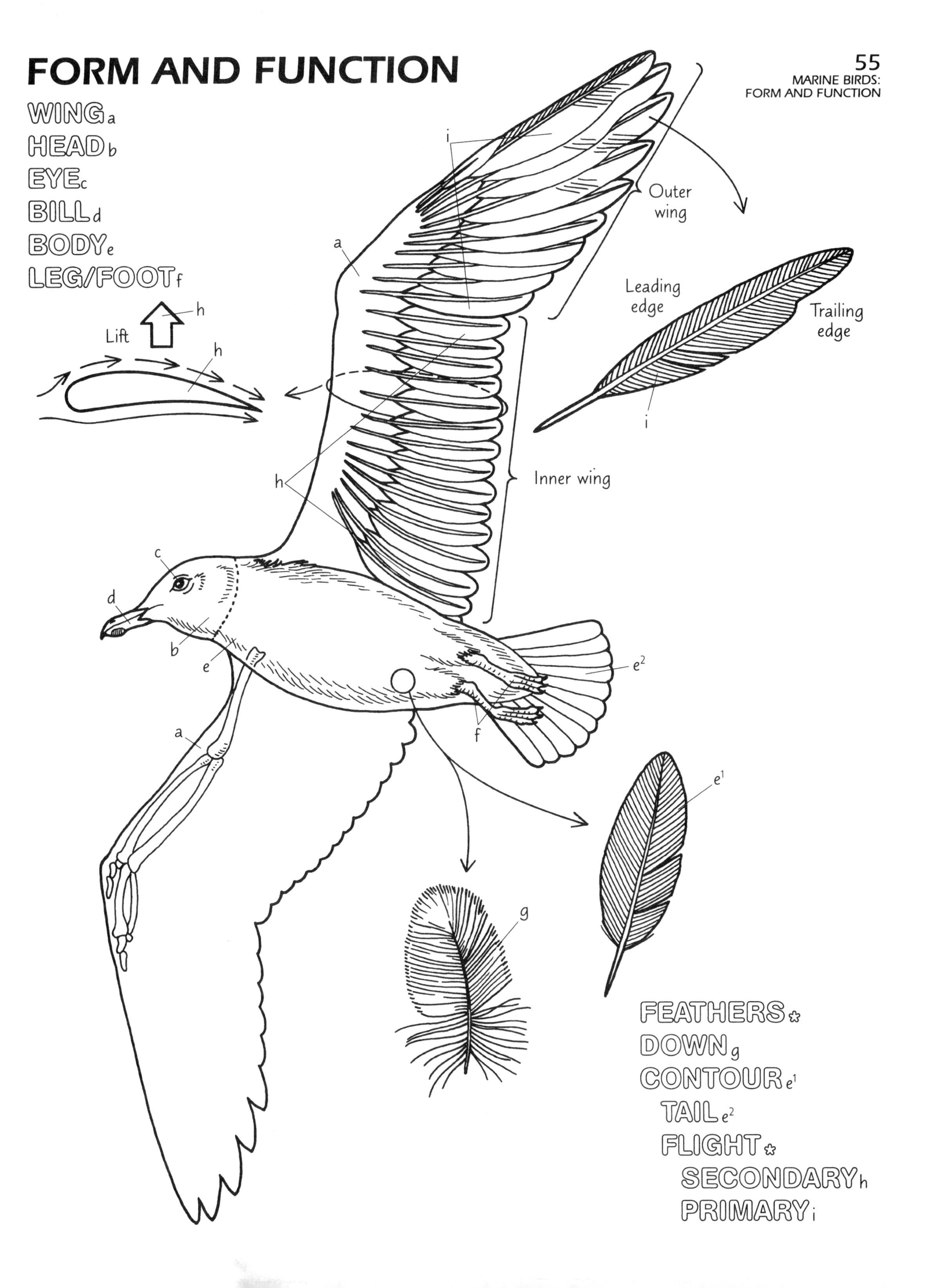

56
MARINE BIRDS: SHORE BIRDS

By looking at birds of specific marine environments, we can get some idea of the many different kinds of marine birds, and the way they utilize marine resources. This plate investigates birds of intertidal habitats, collectively called shore birds.

Color each bird as it is discussed; note the outstanding features highlighted in the text. Not all structures on all birds may be visible to color.

Herons and egrets are large solitary birds commonly seen near the shoreline. The great blue heron is one of the most striking, with its blue-gray *body* and pale yellow *legs* reaching over a meter in height. Great blue herons can be seen foraging in shore environments, such as tidepools and estuarine salt marshes, standing motionless with their long necks cocked in position ready to strike. They are also found inland as well, in lakes, streams, pastures, and even vacant lots! Unmistakable because of its large size and streaming black head plumage contrasted against its white head, the heron wades into shallow water on its long legs and snaps up small fish, crabs, and other prey with lighting thrusts of its long neck and sharp, straight, pale yellow *bill.* When disturbed, the heron lets out an annoyed squawk, lifts itself on huge blue-gray *wings* and flies away, folding back its neck and trailing its long legs behind.

Many birds share the estuary environment with the heron and forage on the exposed mud and sand flats for the burrowed worms, clams, and crustaceans. The birds very effectively divide up the buried prey of the tidal flats by probing to different depths with their bills and foraging in different parts of the tidal flats. Birds with short legs are limited to the exposed flat, while others with longer legs can wade into shallow water. Many of these shore birds belong to the sandpiper group which consists of over 80 species. Sandpipers are also noted for their long distance migrations, many spanning two hemispheres. Their narrow, tapered wings are designed for these sustained migratory flights.

The long-billed curlew can penetrate well into the sand with its long, dark brown, sickle-shaped bill and extract deeply buried shellfish and crustaceans. The long bluish-gray legs of this mottled chestnut-brown, pigeon-sized sandpiper also allow it to patrol the shallow water of a retreating tide and snatch up prey that have not completely covered themselves against the exposure of low tide.

While its other sandpiper cousins are busily probing for buried prey, the ruddy turnstone takes an entirely different approach to dinner. Roughly the same size as the curlew but with shorter, orange-red legs, the turnstone can be seen moving about an exposed estuary flat, systematically flipping over every rock and shell it encounters on the surface with its short, black, pointed bill. It quickly gobbles up the small crabs and other crustaceans that have retreated under these "safe havens" to avoid exposure by the low tide. Turnstones can also be seen performing their act in rocky intertidal habitats. They are unmistakable in breeding season as new feathers grow in to give them white bodies and harlequin-patterned heads and wings.

Turnstones are usually seen in small numbers, either alone or in twos or threes. Sanderlings, in contrast, occur in flocks of 10 to over 100. Sanderlings are also members of the sandpiper family. These robin-sized birds have short, black bills and short, black legs and forage along sandy beaches. Sanderlings look like wind-up toys as they race up the beach to avoid the incoming waves only to turn and quickly follow the wave's retreat to hunt for sand crabs, worms, and small clams in the still wet sand. When disturbed, the sanderling flock takes flight as one, flashing their bold white wing stripes on their dark gray wings, and putting on a show of aerial precision that leaves one slack-jawed with wonder.

The final shore bird considered here is the oystercatcher. As the name implies, these black, stocky, short-legged birds feed on shellfish, including oysters, clams, and mussels as well as limpets and other shelled prey. Shown here is the black oystercatcher of the Pacific coast of North America. Its most notable feature is its bright red-orange bill. The bill of the oystercatcher is long and relatively heavy, coming to a flat, blunt chisel point. With a quick hammer, the oystercatcher's bill smashes open the shells of its prey to expose the soft body within. The behavior of European oystercatchers was studied by pioneer animal behaviorist Conrad Lorenz. Newly fledged chicks must be taught how to use their bills to open shellfish. Opening the shell of a particular prey item is learned from the bird's parents. Lorenz determined that specific lineages of oystercatchers had unique ways of opening a given species of shellfish, and these behaviors were passed down from one generation to the next. For example, oystercatchers in one family lineage opened a clam by hammering through the dorsally-located hinge, while another familial group opened the same species of clam by chipping away the ventral (bottom) edge of the shell.

SHORE BIRDS

BODY a
WING b
EYE c
BILL d
LEG/FOOT e

57
MARINE BIRDS: NEAR-SHORE BIRDS

Once you leave intertidal habitats and shore birds, the lines between the marine bird groups blur. The arbitrary division selected here is between near-shore birds and oceanic or pelagic birds. Near-shore birds are more typically visible from shore, or are at least found on the continental shelf and maintain relatively frequent contact with land. Oceanic or pelagic birds are usually found beyond the continental shelf and stay at sea for extended periods, perhaps coming ashore only to reproduce.

Color each bird as it is discussed, and note the outstanding features highlighted in the text. Not all structures on all birds may be visible to color.

The ubiquitous gull is the most common seabird of the near-shore in the northern hemisphere. Sea gull species range significantly in size from the 25 cm (10 in) long, 4.5 kg (10 lb) little gull to the great black-backed gull of Europe which is over 75 cm (30 in) long and weighs over 22 kg (48 lb). Most gulls are white underneath with gray or black plumage on the back and *wings*. There is also a group of small gulls that have black heads. Juvenile gulls are often almost as big as their parents, but have mottled gray coloration. Gulls use a variety of marine environments and forage in intertidal habitats along with shore birds as well as on the open ocean. Although some species are fearsome predators, the typical gull species is an all-purpose scavenger, seeking live prey, carrion, fish entrails, or unguarded picnic lunches with equal panache. The relatively long *bill* ends in a hook for tearing flesh. Gull wings are medium length and tapered for sustained flight and good gliding ability. They have stout *legs* and webbed *feet* for strong swimming on the water surface. Gulls will surface dive or make shallow plunge dives for small fish near the surface. The gull is not above frequenting land fills, picking through garbage, or following a farmer's plow as it cultivates a field.

The pelican is a much more tasteful diner than the gull and certainly more specialized. Shown here is the brown pelican which feeds by plunge diving. It has a long, flat bill, gray on the top and dark underneath, with a hook at the end. Attached to the lower portion of the bill is a fold of elastic skin which can expand into a deep pouch. After spotting prey from aloft, the pelican makes a spiraling dive and, at the last second, thrusts its legs and great wings backward as it enters the water, bill first. It dives down a few feet in pursuit of fish and scoops them up with its large pouch capable of holding 10 liters (2.6 gal) of water. The skin of the pouch contracts and forces the water out, leaving the fish behind. Once on the surface, the pelican tips back its head and swallows the catch. With a few running hops to generate sufficient lift, the pelican is airborne for another pass. Not all pelicans are plunge feeders. Small groups of white pelicans swim in horseshoe-shaped "scare lines," herding schools of small fish on the surface into a ball where they are scooped up with simultaneous dips of the birds' great pouches.

There is no more tranquil a sight at sunset than a line of pelicans soaring just above the water on wings that can span 3 meters (10 ft) as they ride the updraft created by a moving wave. The majestic birds are no doubt on their way to a nocturnal roosting site, such as a quiet pier or breakwater.

Dusk is also the time to look for another very interesting bird, the skimmer. Skimmers feed at dusk and also at night, the times when the sea tends to calm and small fish come to the surface. Skimmers fly and glide along on long, sharply-tapered, black wings with their lower bill dipping into the water. Skimmers are unique in that the lower portion of their orange, black-tipped bill is longer than the upper portion. When a skimmer comes in contact with a small fish on the surface, its black head quickly snaps down and closes the bill over the fish. Skimmers are not found on all coasts; they tend to be more semi-tropical in their distribution.

The cormorant is a bird familiar to many sea coast visitors. These evolutionarily primitive birds are frequently seen on man-made platforms, drying their outstretched black wings in the sun. Unlike other sea birds that waterproof their feathers with special oil glands, the cormorant's feathers are not waterproof. The cormorant is a master diver, reaching recorded depths of 50 meters (165 ft). Their sleek black bodies are propelled beneath the surface with a crisp surface dive. Submerged, they swim after their fish prey using their large, dark, webbed feet. The fishing ability of the cormorant has long been utilized by humans. Fishermen in Southeast Asia fish using cormorants controlled with leashes and throat rings to prevent them from swallowing their catch. Like many seabirds, cormorants seek inaccessible places such as the steep faces of sea cliffs or oceanic islands to breed unmolested, often occurring in huge numbers on these rookeries.

NEAR-SHORE BIRDS

BODY a
WING b
EYE c
BILL d
LEG/FOOT e

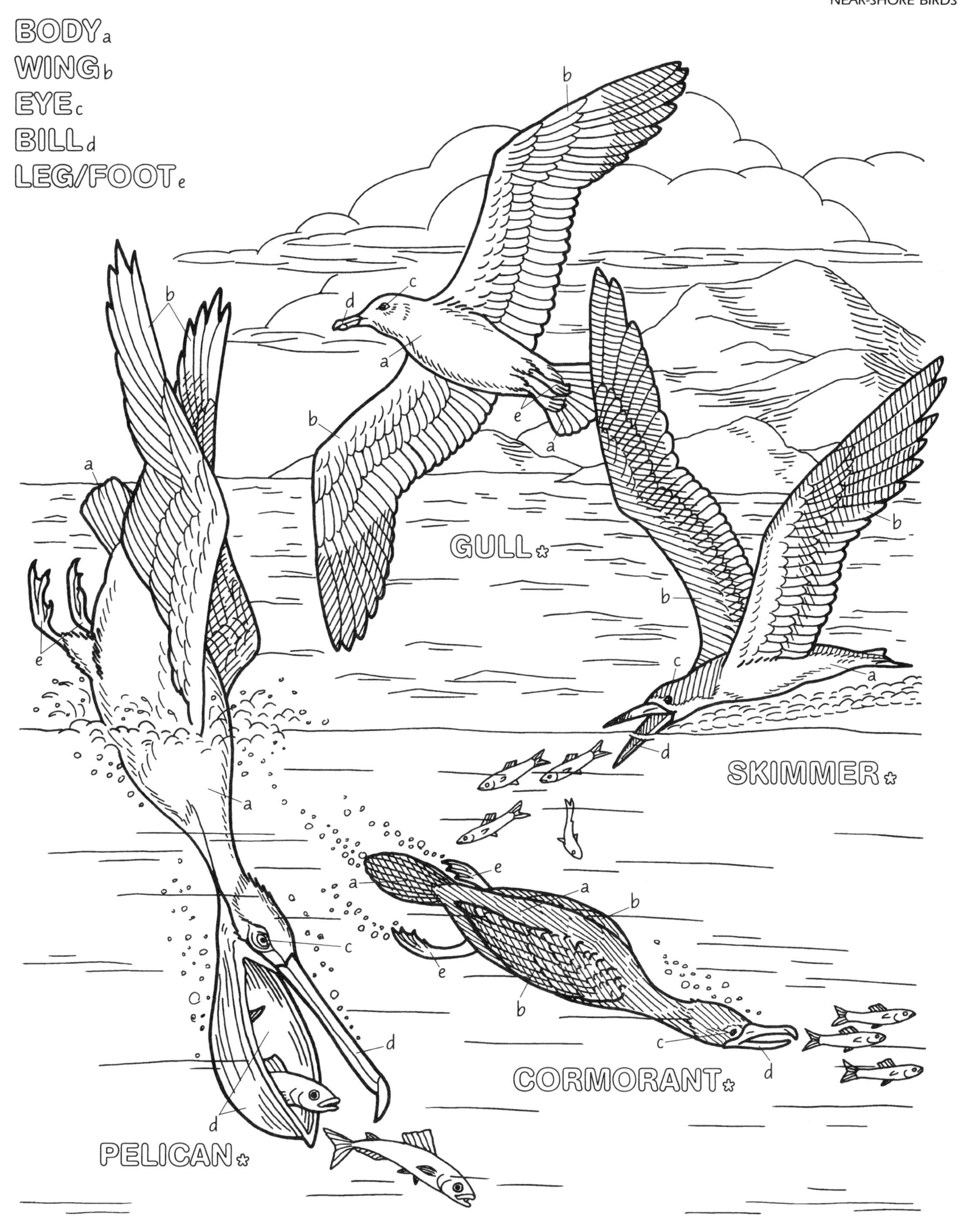

58
MARINE BIRDS: OCEANIC BIRDS

You would not see most oceanic birds unless you went to sea or lived on an oceanic island. These birds take to the sea for extended periods and may not see land for months at a time.

Color each bird as it is discussed, and note the outstanding features highlighted in the text. Not all structures on all birds may be visible to color. The gull being pursued by the skua is not to be colored.

The classic oceanic bird is the royal albatross of the south Atlantic Ocean. Gliding majestically high above the sea on long, narrow, tapering *wings* that may span over 3 meters (10 ft), the albatross epitomizes the oceanic bird. Albatrosses land on the sea to feed, usually at dusk, where they take small fish and squid from the surface or by shallow diving. They may also stay afloat at night to sleep, but by day they are aloft, soaring high on the stiff winds blowing above the sea. The *body* of the albatross is white. Its wings are dark above and white below, and its large, hooked *bill* is colored light pink. Tubular nostrils run along each side of the upper bill, and are the sites for the secretion of salt taken in with food and when the bird drinks sea water. Albatrosses are very long-lived birds, some reaching 50 years old.

Silhouetted against a tropical Atlantic or Pacific sky with huge black, pointed wings and a long, scissor-like tail, the magnificent frigatebird is an unmistakable oceanic bird. Frigatebirds are an enigma. Their black plumage is not waterproof. Their feet are weakly webbed and they cannot swim, yet they are found far out to sea. How do they make a living? Perhaps their other common name, man-of-war bird, will provide a clue. Both common names refer to their pirating lifestyle. Frigatebirds rob other birds of their prey by giving high speed chase until the bird either drops its prey or regurgitates its latest meal which is quickly snatched up by the harassing pirate. Frigatebirds will also fly just above the surface and snap their heads down to make a catch with their strong hooked bills, frequently taking flying fish in midair. Frigatebirds have slim bodies and small weak legs that will not support their weight on land. They nest in noisy colonies in trees.

One of the most formidable seabirds is the great skua of the south Atlantic. Like the frigatebird, this brown, gull-sized bird means trouble because it makes its living by pirating the prey of other seabirds. Skuas may be seen at sea chasing other seabirds, sometimes grabbing them by the wing or tail. They are excellent at aerobatics. Skuas are also very agile on land which makes them bad news on seabird rookeries, where they steal eggs and snatch chicks out of the nests of breeding gulls, terns, and penguins. Some skua species also hunt over land and feed on small mammals such as lemmings. Their large, hooked bills and short legs with webbed feet and sharp claws make short work of prey.

The horned puffin of the north Pacific is sometimes the victim of a harassing skua. At rest on land, this decidedly squat seabird with short, black wings and a large yellow, triangular bill, brightly tipped with red, looks awkward standing stiffly upright on rear-positioned legs. However, once it settles on the sea and dives, its true uniqueness can be appreciated. The puffin pursues its baitfish prey by literally flying through the water using its wings and feet. It is not unusual to see a puffin surface with several small fish lined up cross-wise in its bill. Puffins are members of a group of 21 seabird species of the northern hemisphere called the auks, which includes the guillemots or murres, razorbills, murrelets, and auklets. All these birds share the puffins' superior swimming ability and underwater feeding habit.

The penguin is another great swimmer. In the case of this bird, flight has been lost and the wings function exclusively as flippers, much like those of a sea lion. The wings no longer fold backward like those of other birds, but remain hanging by the penguin's side. They use their webbed feet as rudders to steer. Penguins are found in Antarctica and the surrounding Southern Ocean as far north as the west coasts of South Africa and South America, and the Galapagos Islands. They are able to withstand the cold water because of a layer of subcutaneous body fat or blubber and a dense layer of oily down feathers covered by short contour feathers. Penguins vary in size from the giant 1.2 meter (4 ft) emperor penguin shown here, to the small, 40 cm (16 in) blue or fairy penguin. Penguins go to sea for extended periods, feeding on fish, squid, and shrimp-like, planktonic crustaceans known as krill (Plate 14). Penguins remain in close association even while at sea. Groups of emperor penguins are known to dive to over 240 meters (800 ft) after their squid prey and stay submerged for more than 15 minutes. Like most of its fellow penguin species, the emperor penguin comes ashore to breed and nest and after the chicks are fledged, takes to the sea again.

OCEANIC BIRDS

59 MARINE MAMMALS: FORM AND FUNCTION

Four or more groups of warm-blooded mammals have invaded the seas during evolutionary history. In a relatively short time (fifty millions years), one of these groups has evolved into the largest creatures in the sea: the great whales. In this plate, a representative of each of these four mammalian groups is discussed.

Color the sea otter and the outline illustration.

The least specialized marine mammal is the sea otter; it is also the least changed from its ancestral terrestrial form. Sea otters are closely related to the smaller river otters, and both are classified in the weasel family, along with badgers, wolverines, and minks.

Sea otters have retained all four limbs although they are somewhat modified for a life spent almost entirely in the water. The *forelimbs* are short and have stubby, rounded paws with poorly developed digits; however, sea otters are able to pick up rocks from the bottom and use them as tools. The *hind limbs* of the otter are short, and the feet are large and webbed. Otters swim using their webbed feet and *tail*, which is flattened dorsoventrally to provide a broader sculling surface. Sea otters usually swim on their backs. They also rest, sleep, and eat on their backs, with all four limbs projecting out of the water, possibly to conserve body heat. Unlike most other marine mammals, sea otters do not possess a layer of insulating blubber and must rely on the air trapped in their fine, dense fur to maintain warmth. This fur, which varies in color from reddish brown to black, is highly prized by furriers, and fur hunters nearly rendered the sea otter extinct. Sea otters are relatively large. Male otters often grow to 1.35 meters (4.5 ft) in length, including tail (25–35 cm, 10–14 in), and weigh over 36 kg (80 lb).

Now color the seal lion.

The California sea lion represents the pinniped order, which includes the "earless" seals, sea lions, and the walrus. Their closest terrestrial relatives are the bear and dog. The sea lion's forelimbs are modified into *flippers*, which it uses to "fly" through the water. The hind limbs are also flipperlike, though much less powerful, and the sea lion possesses a short stub of a tail. The sea lion's tawny brown, sleek form is evidence of the streamlining that evolved for ease in swimming. A layer of blubber beneath the skin provides insulation against the cold ocean water.

Color the dugong.

The dugong of Africa and the South Pacific, and its close Atlantic relative, the manatee, are representative of the sea cows. These animals, which grow to 3 meters (10 ft) and can weigh over 400 kg (880 lb), have little hair and rely on a layer of blubber for insulation. Supposedly, sailors mistook them for mermaids, and thus their scientific name (Sirenia) derives from the comely Sirens who lured Odysseus' crew onto the rocks in Greek mythology.

The dugong, whose closest terrestrial relative is the elephant, is the only herbivorous mammal restricted to the sea. It spends its entire life in the water, where its chief food is sea grass. The dugong uses its bristled muzzle to uproot tender young sea grass plants from shallow protected bays and estuaries. In Florida, manatees perform a service to navigation by eating the troublesome water hyacinth that clogs slow-flowing river channels, but have suffered high mortality from motor boats. Manatees are a severely endangered species.

The forelimbs of the dugong are like flippers and the hind limbs are absent. The tail has grown into broad, notched *flukes*, which the dugong moves in powerful beats, propelling itself along at maximum speeds of 21 km/h (13 mph). Although they may appear sluggish, dugongs are actually quite alert and active, and are said to have an intelligence comparable to that of a deer. Although the dugong lacks ear flaps and has only a small opening to the outside, it is able to hear quite well.

Finally, color the dolphin.

The familiar bottlenose dolphin is a representative of the order Cetacea which includes the whales and dolphins. These animals exhibit the most thoroughgoing adaptation to marine existence of all the sea mammals. Like the dugong, the Cetaceans lack external ear flaps, and have lost their hind limbs, relying on horizontally positioned tail flukes for propulsion. The dolphins also possess a rigid, permanently erect *dorsal fin* for swimming stability. The sleek *body* of the dolphin is extremely efficient hydrodynamically; the animal is known to attain swimming speeds of 39 km/h (24 mph). Like all marine mammals, both dolphins and whales must breathe air, and their *nostrils* are equipped with muscular plugs that close when they dive.

FORM AND FUNCTION

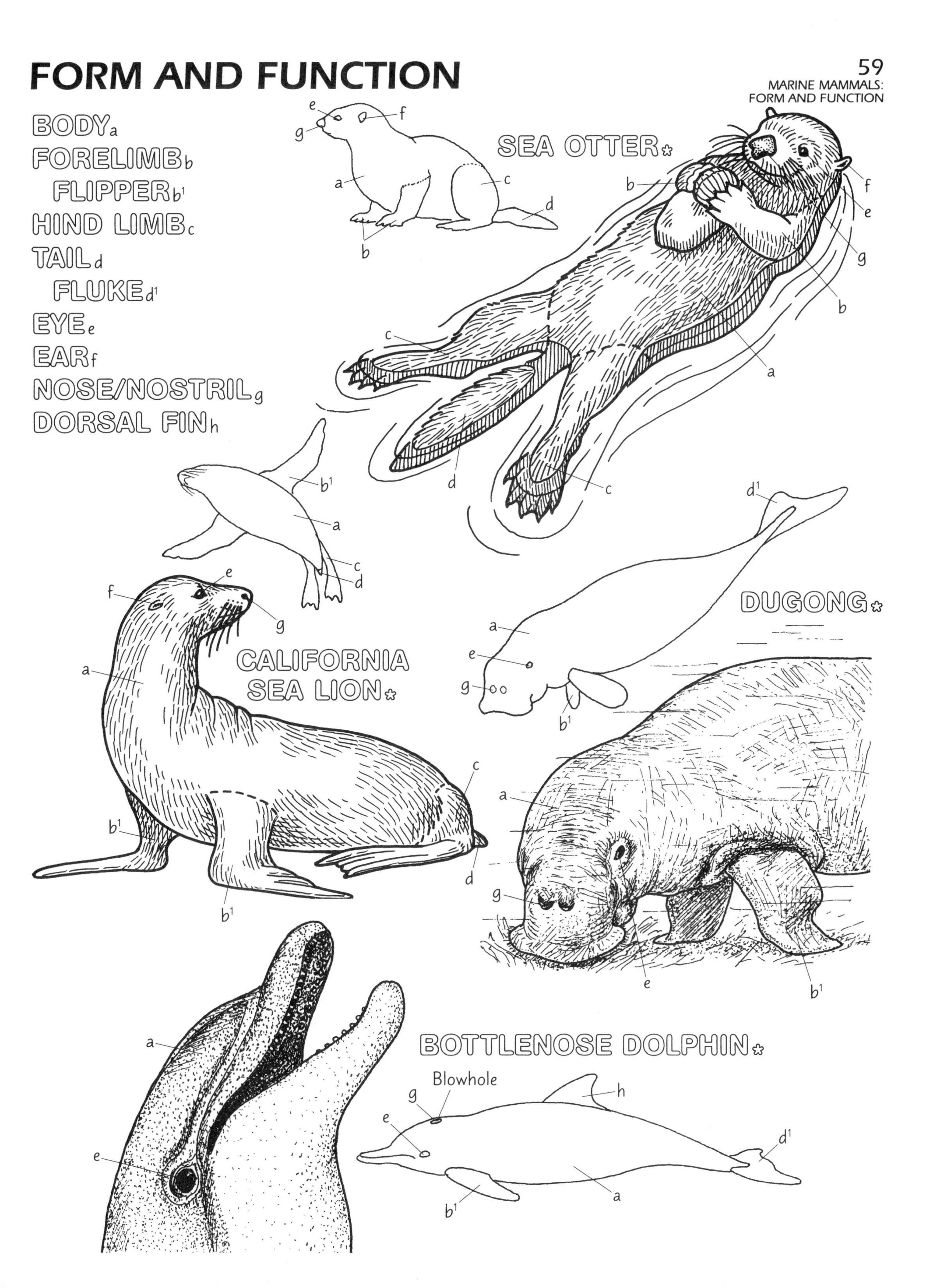

60 MARINE MAMMALS: PINNIPEDS

The pinnipeds ("feather-foot") are the most visible of the marine mammals because they are large, and they frequent coastal areas to rest and sun themselves. In this plate, the fur seal, harbor seal, elephant seal, and walrus are introduced.

Color each animal separately as it is discussed.

In the illustration, a male and female northern fur seal are shown "hauled out" on a rocky platform. These animals are representative of the "eared" seals (including the sea lions) so called because they possess an obvious *ear* flap, or pinna. Fur seals can rotate their *hind limbs* forward to rest underneath them and support their *bodies*. Their *forelimbs* are relatively large and strong enough to allow them to hold their upper bodies erect; they can move fairly well on land.

Northern fur seals have very fine, rich fur that keeps them dry and warm. The large male is dark brown, and the smaller female is dark gray. The northern fur seal is found from Baja California to the Bering Sea, and rarely comes on land, except to breed.

The upright mobility of the fur seal on land contrasts strongly with the harbor seal's awkwardness out of water. Harbor seals belong to the family of true seals, which also includes the elephant seals. These silver gray seals lack the external ear pinna and have fairly restricted movement on land. The hind limbs do not protrude from the body trunk above the ankle; therefore they cannot rotate under the body when on land, but instead drag behind. The forelimbs are relatively small and do not support the upper body well. As a result, a seal's movement on land consists of flopping along on its belly. In the water, the hind limbs propel the harbor seal. When held close to each other, the hind limbs may act in tandem in a sculling motion.

Harbor seals are found in both the Atlantic and Pacific oceans and appear be less nomadic than other pinnipeds. They stay close to shore and often haul out on sandbars in bays and estuaries. Harbor seals feed mainly on small fish, molluscs, and crustaceans in near-shore water.

The dark gray elephant seal is the largest of the pinnipeds. Males may be over 6 meters (20 ft) long. Females are much smaller, reaching 3.5 meters (11 ft). The most prominent feature of the elephant seal is its enlarged nose, or *proboscis*, which is a secondary sex characteristic. As the male becomes sexually mature, the nose begins to grow, achieving full development at, it is thought, eight to ten years of age. During mating season, the nose is used as a display organ while adult males (in the posture shown) bellow challenges at each other. For many of the pinnipeds, mating season involves very complex social interactions (Plate 90).

The brownish tan walrus is found in the cold, northern waters of both the Atlantic and Pacific, near the edge of the Arctic ice pack. The male walrus may reach a length of 3.6 meters (12 ft), and the female is only slightly smaller. Both sexes possess prominent ivory tusks. These enlarged teeth were once thought to serve as digging devices for rooting out shellfish from the bottom. However, underwater video of feeding walrus shows that the broad muzzle is used for this chore, with the stiff whiskers serving as tactile sensory devices that detect prey. The walrus creates a strong suction with its throat and muzzle that not only sucks the clams from their burrows but shucks them completely out of their shells as well! It now appears that tusk length may be related to social status.

At birth, the walrus has a thin coat of reddish hair; the skin of adults is nearly smooth. The walrus relies on a blubber layer for warmth; it spends much of its time hauled out on ice floes where it sleeps and rests. These large animals have only one major predator besides humans—the polar bear, which eats mainly young walrus. The Pacific walrus population has enjoyed a remarkable recovery from its slaughter by hunters in the nineteenth and early twentieth century, but it is still very vulnerable to human predation.

PINNIPEDS

BODY a
FORELIMB b
HIND LIMB c
TAIL d
EYE e
EAR f
PROBOSCIS g
TUSK h

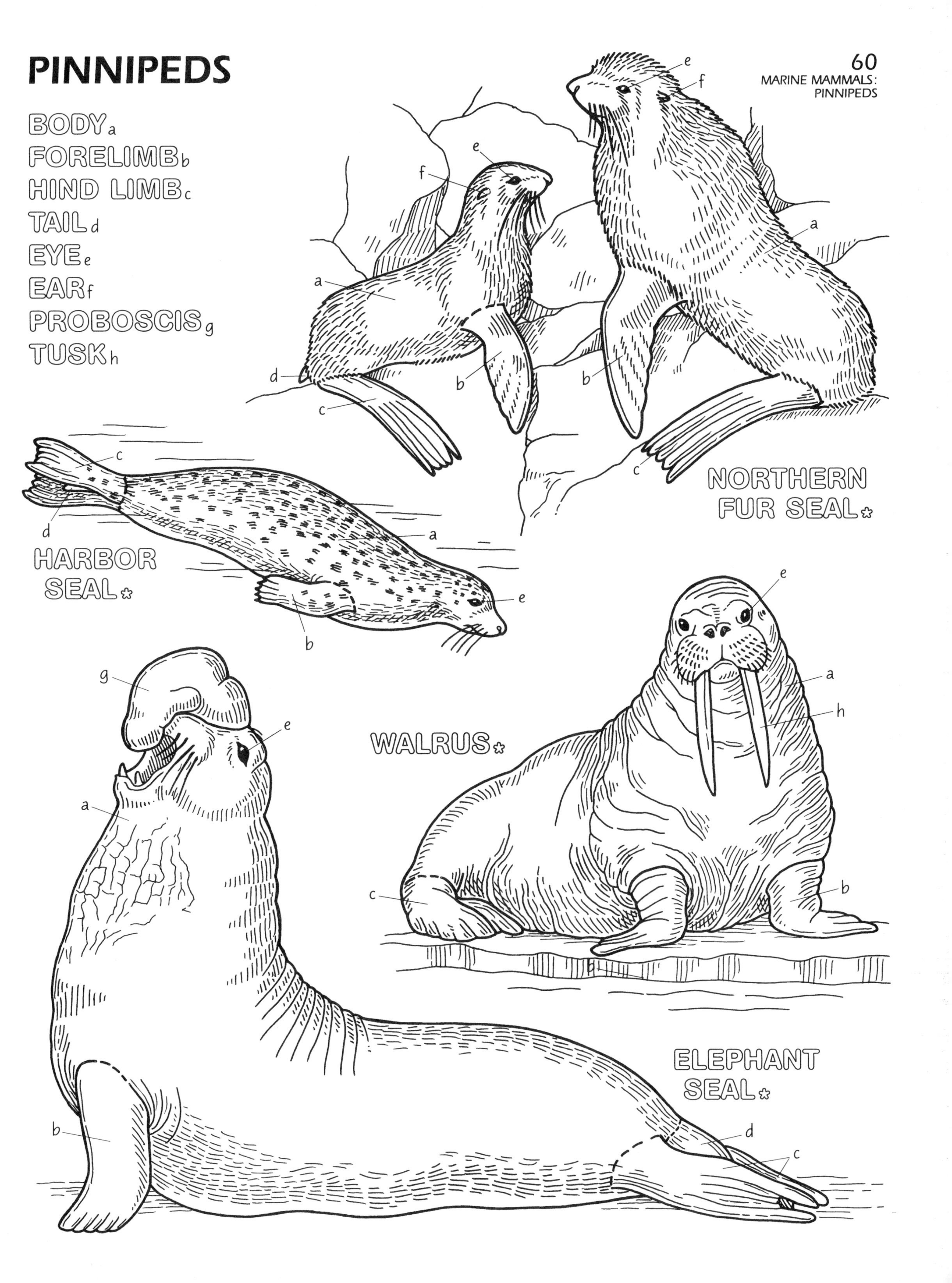

61
MARINE MAMMALS: TOOTHED WHALES AND ECHOLOCATION

There are approximately 74 species of toothed whales, including porpoises, dolphins, sperm whales, beaked whales, and a number of other cetaceans (order Cetacea), all of which bear teeth and have a single blowhole. The toothed whales are all predators and eat fish, squid, and in one or two cases, other marine mammals. They have lost their sense of smell and have fairly good vision, although sight is of little use to the deeper-diving forms that probe the murky depths for prey. The best-developed sense in the toothed whales is hearing, which is used by some species for a very sophisticated behavior known as echolocation.

Color the illustration of echolocation in the dolphin.

Echolocation has been most thoroughly studied in the smaller toothed whales, especially the dolphins. Underwater, these animals emit a wide range of vocalizations from squeals, chirps, and moans to pulses, or trains, of very short "clicks." The former vocalizations are perhaps used in communication, while the "clicks" are used in echolocation. The dolphin is believed to generate the clicks by squeezing a tiny jet of air under high pressure between small sacs in the *blowhole* passage. The moving air vibrates the tightened nasal passage to produce a click in much the same way as air squeezed through the pinched neck of a balloon causes a squeak. As few as five or as many as several hundred of these clicks may be produced each second. The clicks are not emitted from the blowhole, but instead are reflected off the dolphin's concave *skull* (cranium) and focused in an *outgoing* directional beam of sound pulses in front of the dolphin by the *melon*. The melon is a large, lens-shaped organ made of fatty tissue, located on the dolphin's forehead. When the sound pulses strike a *target*, a portion of the signal is *reflected* back. The bony *lower jaw* receives and transmits the impulse via bone conduction to the bone-enclosed *inner ear*. Here the sound impulses are converted to nerve impulses, and these are directed to the brain. The dolphin is able to determine the distance to a target on a continuing, moment-to-moment basis apparently by measuring the time between emission of the clicks and their return. The rate of click production is regulated to allow the returning echo to be heard between outgoing clicks. Using this highly-evolved skill, the dolphin can determine the size, shape, direction of movement, and distance of an object in the water. It scans the water column with low frequency clicks and uses higher frequency clicks to make finer discriminations.

Color the sperm whale. The spermaceti organ is an internal structure exposed for identification. Also color the squid prey in the lower left corner.

The dark brownish-gray sperm whale is the largest of the toothed whales, with males exceeding 17 meters (56 ft) and 47,000 kg (52 tons). Their outsized snouts make up one-third of their length and contain the huge *spermaceti organ* (weighing up to 11,000 kg or 12 tons). This organ is physically very complex, consisting of a mass of oil-filled connective tissue surrounded by layers of muscle and blubber, and is probably analogous to the dolphin's melon. As in the dolphin, the origin of the sound pulses used in echolocation is thought to involve the blowhole passage. Likewise, the mode of focusing the outgoing sound is thought to utilize the sperm whale's concave skullbone as a reflector and the spermaceti organ as a lens. The sperm whale produces very loud, low-frequency clicks that travel for many kilometers. Some believe these loud clicks allow the sperm whale to scan the depths for its squid prey. Once located, the sperm whale dives for its prey. Dives have been recorded that reached depths of 1134 meters (3700 ft) and lasted 90 minutes.

Finally, color the killer whale and its prey. In the lower right corner, the toothed whales are drawn in relative scale to the blue whale. This endangered baleen whale (Plate 62) reaches 33 meters (108 ft), and is considered the largest animal ever to have existed on earth.

The striking black-on-white killer whale is the largest of the dolphins (9 meters, 30 ft) and, as with the sperm whale, the male is considerably larger than the female. The killer whale sports a mouth full of large, conical teeth that are closely spaced and interlock when the mouth is closed. They are skillful hunters that forage cooperatively in tightly knit matrilineal groups (pods). Two distinct types of killer whale pods are recognized along the west coast of North America. The first type remains resident in a defined geographic locale and feeds primarily on fish such as salmon. The second pod type is nomadic and ranges widely along the coast from the tropics to the polar regions, feeding primarily on marine mammal prey such as seals, sea lions, porpoises, and even the great baleen whales.

TOOTHED WHALES AND ECHOLOCATION

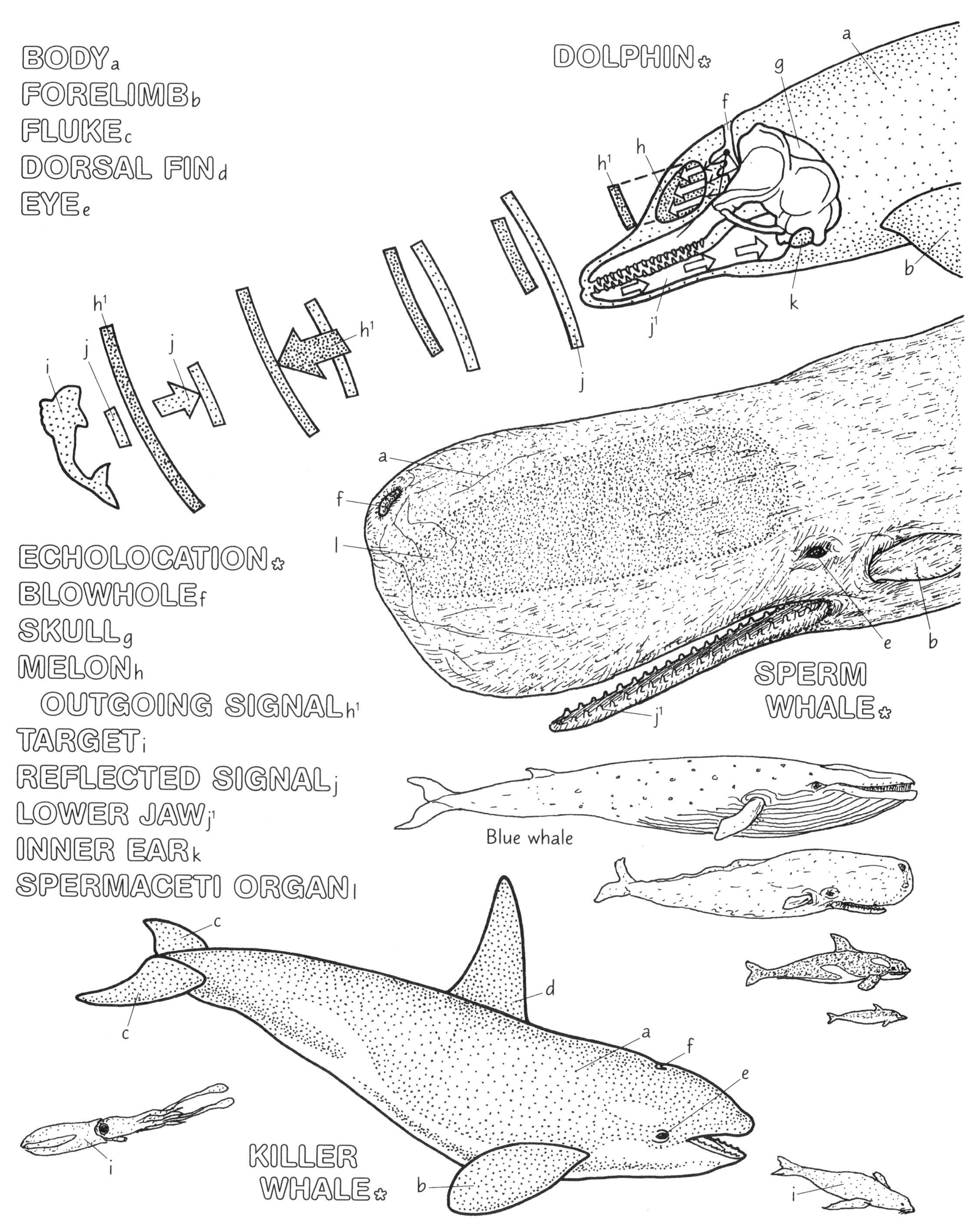

62
MARINE MAMMALS: BALEEN WHALES

The baleen whales are the largest animals that have ever lived on earth, yet little is known about them. Due to past whaling activities, some species are severely depleted and may well disappear before we ever have a chance to learn more. Baleen whales differ from the toothed whales in several ways including: shape of the head; possessing two side-by-side blowholes; presence of baleen plates; lack of teeth; and absence of the large melon in the forehead.

Begin with the upper drawing. Note the bones of the forelimb. A single detached baleen plate is shown to the left of the upper jaw.

Baleen whales are named for the fibrous, horny *baleen plates* hanging from their *upper jaws*. These plates are entirely different from teeth and are derived, like hair, from the epidermal tissues. The individual baleen plate is about 6 mm (0.24 in) thick and consists of long, coarse, fibrous bristles held together by a horny substance. These plates are lined up at intervals along the upper jaw. The outer surface of each plate is smooth, while the fibrous bristles of the inner surface are frayed and cover the spaces between the plates. These frayed bristles make an effective filter for straining sea water. The color, shape, size, and spacing of the baleen plates and the coarseness of the bristles vary among whale species.

Color the two illustrations of the right whale. Use a dark gray for the whale's body, as this whale is nearly black in color.

The right whale (so designated by early whalers as the right whale to kill because it did not sink) has the most elaborate baleen apparatus. It has a significant arch in the upper jaw from which hang baleen plates up to 4 meters (13 ft) in length. The *lower jaw* is not similarly arched, but the large lower lips extend well above the lower jaw to enclose the baleen. The dark colored, flexible baleen plates fold backward toward the throat when the mouth is closed and spring forward when the mouth opens. The head of the right whale takes up nearly one-third the length of the animal's body.

Right whales feed on copepods and other planktonic crustaceans that accumulate in large patches or shoals on and beneath the ocean's surface. To feed, the whale swims through the plankton patch with its scooplike mouth open, forcing the water through the baleen plates and trapping the plankton on the bristles. The black right whale is found in all oceans between the Arctic and Antarctic circles, and reaches a length of 20 meters (66 ft).

Color the humpback whale illustrations. It is dark gray above, white below, and the forelimbs are white as well.

The humpback whale and its fellow great whales, or rorquals (fin, sei, blue, minke, Bryde's whales), are distinguished from other baleen whales by the presence of many long, pleated grooves in the throat region. Humpbacks are characterized by their long, knobby forelimbs, and are known for the elaborate songs that are sung each year by the males on the breeding grounds (Plate 71).

The baleen of the humpback and other rorquals is black, short, and broadly based in the roof of the mouth. These whales feed on a variety of planktonic crustaceans; some species also take fish and squid. The whale swims through the plankton patch, and its capacious mouth engulfs several tons of seawater. Its huge tongue is used as a piston to force water out over the baleen where the small crustaceans are trapped and drawn into the throat to be swallowed. Observations of feeding behavior suggest the humpback swims upward in a slow spiral while emitting air from its *blowholes*. This forms a circular curtain of bubbles which apparently startles the planktonic crustaceans. In this manner, they are driven to the circle's calm center, where they are then engulfed by the humpback lunging upward through the concentrated crustaceans with its mouth opened widely. Groups of humpbacks are known to hunt cooperatively by driving prey into a circled area and all rising together to feed.

Color the gray whale illustrations. The animal is illustrated swimming on its side while feeding just above the bottom. The patchy areas on the body are barnacle encrustations and should be left white.

The gray whale is the only member of its family. Gray whales are medium-sized (15 meters, 50 ft), active whales found in the north Pacific (Plate 89). Their baleen is the shortest of all the whales, and their mode of feeding is unique. Gray whales feed primarily on benthic crustaceans, especially amphipods. They swim just off the bottom, turn on their sides, and sweep their heads back and forth; this disturbs the amphipods and causes them to rise off the bottom. Then the gray whale sucks the amphipods in under the yellowish baleen by pushing its tongue against the floor of its mouth while expanding the throat grooves to create suction. The gray whale's baleen is usually shorter on one side of its mouth, and that same side is also relatively free of the barnacles that encrust the hides of these gentle giants, suggesting that individuals consistently feed from the same side of the mouth.

BALEEN WHALES

BODY a
FORELIMB b
FLUKE c
EYE d
BLOWHOLE e
DORSAL FIN f

HEAD *
UPPER JAW g
BALEEN PLATE h
LOWER JAW i
TONGUE j

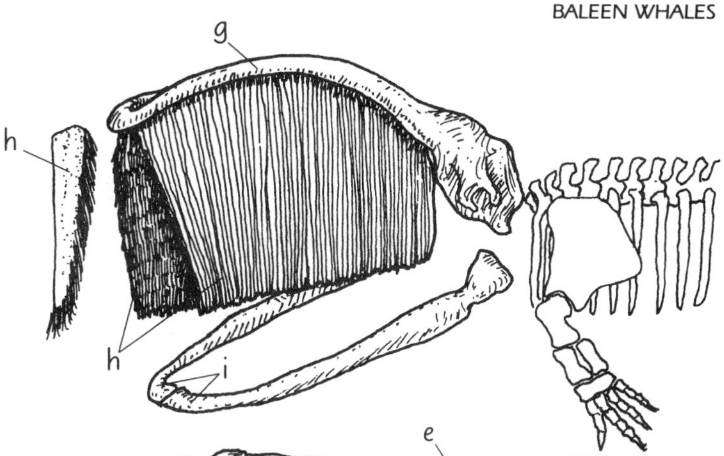

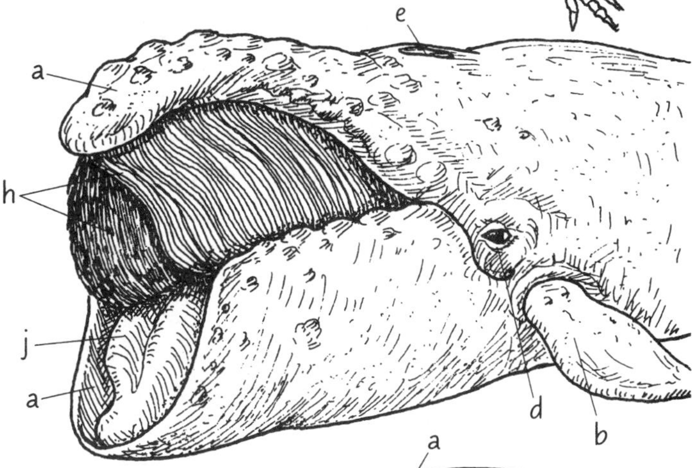

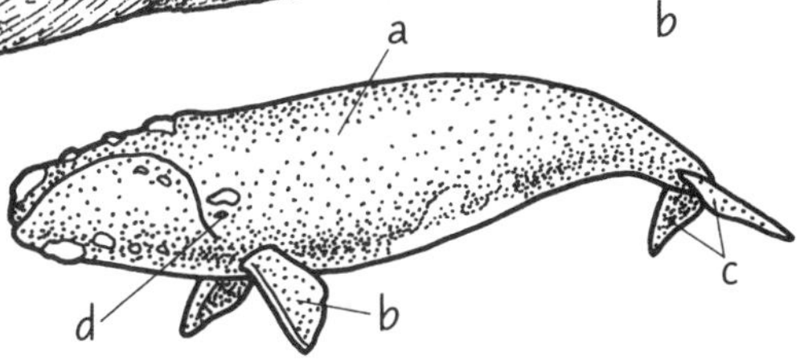

RIGHT WHALE *

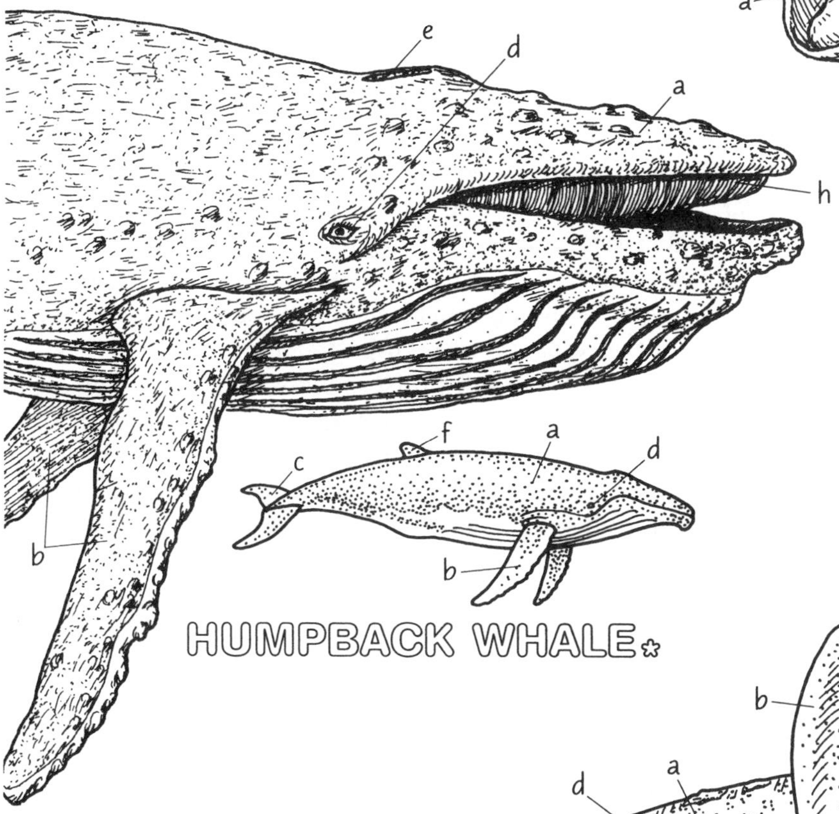

HUMPBACK WHALE *

GRAY WHALE *

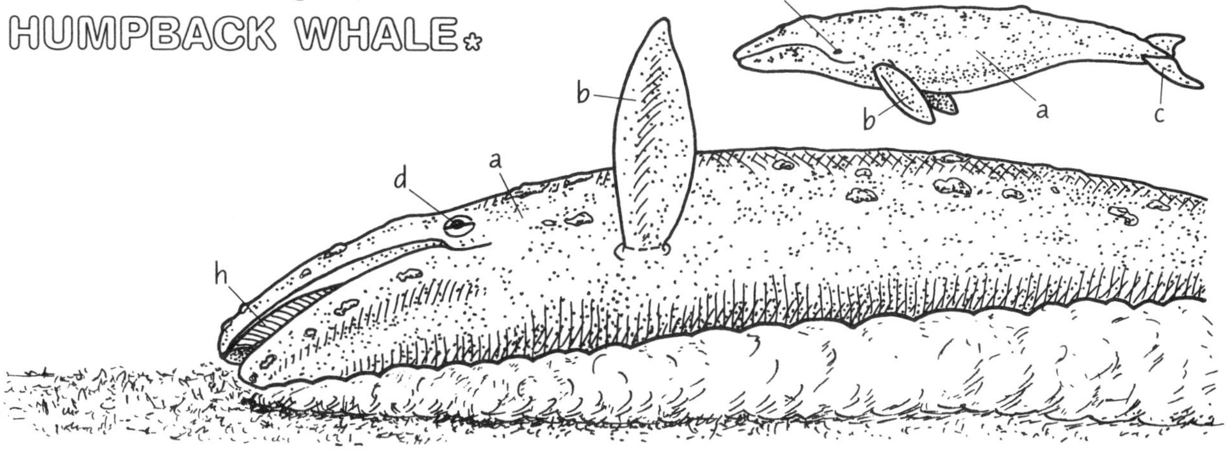

63 COLORATION IN FISH: THE ADVERTISERS

Coloration in fish (and in many other animals) serves a variety of important functions. These will be investigated in the following seven plates, which offer an opportunity to employ some of your brightest colors. This plate is concerned with fishes that call attention to themselves: that is, they advertise.

Color the garibaldi and lionfish. Color the entire garibaldi, including fins, a golden orange or golden yellow, also color the center portion of the traffic light on the fish's right. On the lionfish, color red the dotted, striped areas only. Don't be too concerned about accuracy in coloring the confusing pattern. Color the skull and cross-bones red.

One type of advertisement is seen in the bright, golden orange *garibaldi* of southern California. This fish uses solid, bright color to advertise its presence, warning other garibaldis not to encroach on its closely guarded territory. Male garibaldis most vigorously defend a small patch of red algae in the center of their territory. It is here that the females are enticed to lay the eggs which the male defends until they hatch (Plate 87). Many brightly colored reef fishes are thought to use their colors in this kind of territorial display.

Warning coloration also functions between species (interspecifically), as in the case of the golden boxfish in Plate 47. The butterfly *lionfish* of the tropical Pacific has showy red and white stripes on its body, pectoral fins, and highly venomous dorsal spines. The contrasting red and white colors give warning to predators, especially those that have felt the sting of those poisonous spines before. Interspecific advertisement is not always negative; one example being that of the cleaner wrasses (Plate 92), which advertise their availability with bright color patterns.

Now color the angelfish and the butterflyfish. The larger angelfish on the left receives a yellow color in the areas marked (c), as does the officer's insignia in the drawing on the right. The areas labeled (d) are given a bright blue color. For the juvenile angelfish, use bright blue (d) in the striped areas that are drawn in light lines. The heavily lined stripes are left blank. The remainder of the smaller fish is colored black. In the copperband butterflyfish, the stripes marked (e) receive a pink- or orange-tinted copper color. The well-advertised (but false) eye is colored black. Leave the rest of the fish uncolored. The false bottom of the bottle represents false advertising in that the apparent amount in the full bottle is not, in fact, the true amount.

Coloration may also play a role in species recognition, especially when many closely related fish species are living near one another, such as on a Pacific coral reef. Many fish have a different color pattern in their juvenile phase, sometimes strikingly so. Illustrated here is the *Koran angelfish*, which as a juvenile is black with blue and white semicircles, and as an adult is light yellow with darker yellow spots and light blue on the fins and operculum. These distinct color patterns allow quick recognition and serve to clarify social behavior within a species. In many species of angelfish, the distinctive coloration of the adults facilitates a very tight bond between mated pairs, which are almost always seen swimming together, rarely out of sight of one another.

Coloration can make a fish appear to be something it isn't. Eye-bars, broad stripes, and other tricks are often employed in such "false advertising." The beautiful *copperband butterflyfish* possesses a black eyespot (ocelus) near its tail which potential predators mistake for an eye. The predator will "lead" its prey (just as a human hunter will lead a moving target) and plan its strike for the head, only to gobble empty water or a small piece of dorsal fin that the butterflyfish can regenerate. The eyespot trick is employed by several types of fish and appears throughout the animal kingdom in groups as diverse as moths and octopuses.

THE ADVERTISERS

TERRITORIAL WARNING*
GARIBALDI a

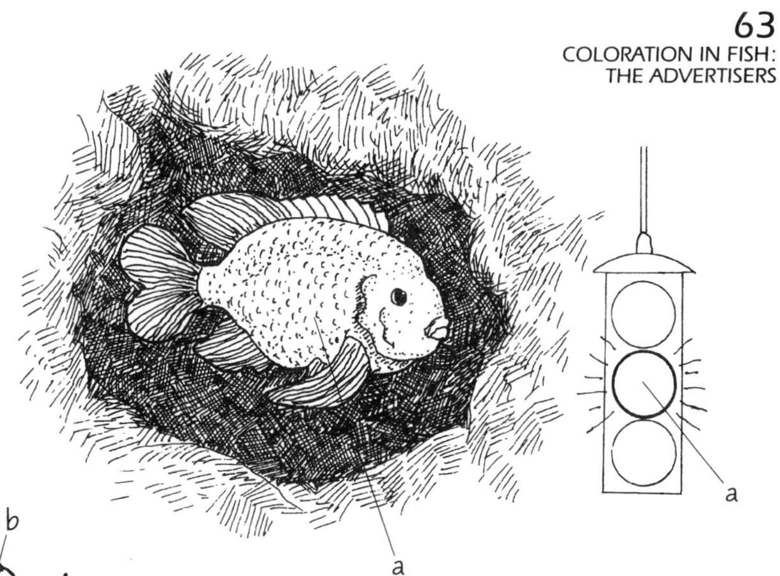

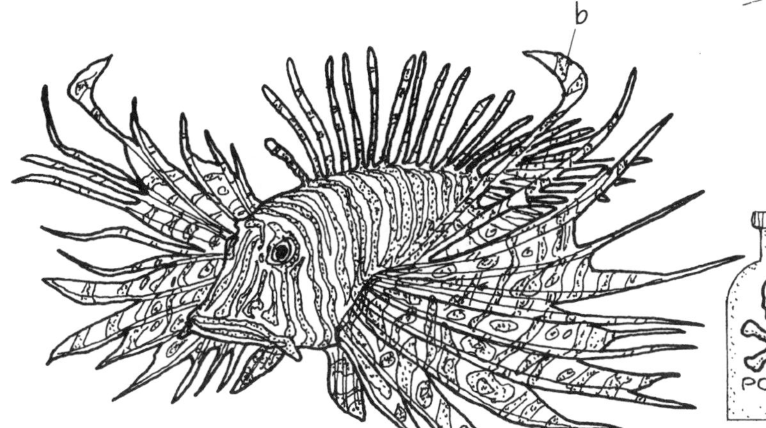

DANGER WARNING*
LIONFISH b

SOCIAL STATUS*
KORAN
ANGELFISH c, d

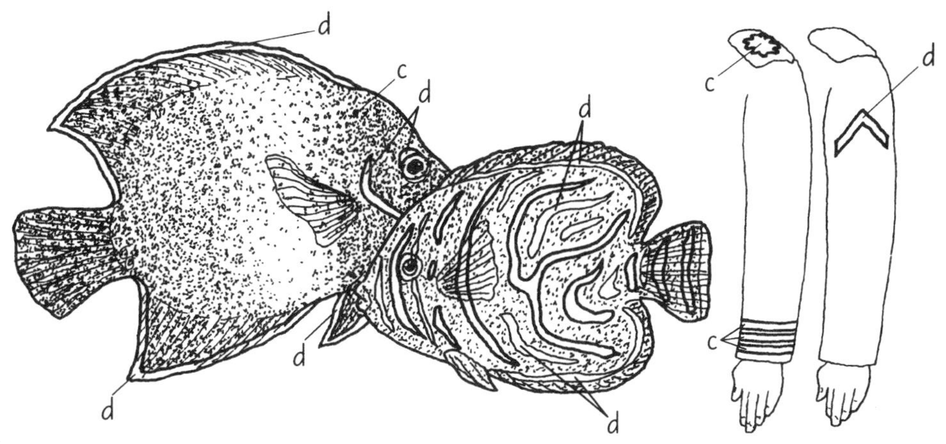

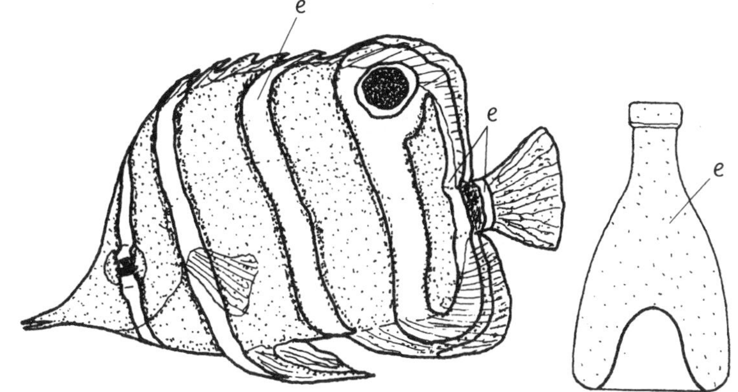

FALSE ADVERTISING*
COPPERBAND
BUTTERFLYFISH e

64
COLORATION IN FISH: THE CRYPTIC ONES

Color the countershading of the mackerel in the drawing at top right (above the rectangle) and the related title dark blue (a). Do not color its white or silvery underside. On the left, color the entire grouper and related title orange (b). Now color the rectangle representing sea water blue-green (d), but do not color the fish. Then color the top half of the grouper (above the dotted line) a light orange (b), or leave small areas within the colored area white. This is to suggest the lightening of the top half of the fish by light shining down from the surface. Color the bottom half solid orange (b), and then color over the bottom half with gray (c) to simulate the darkening of the bottom side that would be in shadow. Color the mackerel the same blue-green as the water on top (d), and shadow the bottom surface (c).

There are several modes of cryptic (hiding) coloration among fish. Many fish of the upper pelagic zone, such as the common mackerel of the North Atlantic seen here, have a coloration pattern known as obliterative countershading. These fish are green to blue-black on their dorsal surface and grade into a silver to almost white underneath. Viewed from above, they blend with the dark void below; seen from below they blend with the bright, sunlit surface water. Viewed from the side (upper illustration), it is apparent that the surface normally directed toward the light is *countershaded* by the dark color on the fish's back, while the ventral (belly) surface, which would normally be in shadow, is counterlighted. The sides have tones that grade between these colors. As a result, the fish is rendered optically flat and reduced in visibility. This type of countershading is employed not only by fish, but also by many birds, mammals, reptiles, and amphibians. Compare the mackerel's color pattern with that of the *non-countershaded* grouper (upper left). One can easily see how the grouper, which inhabits coral reefs, would stand out in open water.

Within the rectangle labeled "disruptive coloration," color the anemonefish (including fins) to the right, bright orange (b^1), but do not color the two stripes on the body of the fish or the tubular coral surrounding it. On this grouper, to the left, color the dotted patches and stripes (e) brown and the body (f) tan or dull yellow. Color the coral marked (g) red, (h) pink, (e) brown, and (i) dull green. Color the sea water blue-green (d).

Another type of color pattern widely employed by fish is disruptive coloration. A familiar object is recognized by a specific contour or outline that shows an obvious surface continuity. The orange garibaldi against a dark bottom is a good example. Disruptive coloration conceals a particular recognizable form by employing contrasting colors in different-sized patches that effectively break up and distract from a recognizable outline, such as the camouflage employed on military equipment. Fish that live in a habitat with a high degree of surface relief, like a coral reef or a kelp forest, take advantage of its many shapes, shadows, and colors to blend into the background.

The grouper on the left has patches of color in irregular shapes and positions all over its body, including its fins and lips. Standing alone against a solid background it would be easily recognized. However, when the fish remains still against a highly colorful and irregular coral reef habitat, the shadows and light areas tend to blend with the fish's coloration, and it no longer presents a clear outline to the viewer. Furthermore, groupers are capable of quickly changing their color patterns and background colors to match the changes in lighting and habitat encountered as they swim.

The clown anemonefish also employs disruptive coloration. When seen against a solid background, the small fish's body, vivid orange with boldly contrasting white stripes, is clearly obvious. However, among the tubular coral and finger sponges of coral reefs, the broad white stripes blend with the background, effectively disrupting the outline of the fish and making it much less recognizable.

Color the stonefish in the middle of the bottom drawing brown (e) and the surrounding rocks (e^1) brown with various red accents (g) that are surrounded by dark lines. Color the bottom (f) tan or dull yellow.

The deadly stonefish mimics and blends into its rocky background with the aid of its coloration. Its body is lumpy and unstreamlined to further conceal its motionless presence among the encrusted stones, from which it cannot be distinguished by humans or prey fish (aggressive mimicry). Like its relatives the lionfish and scorpionfish, this fish has extremely venomous spines; they have caused death to humans who inadvertently stepped on them while wading on shallow Pacific coral reefs.

THE CRYPTIC ONES

OBLITERATIVE COUNTERSHADING✿

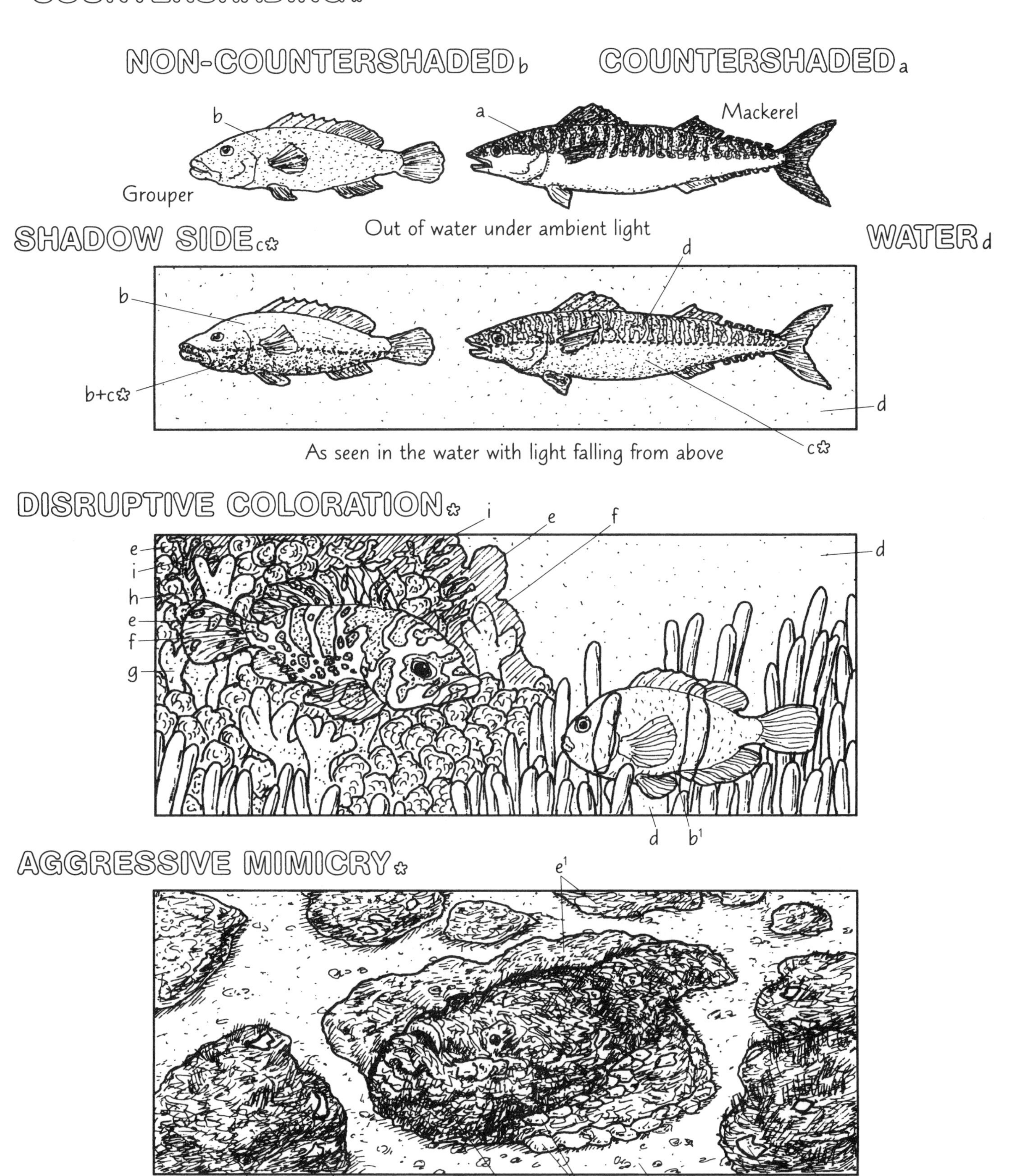

65 COLORATION IN FISH: CHANGING COLOR

This plate explores the source of color in fishes and how coloration can be changed in certain situations.

Begin at the bottom of the plate with the patches of skin. Both squares are of the same magnification and represent enlarged views of skin chromatophores. Choose pale yellow, tan, or light grey for the base skin color (a). Then use dots of color for the chromatophores as follows: (b) red, (c) orange, (d) yellow, and (e) black.

In the large illustration of the flatfish, begin with the left half, coloring the skin of the fish with (a) only. This is usually very close in color to that of the sandy bottom, which receives the same color. This base color is also the base color of the fish in the pebbly environment, and, since there is sand between the pebbles, color over the entire area with the base color first. Then use (b), (c), (d), or (e) in any combination you wish on both the fish and the pebbly substrate. To get the optimum effect, if you do not want to color the entire illustration, color small adjacent sections of each illustration completely. The pigmented areas in the fish on the sandy bottom are so highly concentrated that they cannot be seen.

Fishes take their color from two types of pigment cells that are located in various layers of the skin. One type of cell, the iridocyte (mirror cell) contains guanin, a substance that reflects light and color from the environment outside the fish. The iridocytes give rise to the pearly white color and the silvery, iridescent blues and greens often seen in fishes.

The *chromatophore* is a second type of color cell, which contains its own pigment particles of red, orange, yellow, and black. The cell body itself is highly branched and, in order for color to be seen, pigment granules must be dispersed throughout the branches (expanded pigment). When the pigment is concentrated in the center of the cell, very little color shows. Fishes can produce other than the pigment colors by activating a mixture of chromatophore types. Green, for example, can be obtained by combining black and yellow chromatophores.

Fishes change color depending on the color of their surroundings, the stage of their life cycle, or even a state of excitement. A startled fish will often blanch, the colors appearing to wash out of its skin as the pigment concentrates in the chromatophores. An angry fish may turn bright red as the pigment disperses in its red chromatophores. Some fishes change color pattern between night and daytime. A courting male fish may put on a "suit" of dazzling nuptial colors to attract a mate. Billfishes, making leaping "tail walks" when hooked, are observed "lighting up" in flashes of chromatophore color.

The best-studied color change in fishes deals with their response to surroundings. Surrounding colors may change due to variations in incoming and reflected light or because the fish moves from one habitat to another. Many fishes are able to assume the color of their background, and some are actually able to mimic its pattern. Flatfish are especially adept at this. As shown in the left drawing, the flatfish concentrates the pigments in its chromatophores, enabling it to blend with the sandy bottom. When it moves to a pebbly bottom, clusters of chromatophores expand their pigments to match the size of the pebbles encountered.

Some fishes change color with startling quickness, while others take minutes or even days. The color change is controlled by the nervous system or by the endocrine (hormonal) system. Rapid changes are controlled by direct neural impulses to the chromatophores, while slower changes are brought about by blood-borne pituitary hormones. Short-term changes in color involve the concentration and dispersion of pigment granules in established chromatophores. Long-term changes, such as those brought about by a permanent or lengthy change in surroundings, are the result of an increase or decrease in the number of chromatophores.

The ability to register and recognize mood, sexual readiness, or social status by color has allowed fishes to develop highly complex behavior patterns that are still only partly understood.

CHANGING COLOR

BASE COLOR OF SKIN$_a$
CHROMATOPHORES$_{b, c, d, e}$

SANDY BOTTOM$_{a^1}$

PEBBLY BOTTOM$_{b, c, d, e}$

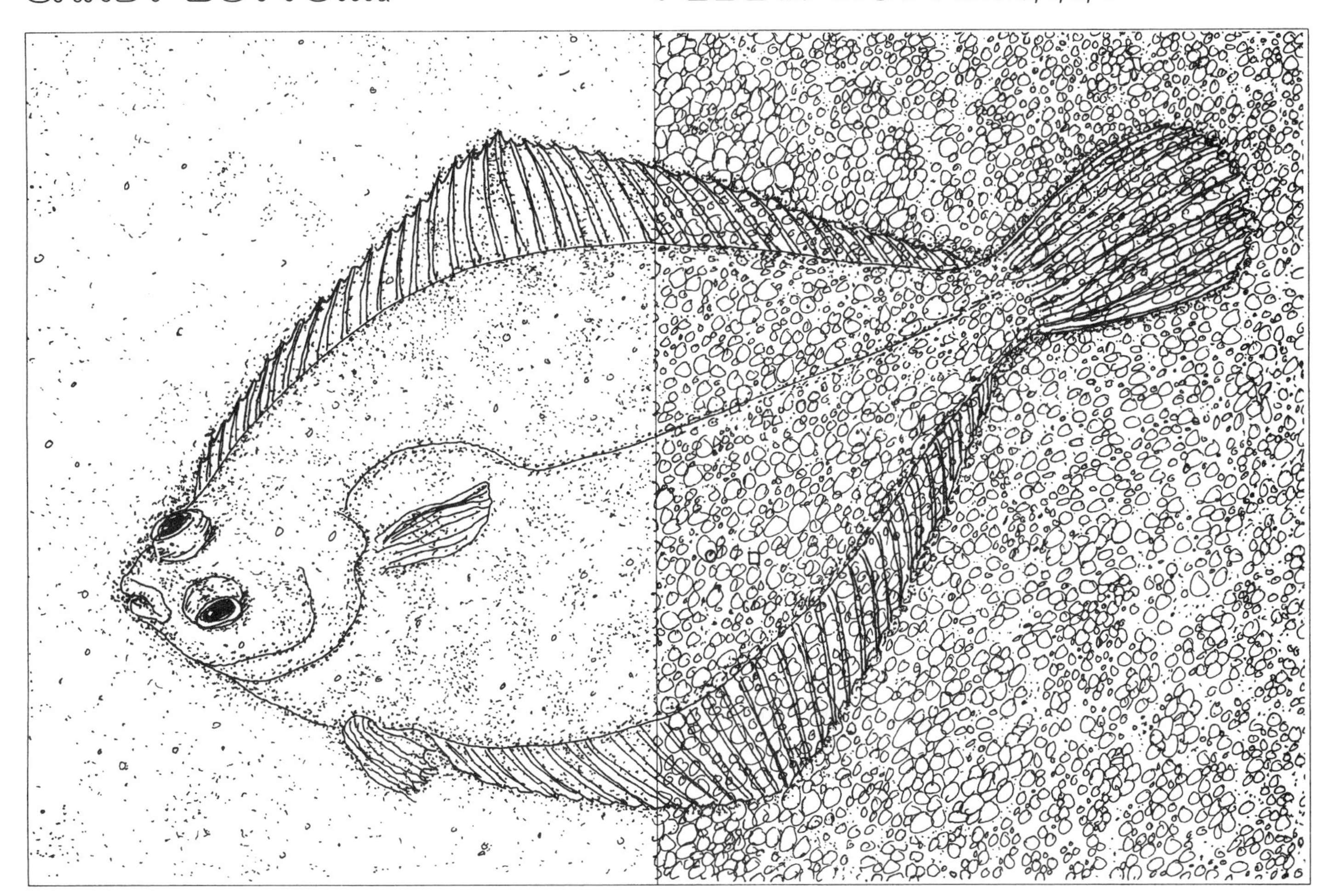

CONTRACTED PIGMENT*

EXPANDED PIGMENT*

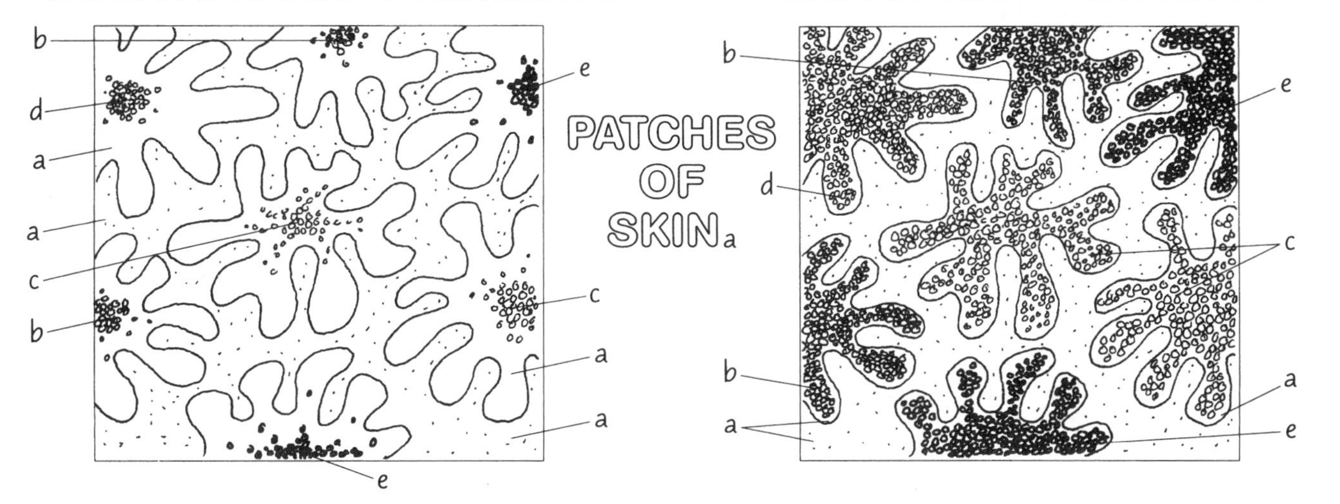

66
COLORATION IN MARINE INVERTEBRATES: THE ADVERTISERS

As we have seen, many fishes display various colors to attract attention to themselves. The same is true of some marine invertebrates.

In this plate, you should use precisely the colors indicated so that the animals shown will look as lifelike as possible. Begin by coloring the peppermint shrimp. The antennae and the stripe down the middle of the back are white and should be left blank. The areas (a) adjacent to the white stripe are bright red, and the underside (b) of the shrimp is golden yellow. You may wish to color the title, peppermint shrimp, alternating red and white (blank).

The *peppermint shrimp* is a cleaner of fishes, and its brightly patterned body apparently advertises this service. This Caribbean shrimp resides in cracks and crevices of coral reefs, where its long white antennae and red-and-white striped body are visible to passing fishes. Once attracted, a fish presents itself to the cleaner, often using complicated patterns of behavior, and the shrimp emerges to pick parasites from the fish's body. Some observers believe that this and other cleaning shrimps (Plate 92) actually establish routine "cleaning stations" which may be frequented daily by their customers.

The female cuttlefish (the left member of the courting pair) in the middle of the plate displays a brown or tan (c) background with dark brown (d) spots. The male on the right shows dark brown (d) stripes on a white (blank) background. Apparently to confuse prey, an attacking cuttlefish can rapidly change colors from dark brown (d), to white, to a mottled pattern of light brown (c) splotches with dark brown (d) centers. The nudibranch at the bottom of the page should be colored blue-purple (e) with yellow-orange (f) stripes. Color the titles with the alternating colors indicated.

The *cuttlefish* is a master of color change. Like its relatives, the squid and octopus, the cuttlefish can change color in a fraction of a second (Plate 68) for a variety of purposes, two of which are described here. Cuttlefish tend to be rather solitary, except during the mating season when the males begin to search out females. Identifying females could be a real problem were it not for courtship behavior involving color changes. The male, sporting a distinctive zebra-striped pattern over its body, approaches another cuttlefish; if the prospective mate is another male, it responds by presenting a similar pattern. A brief skirmish may ensue until one swims off. However, when the approached cuttlefish is a female and does not assume the striped pattern, the male initiates mating by grasping its mate head-on, as shown in the illustration.

The cuttlefish also uses color changes to confuse or dazzle potential prey. As the cuttlefish moves toward its intended victim, it rapidly changes color patterns from dark to light to mottled. Thus distracted and taken off guard, the prey animal is captured by the cuttlefish's two long tentacles which are quickly shot out ahead of the body.

The warning colorations displayed by various *nudibranches* (sea slugs) have been discussed earlier (Plate 32). The animal shown here is brilliant purple with contrasting yellow stripes, seemingly advertising itself as a potential meal. Many nudibranchs possess some noxious attribute that a predator might well learn to associate with their distinctive color patterns: some species (like the one illustrated here) secrete distasteful or irritating chemicals on their bodies; some have stinging nematocysts imbedded in their surface (Plate 100); and still others may accumulate noxious substances from their prey. However, in some species, no such explanations for their bright colors have been found; perhaps they are mimicking some distasteful look-alike.

THE ADVERTISERS

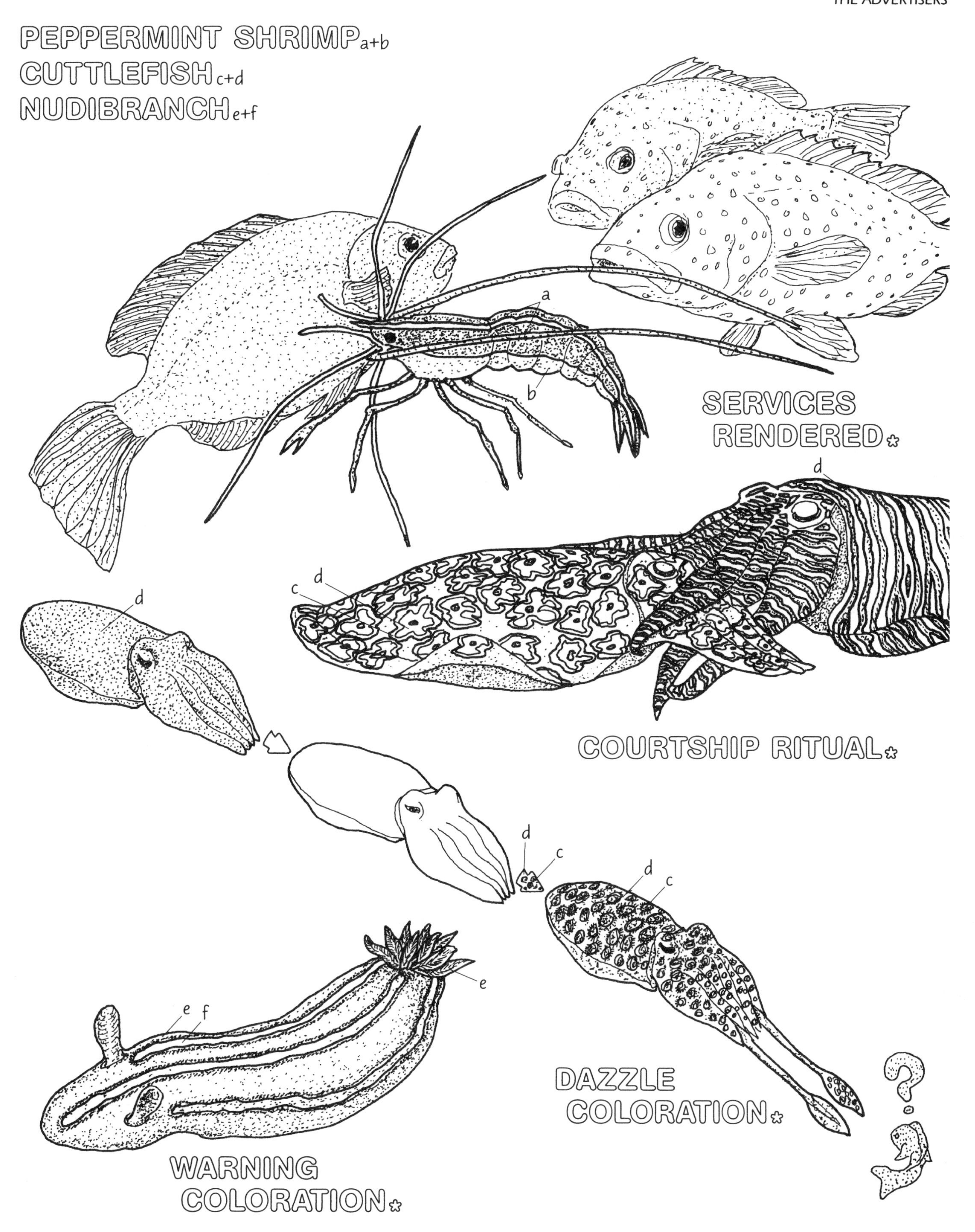

67
COLORATION IN MARINE INVERTEBRATES: THE MIMICS

Many marine invertebrates use color as camouflage, mimicking their environmental backgrounds for protection from predation. As examples of such mimicry, this plate depicts four animals of the rocky intertidal zone, found along the Pacific coast of North America.

In coloring this plate, you may wish to use the natural colors of the invertebrates and their substrata as described in the text. As it is discussed, color each animal both in the large illustration and the smaller drawing in the center panel. Use red, green, or brown for algae (f) on the decorator crab.

When seen against any other than its normal background, the red sponge *nudibranch* stands out vividly. However, this small (10 mm, 0.4 in), bright red sea slug blends almost perfectly with the red *sponges* on which it lives, feeds, and lays its eggs. The nudibranch feeds by scraping away bits of the sponge with its radula (Plate 106). The pigments of the sponge apparently are incorporated into the tissues of the nudibranch, resulting in the close color match between predator and prey. The red pigments are also deposited in the nudibranch's jellylike egg mass, which is laid on the sponge in a characteristic *egg spiral*. Many sponge-feeding nudibranchs not only utilize their prey's pigment but also incorporate noxious defensive chemicals produced by the sponge into their integument and egg spirals.

Also shown on this plate is a particular species of small (15–20 mm, 0.6–0.8 in) crustacean known as an *isopod*, another invertebrate that mimics the color of the substrata on which it lives. This isopod feeds on both the green *surf grass* in the wave-swept lower intertidal and on various *red algae* (Plate 20) in the upper zones; in either case, its coloration matches the plant it is feeding on. Dr. Welton Lee has studied these color changes and their significance. *Adult* isopods live and feed on surf grass, and they are green. When the adult females release the young isopods in this habitat they are unable to cling to the plants because of the strong wave surge. They are carried by the waves into higher tide zones where they cling to various red algae. Here the *juvenile* isopods molt (shed their old outer skeleton) revealing a new, red skeleton below. In the living tissues beneath the reddish skeleton are chromatophores (Plate 65) which are used to adjust the color to the precise shade of red to match the particular red alga on which the animal is located. These animals feed on the red algae, grow, and upon reaching adulthood, migrate back down to the surf grass. Once again they molt, and the new exoskeleton is green, matching their new substratum. This color-change ability allows the adults and juveniles to utilize different plants for food and attachment while both remain camouflaged as protection from predators. In contrast to the red sponge nudibranch, which utilizes pigment from its prey, the isopod produces the mimicking colors on its own.

The ribbed or digit *limpet* is a common gastropod (Plate 5) of the upper intertidal zone. Most of these limpets are found on bare rock and are generally dull brown to gray. However, another color variety is found in the middle intertidal zone associated with the common goose or *stalked barnacle* (Plates 35, 97). These limpets have light gray shells with black stripes. When clinging to the stalked barnacles, the limpet's black stripes blend with the dark areas between the barnacle's gray plates. Thus camouflaged, the limpets are very difficult to detect, even when viewed at close quarters.

Several species of long-legged spider crabs "decorate" their bodies to blend with their backgrounds. The small (3–4 cm, 1.2–1.6 inch wide) *decorator crab* shown here lives among rock rubble in the low intertidal zones, or subtidally around kelp holdfasts. These habitats are often characterized by dense, low growths of attached plants and animals. The decorator crab's greenish-brown exoskeleton bears stout hooked "hairs," or setae, to which the animal attaches bits of local plants and animals (seaweeds, sponges, hydroids, and so on). These "decorations" grow and become living camouflage for the crab. When the crab remains motionless, it is extremely difficult to discern against the background of its habitat.

THE MIMICS

NUDIBRANCH$_a$
EGG SPIRAL$_{a^1}$
SPONGE$_{a^2}$

ADULT ISOPOD$_b$
JUVENILE ISOPOD$_c$
SURF GRASS$_{b^1}$
RED ALGA$_{c^1}$

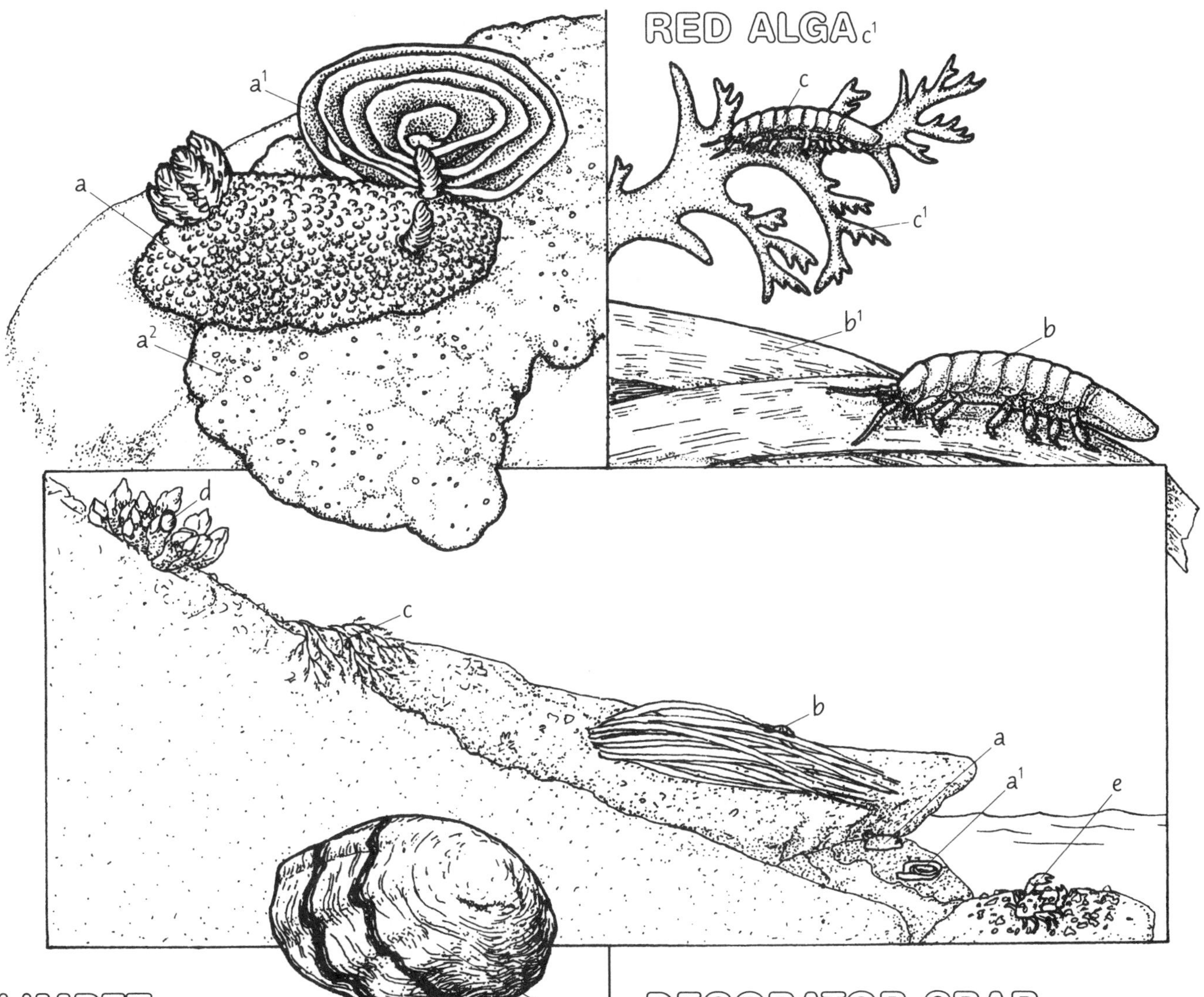

LIMPET$_d$
STALKED BARNACLES$_{d^1}$

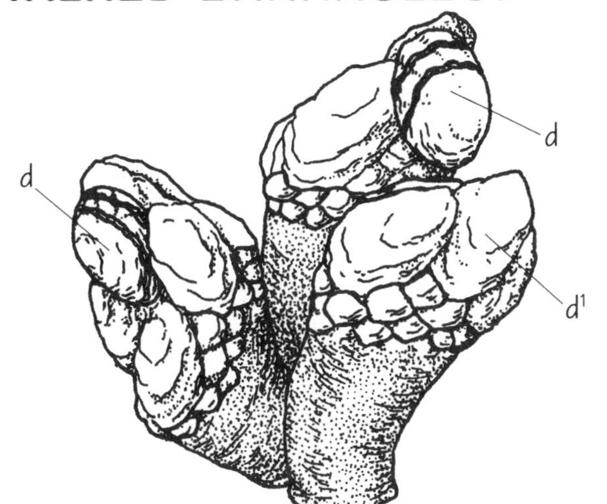

DECORATOR CRAB$_e$
ALGAE$_f$

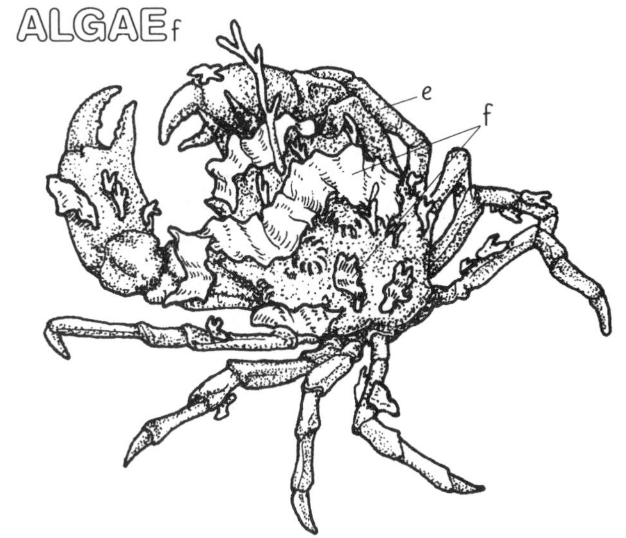

68
COLORATION IN MARINE INVERTEBRATES: CRUSTACEANS AND CEPHALOPODS

Marine invertebrates employ color changes both to advertise their presence and to mimic their surroundings. Crustaceans (crabs, shrimps, and others) and cephalopod molluscs (squids, octopuses, etc.) produce these color changes through the activities of the special pigment-containing cells called chromatophores. However, the structure and methods of controlling the chromatophores differ greatly between these two groups of animals, as explained in this plate.

First, read the text concerning crustacean chromatophores. Next, color the chromatophore diagram at the upper right, coloring only the pigment granules in their concentrated and dispersed conditions. Now color the pathway of the hormone that controls the movement of pigment within the cell. Finish by coloring the left half of the fiddler crab as it appears in daylight with the same color used for the pigment granules.

The color of many crustaceans is often due, at least in part, to pigments deposited in the exoskeleton; these colors are relatively permanent. In some crustaceans, however, the exoskeleton is thin or transparent, and the colors of the *chromatophores* in the underlying tissue can be seen. A crustacean chromatophore has highly branched processes. Each chromatophore is tightly bound with a number of similar chromatophores to form a multicellular organ with radiating processes. Each cell may contain one or several colors of chromatophore *pigment granules* (black, white, blue, yellow, red, or brown). When the pigment granules are concentrated in the center of the chromatophore, they are more or less inconspicuous, but when the granules are dispersed and spread into the cell's branches, their color is plainly visible.

The movement of pigment granules within crustacean chromatophores is controlled by special *hormones* called chromatophorotropins. These hormones are usually secreted by special cells in the animal's *eyestalks* or *brain*. Apparently, each pigment color is individually regulated by particular and separate concentrating and dispersing hormones. Because these hormones must be carried through the blood from their sites of production to their sites of action, color change in crustaceans is relatively gradual.

Shrimp and other crustaceans that have a variety of colors of pigment granules can adapt to blend with nearly any background within a few hours. However, most crustacean color changes are much simpler, as in the example of the fiddler crab. This animal undergoes a daily color change involving very dark pigment granules. These granules are dispersed during the daylight hours, darkening the crab to blend with the muddy sand background of its normal habitat. At night, the pigment granules are concentrated and the crab appears pale in the moonlight. This color change is especially critical in the fiddler crab as it is active during low tides when it is exposed to sight-feeding terrestrial predators like birds. Fiddler crabs remain in their burrows at high tide, most likely to avoid predation by fish.

After reading the text, color the pathway illustrating nervous and muscular control of the cephalopod chromatophore. Next color the boxed illustrations of the pigment granules in the retracted, partially stretched, and completely stretched pigment sac. Color only the labeled, active nerves and their associated contracted (shorter and thicker) muscles. Finally, color only the concentrated and dispersed chromatophores in the young octopuses illustrated at the bottom of the page.

In contrast to the relatively slow-acting, hormonally-controlled chromatophores of crustaceans, those of cephalopods are regulated directly by the nervous system and can react in a fraction of a second. These fast-acting chromatophores are also very different structurally from those of crustaceans. The pigment granules are contained within an elastic *pigment sac*. *Muscles*, each with their own individual *nerve* for separate control, are attached to and radiate from the pigment sac. When some of these muscles contract, the pigment is partially dispersed by stretching the pigment sac. Complete pigment dispersion occurs when all of the muscles are stimulated to contract simultaneously, stretching the sac into a flat sheet of color. When the nerve stimulus ceases, the muscles relax, and the elastic pigment sac draws the pigment back into a condensed, inconspicuous spot. The difference in color achieved between dispersed and condensed chromatophores is clearly seen on the small (3 mm, 0.12 in) bodies of very young animals.

Adult cephalopods may have millions of variously colored chromatophores (yellow, orange, red, blue, black) arranged in patches and layers in their skin. By precise regulation of different chromatophores, these animals can quickly change colors and color patterns to match their background, or flash various signals that play a part in their courtship, defense, and aggressive behaviors (Plates 66, 80, 104).

CRUSTACEANS AND CEPHALOPODS

EYE STALK$_a$
BRAIN$_b$
HORMONES$_c$
BLOOD STREAM$_d$
CHROMATOPHORE$_e$
PIGMENT GRANULES$_{e^1}$

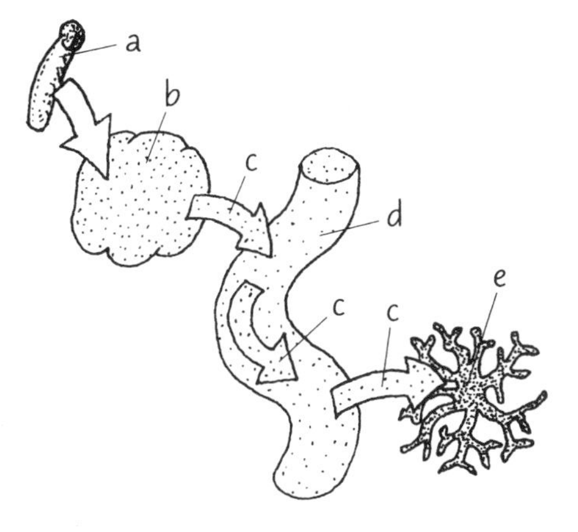

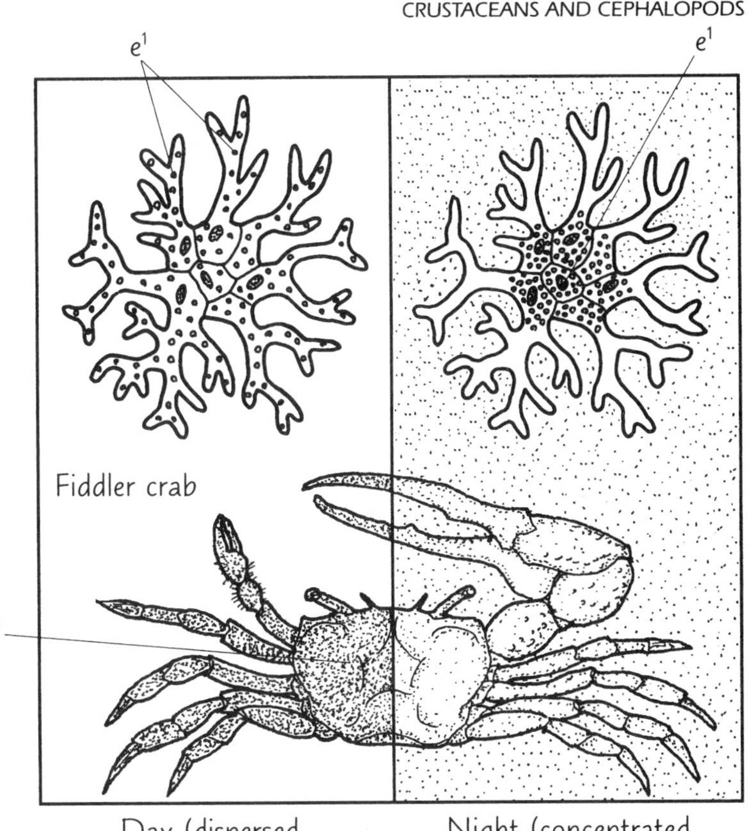

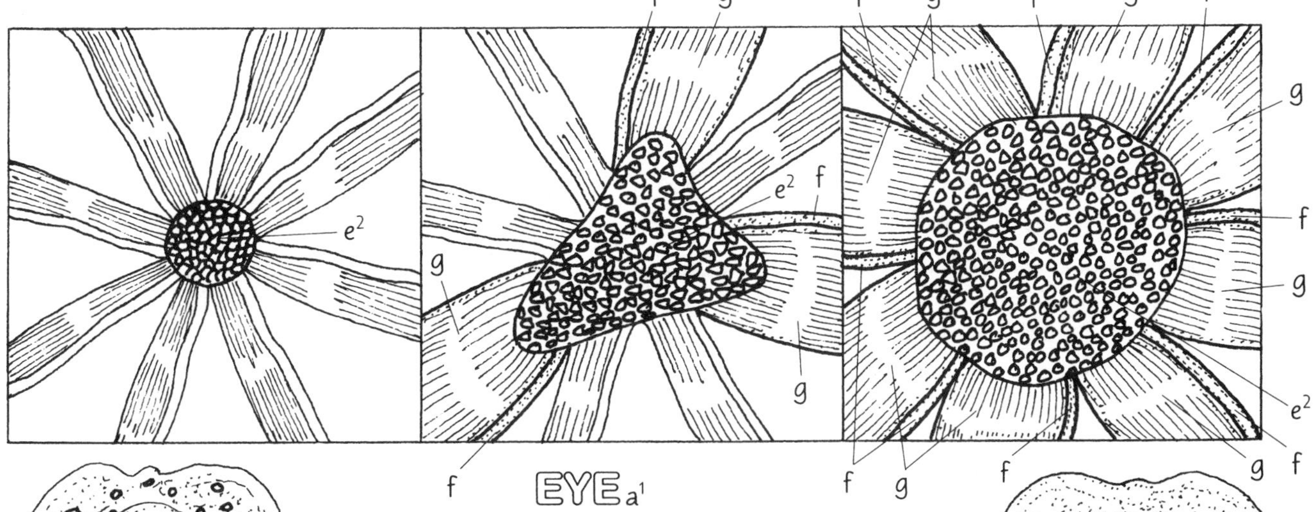

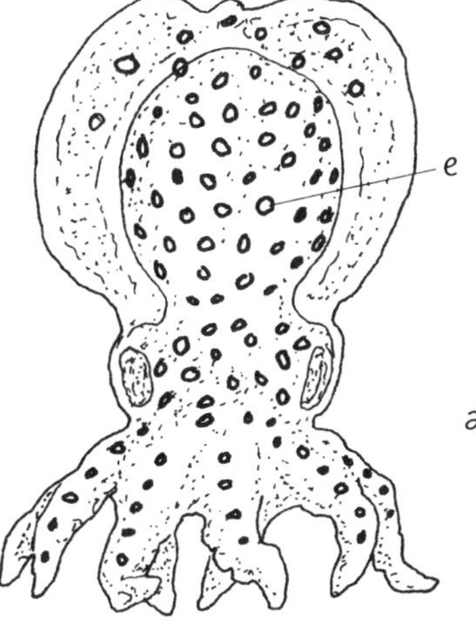

EYE$_{a^1}$
BRAIN$_{b^1}$
NERVE$_f$
MUSCLE$_g$
CHROMATOPHORE$_e$
PIGMENT SAC$_{e^2}$

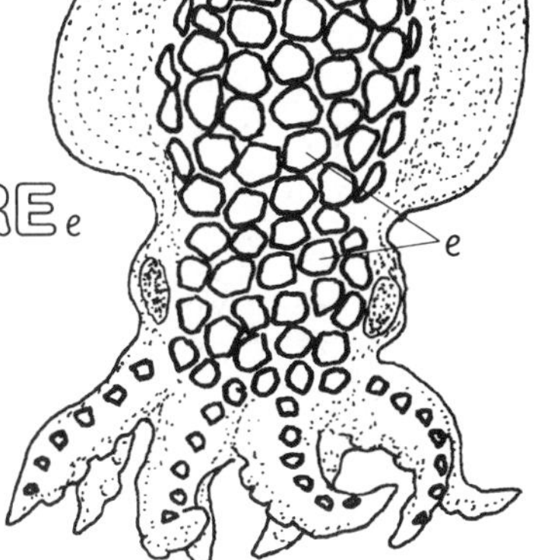

69
BIOLUMINESCENCE IN MARINE ORGANISMS

The ability to produce light is a characteristic of many marine organisms. In the shallow, lit ocean (photic zone, Plate 14) relatively few organisms, 5 to 10 percent of a given group, exhibit bioluminescence. This percentage jumps quickly to 70 to 80 percent in the mesopelagic zone (Plate 15). Animal light is the product of a chemical reaction that requires oxygen, a substrate called luciferin, and an enzyme called luciferase. The luciferin substrate is oxidized in the presence of the luciferase enzyme and releases a photon of light. The chemical structure of the substrate and enzyme molecules varies considerably from group to group, suggesting that the ability to produce light has evolved independently many times. The light produced may be a short flash or a sustained glow. Some of the luminescent creatures of the sea are discussed below and in the next plate.

Begin coloring with the illustration of *Noctiluca*. Use blue-green for the luminescent wake created by the boat. Then color the enlarged view of the planktonic organisms that create the glow. Use shades of green to color the enlarged view of the fireworm and the group of male worms swimming toward the ring created by the female.

Many people who live near the ocean can recall dark summer nights when the sea appeared to glow with light. Each breaking wave had an eerie bluish shine, and boats left a luminous wake as they cut through the water. This commonly observed example of bioluminescence is caused by tiny single-celled dinoflagellates, such as *Noctiluca*. When *Noctiluca* is agitated (as by a breaking wave or the propwash of a boat) it emits a brief blue-green flash of light. If present in large numbers, the combined luminescence of these organisms causes the glow observed at night.

As Columbus approached the coast of North America for the first time, he reported seeing what appeared to be "candles moving in the sea." Some biologists suspect that Columbus was witnessing the mating ritual of the polychaete worm called the *Bermuda fireworm*. This small, bottom-dwelling worm swarms near the ocean surface on summer evenings for a few days following a full moon. The *females* swim in individual tight circles, emitting a green *luminous secretion*. The *males* are attracted to this circle of light and swim toward it, emitting short bursts of bright light as they move through the water. Several males may be attracted to a single circle; soon both sexes release their gametes (sperm and eggs) into the water, where fertilization occurs.

Color the comb jelly shades of red. Use bright blue for the group of small euphausiids and the photophores in the enlarged view.

The *comb jelly* shown here (phylum Ctenophora) is a marble-sized planktonic animal. The illustration shows the animal's *comb rows*, which are bands of cilia used for locomotion, and the long *tentacles* used in capturing prey. The comb jelly is highly luminescent and uses special light-producing cells, called photocytes, located beneath the comb rows and on the tentacles to produce a fiery, scarlet-red when agitated. The light combines with the beating of the comb rows to create a truly beautiful and startling effect. The function of this animal's bioluminescent activity is not understood, but it may serve to confuse, frighten, or temporarily blind a potential predator.

Another bioluminescent zooplankter is the small (2–4 cm, 0.8–1.6 in) crustacean known as a *euphausiid* (also called krill; Plate 14). Euphausiid is a Greek word meaning "true shining light." The name refers to the blue bioluminescence produced by photocytes organized into distinct light organs (*photophores*) on the animal's body. These photophores tend to face downward (ventrally) and are located on the abdomen, the thorax, and near the eyes. Many species of euphausiids occur in massive groups or schools that remain in deep, dimly lit water by day, and migrate into the upper layers at night. The luminescent activity of these crustaceans is thought to help keep the group together at depth in the daytime, and during their upward migrations at night. In some species, the production of light may function in mating behavior.

The photophores of the firefly squid emit white light. You may wish to leave them blank and color around them in dark gray.

The firefly squid also employs bioluminescence in its mating behavior. In late spring, this small (10 cm, 4 in), deep-sea squid migrates towards the surface to breed. This squid has a large number of photophores located on the mantle surface, around the eyes, and on the ventral arms. These photophores give off a brilliant white light that the squids flash during their nocturnal mating activities.

The photophores of most squids can be turned on or off and their intensity varied by direct nervous control. Some deep-sea squids that form schools probably use their bioluminescence to maintain contact with members of their species. Species that migrate to the surface at night to feed may employ photophores to produce "counterlighting," as described in the next plate.

BIOLUMINESCENCE IN MARINE ORGANISMS

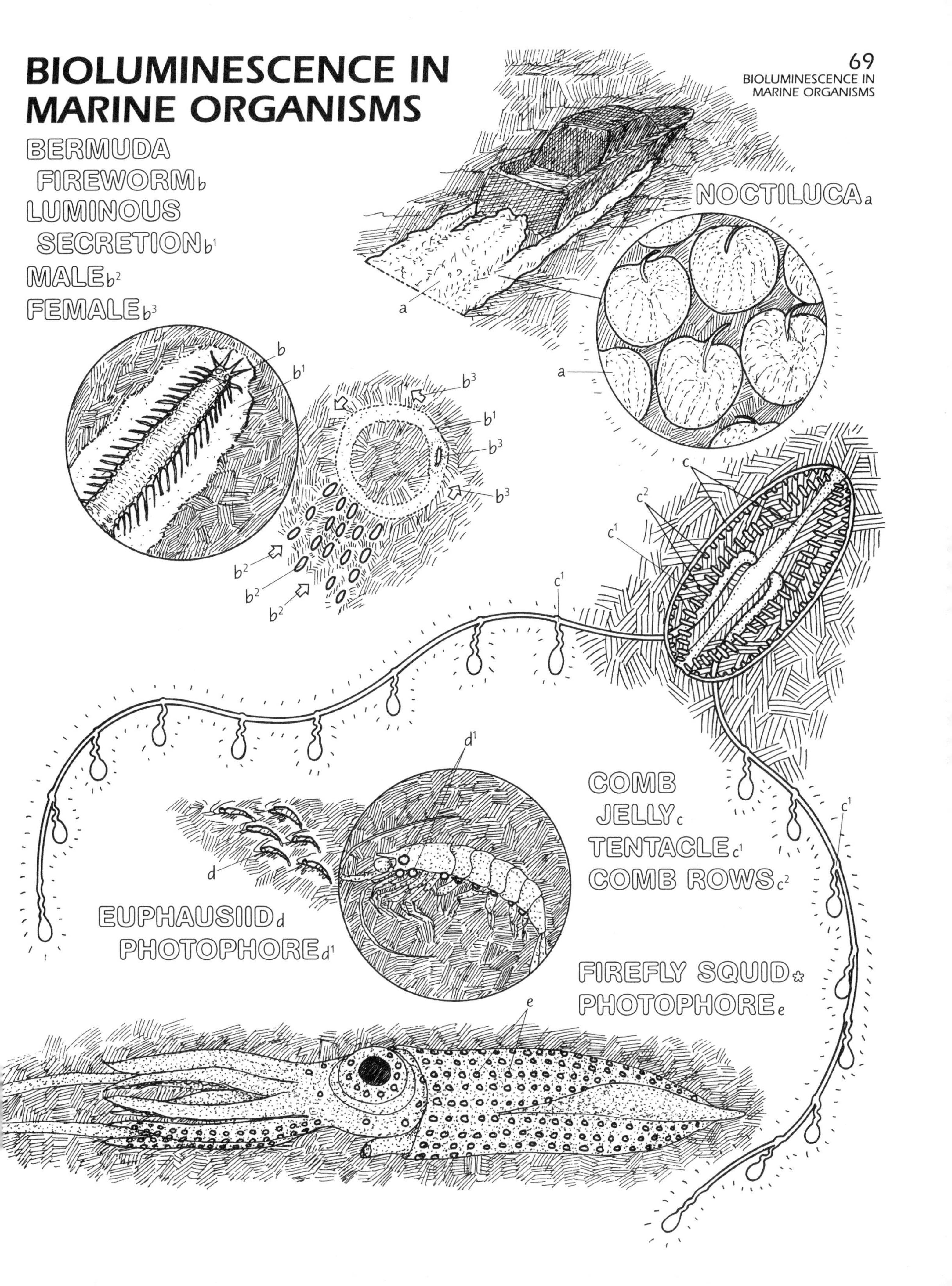

70
BIOLUMINESCENCE IN MARINE FISHES

Sunlight penetrates only the upper layers of the ocean. In the dim midwater realm and the black abyss, many fishes produce their own light with special light-producing structures, called photophores.

Color the enlarged photophore at the upper left. Use a light blue for the photocytes. The bars (a^2) represent the bioluminescent glow.

Photophores are usually cup-shaped and may have elaborate focusing *lenses* and *reflectors* to concentrate and direct the *light* produced by the *photocytes*. Photophores are generally controlled by *nerves,* and are supplied with *blood vessels* that deliver the oxygen and energy sources needed for the light-producing chemical reaction.

Color each fish as it is discussed in the text. Note that the bright flash (a^2) of the caudal photophores of the lanternfish is represented as a burst of light. When coloring the flashlight fish, note that one is "blinking" and has its photophore covered by the skin fold.

Many midwater fishes (Plate 48) utilize bioluminescence for purposes that are often not fully understood, due to the difficulties in studying these animals alive. The *lanternfishes* have photophores along their ventral surface. At night, these fishes migrate to the upper layers of water where they feed on zooplankton. Seen from below, an unlighted fish would be clearly silhouetted against the moonlit water surface, and thus be vulnerable to attack from below by predators using visual cues. Lanternfishes use their ventral photophores to erase this silhouette and blend with the moonlit upper waters. This phenomenon is known as counterlighting and occurs in a number of fishes, squids, and crustaceans (Plate 69). The numbers and patterns of photophores vary among different species of lanternfishes and between sexes of the same species, thus providing mechanisms of recognition for schooling and mating.

To evade predators, many lanternfishes emit a bright *flash* of light from the caudal (rear) photophores and, at the same time, swim rapidly away. The predator is startled and confused, and focuses its attention on the spot where the flash occurred, giving the lanternfish an opportunity to escape.

The *hatchetfish* uses its ventral photophores for protective counterlighting in the manner just described. It also possesses unique, upward-facing eyes equipped with *yellow lenses*. The lenses serve as filters that allow the fish to distinguish the narrow color range of bioluminescent light from the broad color range of normal background light. By looking up onto the moonlit water, the hatchetfish can discern potential prey animals that are using photophores in counterlighting behavior, thus actually capitalizing on the defense mechanisms of its prey.

The deep-water predatory fish shown here is a *stomiatid.* It attracts potential prey with a photophore lure located on the fish's "chin whisker" or *barbel.* Of the species illustrated here, only *females* possess the barbel. *Males* are smaller and have a large photophore beneath their eyes. This photophore probably aids recognition between males and females in their dark environment.

The small (7–8 cm, 3 in), shallow water, *flashlight fish* of the Red Sea bears photophores that are among the brightest and largest found in any bioluminescent organism. However, the blue-green light emitted from these photophores is not produced by the fish themselves, but rather by billions of luminescent bacteria harbored within the photophore by the host fish. *Bacterial photophores*, also called glandular light organs, are quite common in fish, and occur in some squid as well. The flashlight fish's blood stream keeps the microscopic inhabitants supplied with nutrients and oxygen. The bacteria glow in a special pouch that is lined with dark skin arranged in such a fashion as to prevent the light from blinding the fish. The fish possesses a *fold* of *skin* that can be raised over the pouch to cover the photophore and essentially "turn off" the light. Flashlight fish remain hidden in the coral reef by day and on moonlit nights. On dark nights, groups from a few to 60 fish congregate near the surface. The combined glow of their bacterial photophores attracts their small zooplanktonic prey. If a larger potential predator is attracted to the light, the flashlight fish executes a strategic defense response known as "blink and run." The fish swims in one direction with its light "on," then covers the light and swims in a different direction. Each fish performs this "blink and run" behavior up to 75 times per minute. The effect of several flashlight fish "blinking and running" is to confuse the predator and permit the escape of the small fish.

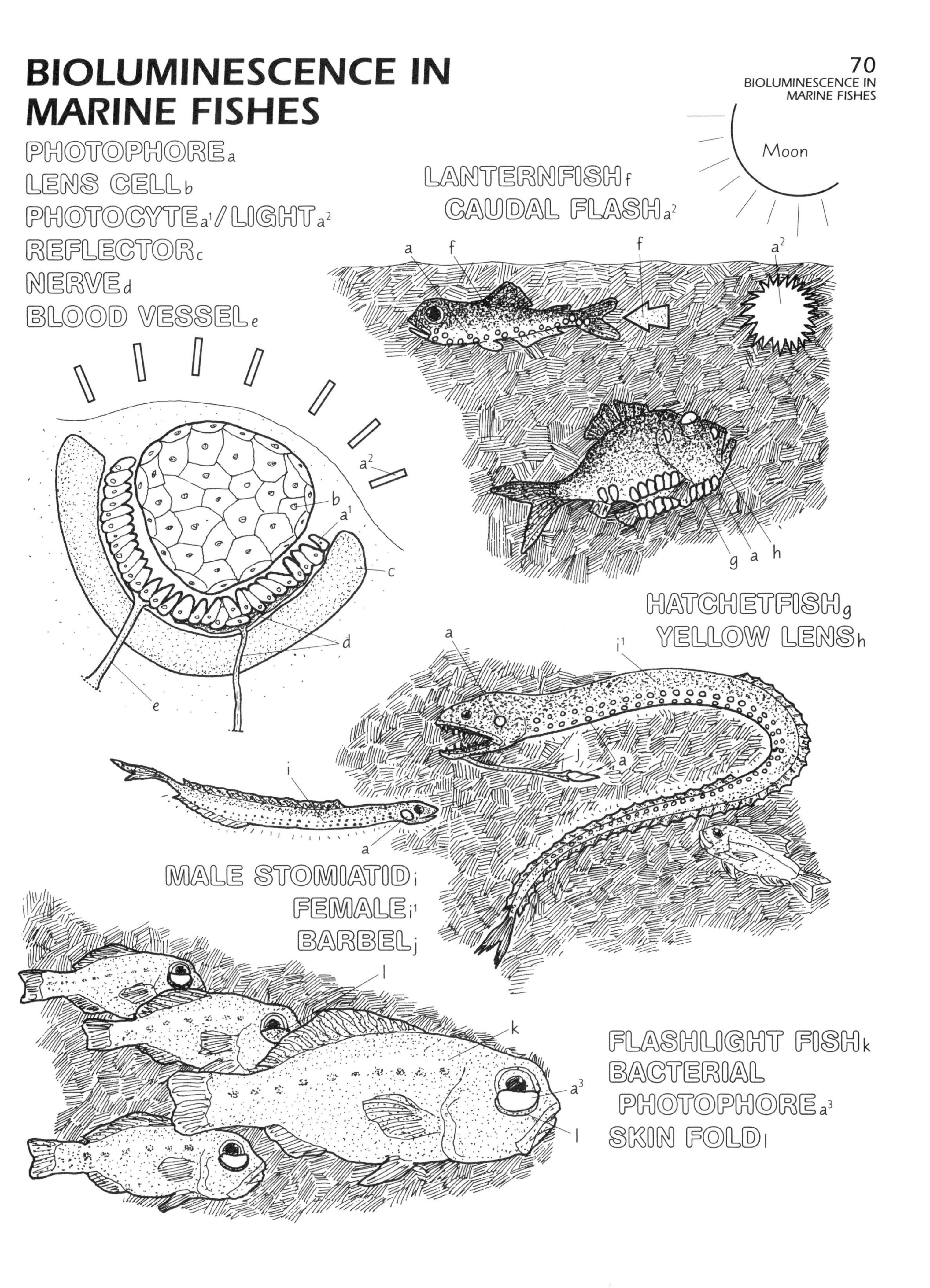

70
BIOLUMINESCENCE IN MARINE FISHES
BIOLUMINESCENCE IN MARINE FISHES
PHOTOPHORE a
LENS CELL b
PHOTOCYTE a1 / LIGHT a2
REFLECTOR c
NERVE d
BLOOD VESSEL e
Moon
LANTERNFISH f
CAUDAL FLASH a2
a
f
f
a2
a2
b
a1
c
d
e
g
a
h
HATCHETFISH g
YELLOW LENS h
a
i1
j
a
i
a
MALE STOMIATID i
FEMALE i1
BARBEL j
l
k
a3
l
FLASHLIGHT FISH k
BACTERIAL PHOTOPHORE a3
SKIN FOLD l

71
SOUND PRODUCTION IN THE SEA

As has been seen in the previous eight plates, the use of color and light by marine organisms is widespread and complex. However, light is quickly attenuated in sea water and visual cues are useful only for relatively short distances. Sound, on the other hand, travels approximately five times faster in water than in air, and can be heard over vast distances. Because light is so limited, sound is more useful and has a greater range for orienting an animal to its surroundings in water. This plate discusses the use of sound in the sea.

Color the singing humpback whale.

Marine mammals are the most sophisticated producers and users of sound in the ocean. The use of sound for echolocation by toothed whales is reviewed in Plate 61. Evidence of echolocation ability is also found in other marine mammals, including sea lions and some baleen whales. Virtually all marine mammals use sound for intra- and interspecific communication. The *song* of the male *humpback whale* is an excellent example of the complex nature of the use of sound for communication by marine mammals. Male humpbacks only sing when they are on the winter breeding grounds. Most researchers believe the song is generated in the nasal passages between the lungs and the blowhole, and that it is part of humpback courtship; males calling attention to themselves and announcing their availability to females. The singing male floats suspended in the water, head down and motionless; his haunting song, a rich repertoire that covers many octaves, lasts ten minutes or longer and may be repeated over and over for many hours. Not all males sing, but those that do all sing the same song on a given breeding ground. The song evolves over the breeding season, and as the theme changes, each male makes the same change at the same time. In addition, the males on breeding grounds in the Pacific Ocean from Japan to Hawaii to Mexico make the same changes in the song at about the same time. How and why this happens remains a mystery. Humpback whale herds in the North Atlantic and the Southern Ocean each have their own distinct song.

Now color the midshipman on his nuptial rock pile. Note that the drawing is cut away to show the swimbladder used by the fish to produce sound.

Love is also the motivation behind the sound produced by the fish known as the singing toadfish or *midshipman*. This big-headed, silvery fish has rows of ventral photophores that reminded an ichthyologist of the buttons on the front of a midshipman's trousers, and the common name persists. Come springtime in the temperate eastern Pacific, the midshipmen enter shallow, still waters of estuaries and embayments and set up a nuptial territory that consists of a pile of likely-looking rocks. After dark, the males begin their serenade in the hope of luring fecund females which they can court and convince to lay their sticky *eggs* on the rocks. The male fertilizes the eggs and then stands guard over them until they hatch some weeks later.

Male midshipmen create a rhythmic, mechanical sound by rapidly contracting the musculature of their *swimbladder*, causing it to vibrate. The vibrations resonate in the air-filled swimbladder like in the sounding board of a base viol, magnifying the sound to surprising levels and broadcasting it considerable distances. The love call of the midshipman is so loud that it can be transmitted through the hull of a boat, causing considerable puzzlement and wonder to the occupants.

Now color the pistol shrimp and the enlargement of the snapping claw.

Finding a mate isn't the only purpose for sound production. The small (2–6 cm, 0.8–2.4 in) crustacean known as the snapping or *pistol shrimp* uses sound as an offensive and defensive weapon (Plate 94). In shallow rocky areas and coral reefs where pistol shrimp are common, their sound can form a continuous wall of noise that accompanies night SCUBA dives. Pistol shrimp produce a tinny, popping noise with an enlarged, modified claw. At the base of the moveable *finger* of the claw is a large *tuberculate process* that fits into a *socket* on the *palm*. The claw is "cocked" when contact is made between two specialized discs, one at the base of the tuberculate process and one on the palm. The adhesive force between the discs, created by surface tension, prevents the claw from closing until the contracting muscle has generated a counteracting pull. The claw then slams shut with great speed and force, creating a snapping or popping sound and a little jet of water. Some species of pistol shrimp use the snap to crack small clams or stun fish. The snap also functions in threat displays between individual shrimp in territorial disputes, and still others use it in defense of their dwelling. Pistol shrimp live in holes and crevices beneath shells, rocks, and coral rubble. They also construct burrows or live within large sponges, tunicates, or coral heads.

SOUND PRODUCTION IN THE SEA

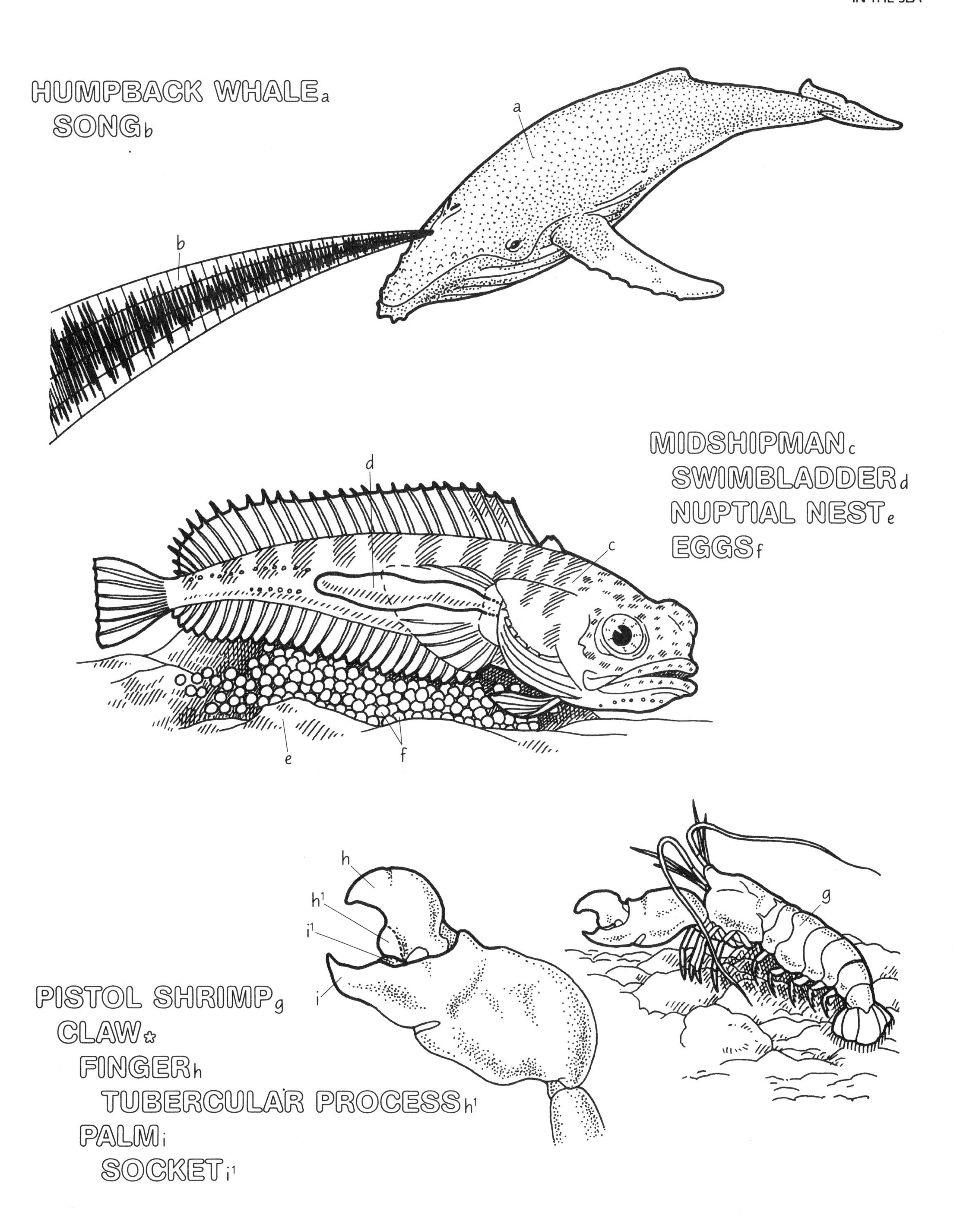

72
PHYTOPLANKTON REPRODUCTION: DIATOMS AND DINOFLAGELLATES

Diatoms and dinoflagellates are among the most abundant large phytoplankters found in temperate coastal waters. Both of these single-celled plants are capable of sexual reproduction. They can also utilize asexual reproduction to take advantage of favorable conditions (sunlight, nutrients) and rapidly increase their numbers.

Color the large illustration of the diatom at the upper left. Then, color the diatoms undergoing asexual reproduction, indicated by arrows (e) in the upper half of the diagram. After reading the text, continue by coloring the sexual reproduction portion, indicated by arrows (f). Choose contrasting colors for (e) and (f) and light colors for (g^1) and (h^1).

A round, or centric, diatom typically found in the plankton has an outer, two-piece frustule composed of silica which surrounds the cell or protoplast (Plate 19). In this diagram, only the two plates of the frustule and the cell *nucleus* are shown. At the onset of asexual reproduction, the nucleus divides and the interlocking pieces of the frustule separate. Each plate of the parent frustule serves as the outer piece of the frustule of a new individual. Thus, with each division, one of the offspring is the same size as the *parent* (the individual with the outer half of the parent frustule plus a new *large* inner frustule). The other offspring, however, is smaller than the parent since it is formed from the parent's smaller inner plate and a new *small* inner frustule (b). In subsequent asexual divisions, descendants of the smaller offspring continue to decrease in size as the small frustule acts as a smaller and smaller parent template. When these asexually produced offspring are 60 to 80 percent smaller than the original parent, the frustule is too small to contain the necessary cell mass for proper functioning, and under favorable conditions, sexual reproduction occurs. If environmental conditions are unfavorable at this time, the small diatoms will die. Note that the other offspring continues asexual reproduction (upper right arrow).

Sexual reproduction in diatoms involves separate *male* and *female* individuals. The frustule of each cell separates, but its halves remain connected by a *membrane*. The male produces and releases flagellated, motile *sperm*, which swim to a female diatom containing a single *egg*. The membrane of the female separates, allowing the entrance of sperm and subsequent fertilization of the egg. The fertilized egg, or *zygote*, swells to a globular shape, and the old frustule is shed. As the zygote develops, a new frustule is formed. The new frustule is large, the size of the original parent diatom, and the asexual cycle begins again. If there is sufficient sunlight, and the required nutrients are abundant, diatoms may undergo asexual reproduction more than once each day.

Color the dinoflagellate illustration.

Dinoflagellates primarily reproduce asexually by simple division of mature individuals. Shown here is an armored dinoflagellate that has a rigid cellulose *theca* (Plate 19). The nucleus divides, and the theca separates. Each new nucleus remains in one of the halves of the theca, and a new second half is then formed. Unlike the diatom, both offspring grow to the same size as the parent. Under favorable conditions, dinoflagellates may divide every 8 to 12 hours. If conditions become unsuitable, the dinoflagellates form *temporary cysts,* which fall to the bottom. When conditions improve, the temporary cysts quickly become active cells again. Dinoflagellates are also capable of forming *resting cysts* by a sexual process which involves the formation and subsequent fusion of motile gametes. These resting cysts persist for months before becoming active again. Cyst formation in dinoflagellates allows these organisms to survive unfavorable environmental conditions.

Some species of dinoflagellates multiply to such incredibly high numbers (10 to 20 million cells per liter) that the water in which they occur becomes strikingly discolored. These *"red tides"* occur along all North American coasts and are triggered by a combination of biological and physical conditions. Certain dinoflagellates produce a neurotoxic substance called *saxitoxin*. When mussels and clams feed on the dinoflagellates by filtering them from the water, this neurotoxin becomes concentrated in their tissues. When bivalves (whether cooked or uncooked) containing high concentrations of this neurotoxin are eaten by humans, a disease called paralytic shellfish poisoning results. In severe cases, the saxitoxin can cause death in 12 to 24 hours. Saxitoxin is over 100,000 times more potent than cocaine and concentrations sufficient to cause poisoning may be present in *shellfish* even when the "red tide" is not noticeable.

DIATOMS AND DINOFLAGELLATES

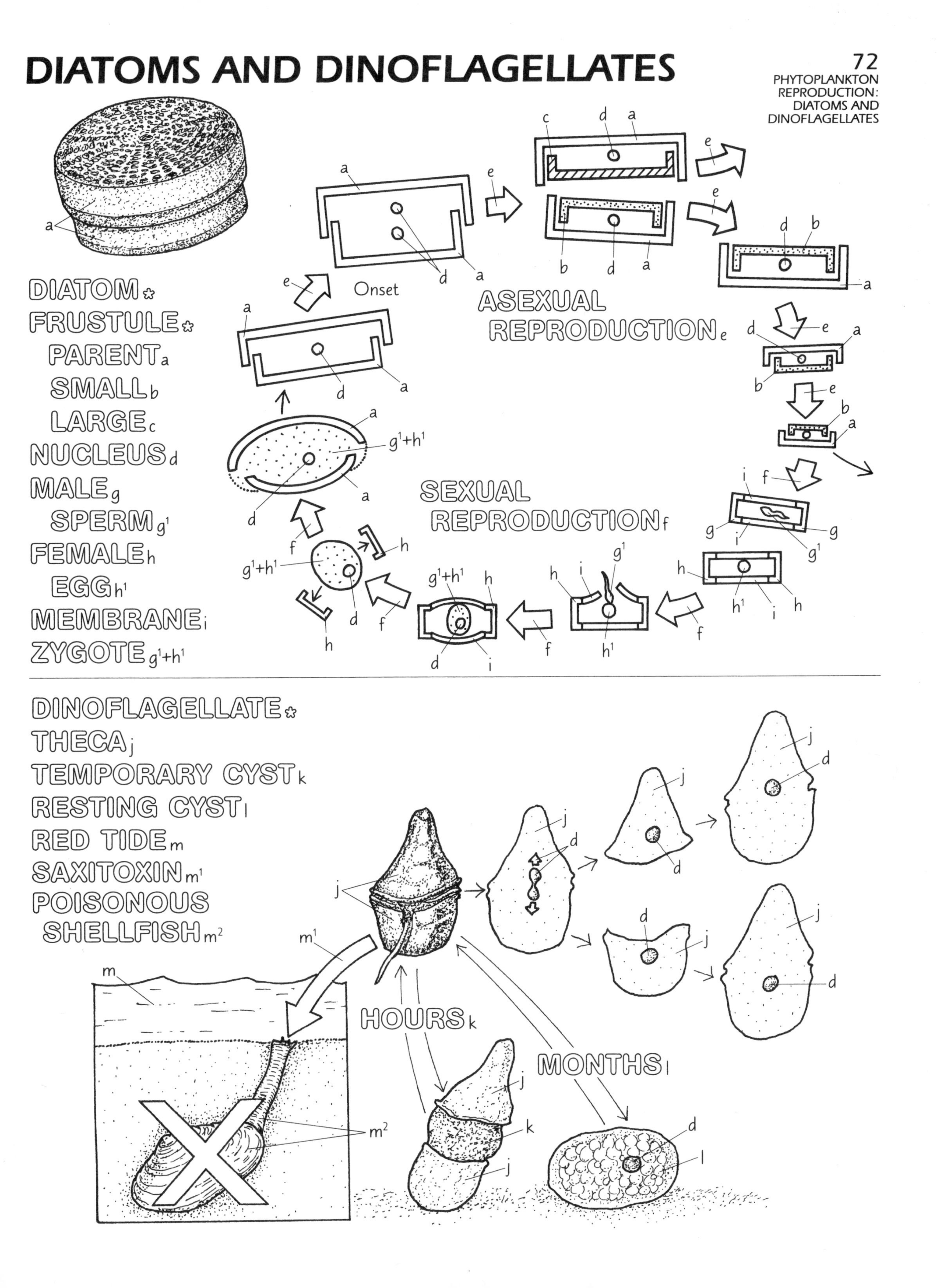

73
LIFE CYCLES OF ALGAE

Much of the growth and proliferation of marine algae takes place by asexual or vegetative means. However, most algae have a sexual phase of their life cycle that is intimately involved in their overall survival strategy.

Begin with the diploid phase in the life cycle of *Ulva*. Locate and color the stages in each life cycle as they are mentioned in the text. The stages drawn within circles are microscopic in actual size.

The common green alga *Ulva* (sea lettuce) undergoes a cyclic alternation between a spore-producing plant (*sporophyte)* and a gamete-producing plant (gametophyte). In *Ulva,* these two types of plants appear identical without the aid of a microscope and detailed examination. Each cell of the sporophyte possesses a full complement of chromosomes (a condition designated as diploid). A cell division process called meiosis (reduction division) produces motile flagellated *zoospores*, each containing half the normal number of chromosomes (a condition designated as haploid). These haploid zoospores swim and attach to a hard substratum, where each can grow into a completely haploid *male* or *female gametophyte* plant. These plants then produce haploid gametes (*sperm* and *egg)* which are released into the water. A sperm and egg combine (fertilization) to produce a diploid *zygote* (two haploids= one diploid). The zygote settles and grows into another diploid sporophyte plant and the cycle is complete.

The large brown alga *Nereocystis (*bull kelp) also undergoes an alternation between diploid sporophyte and haploid gametophyte. In this case, however, there is a great difference in size between the two generations. The huge bull kelp with its 30-meter (100-ft) stipe is the sporophyte phase; it produces haploid flagellated zoospores in special regions of its blades. A single bull kelp might produce over 3.5 trillion zoospores in one year. The zoospores settle to the bottom and grow into separate, microscopic, male and female gametophytes. The male gametophytes produce motile sperm, which swim to the female gametophytes and fertilize the eggs in place, thus restoring the diploid chromosome number. The zygote grows as a sporophyte to eventually become the huge bull kelp.

Why the great difference between the sporophyte and gametophyte generations of this alga? Consider the timing involved. The bull kelp (sporophyte) grows rapidly during the spring and summer months when optimal conditions prevail. It matures in late summer and produces zoospores. The microscopic gametophytes that grow from these zoospores produce their gametes and achieve fertilization during the winter, and the young sporophytes can usually be found in early spring. Despite its large size, the bull kelp is an annual plant, completing its life cycle in one year, and sporophyte plants are torn loose from the bottom during fall and winter storms. The small gametophytes and young sporophytes are safe in cracks and crevices on the bottom during this inhospitable time, but are ready to take advantage of calmer weather and spring sunshine to start the cycle anew.

The red alga *Porphyra* is known as "nori" in Japan, where it is dried and used extensively in cooking. This seaweed represents a food industry of several million dollars annually. The details of sexual reproduction in this alga are not fully understood. However, a general knowledge of the life cycle is the basis for one of the most successfully managed marine cash crops in the world. (The raising and harvesting of marine organisms is called mariculture.) The stage of the life cycle that is dried and eaten is the *foliose* (leafy) *phase.* The foliose phase produces two types of spores, one that results in the growth of more foliose phase plants (the loop in the illustration), and one that generally settles on and bores into an empty mollusc shell. This shell-boring stage is known as the *conchocelis phase.* "Conch" means shell, and the name of this phase comes from *Conchocelis rosea,* a scientific name given to the shell-boring plant before its relationship to the foliose phase was understood. The conchocelis phase can also produce two types of spores. One type grows into more conchocelis phases, and the other type (called a *conchospore*) grows into the foliose phase. In nature, the foliose phase appears during the winter months in the high intertidal zone, where conditions are conducive to its growth: it is kept moist by rain and spray from storm waves; there is little competition from other plants; and the grazers of this area (limpets and littorines) are not particularly active. In the spring and summer, the high intertidal is dry from exposure to the air and sun during daytime low tides, and grazing can be severe. *Porphyra* survives these adverse conditions in the conchocelis phase, safe in a shell on the bottom. Japanese mariculturists take advantage of the conchocelis phase by inducing it to release conchospores onto special ropes in closed tanks. The ropes are then suspended in quiet water marine habitats to allow the foliose phase plants to grow and be harvested as nori.

LIFE CYCLES OF ALGAE

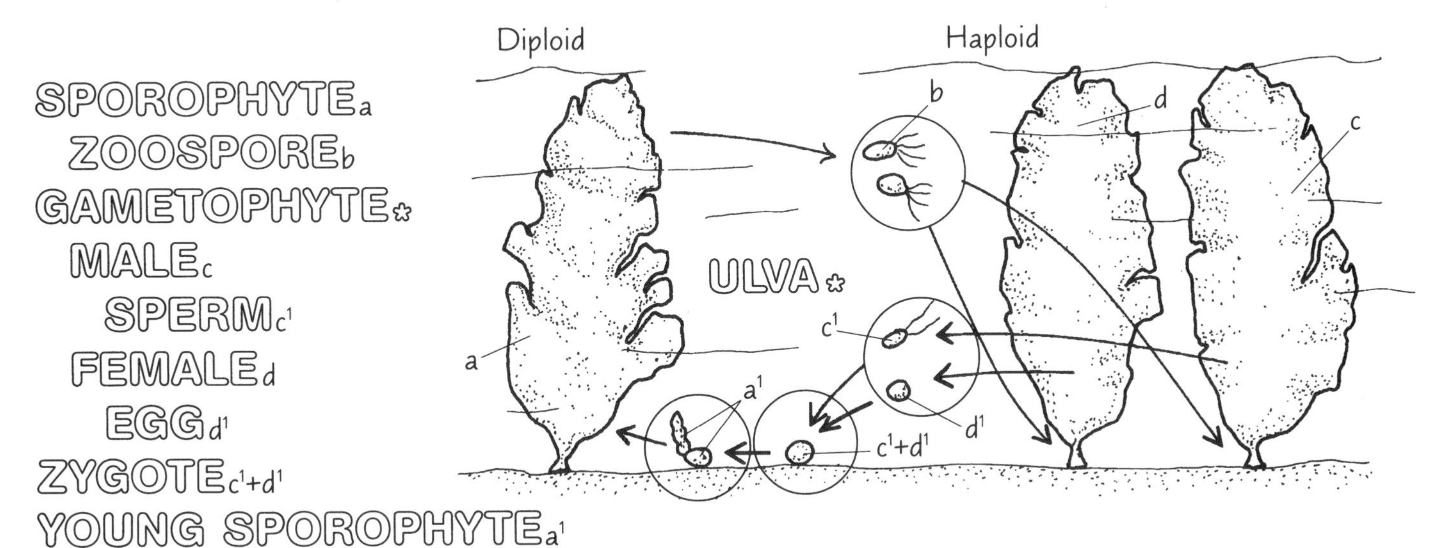

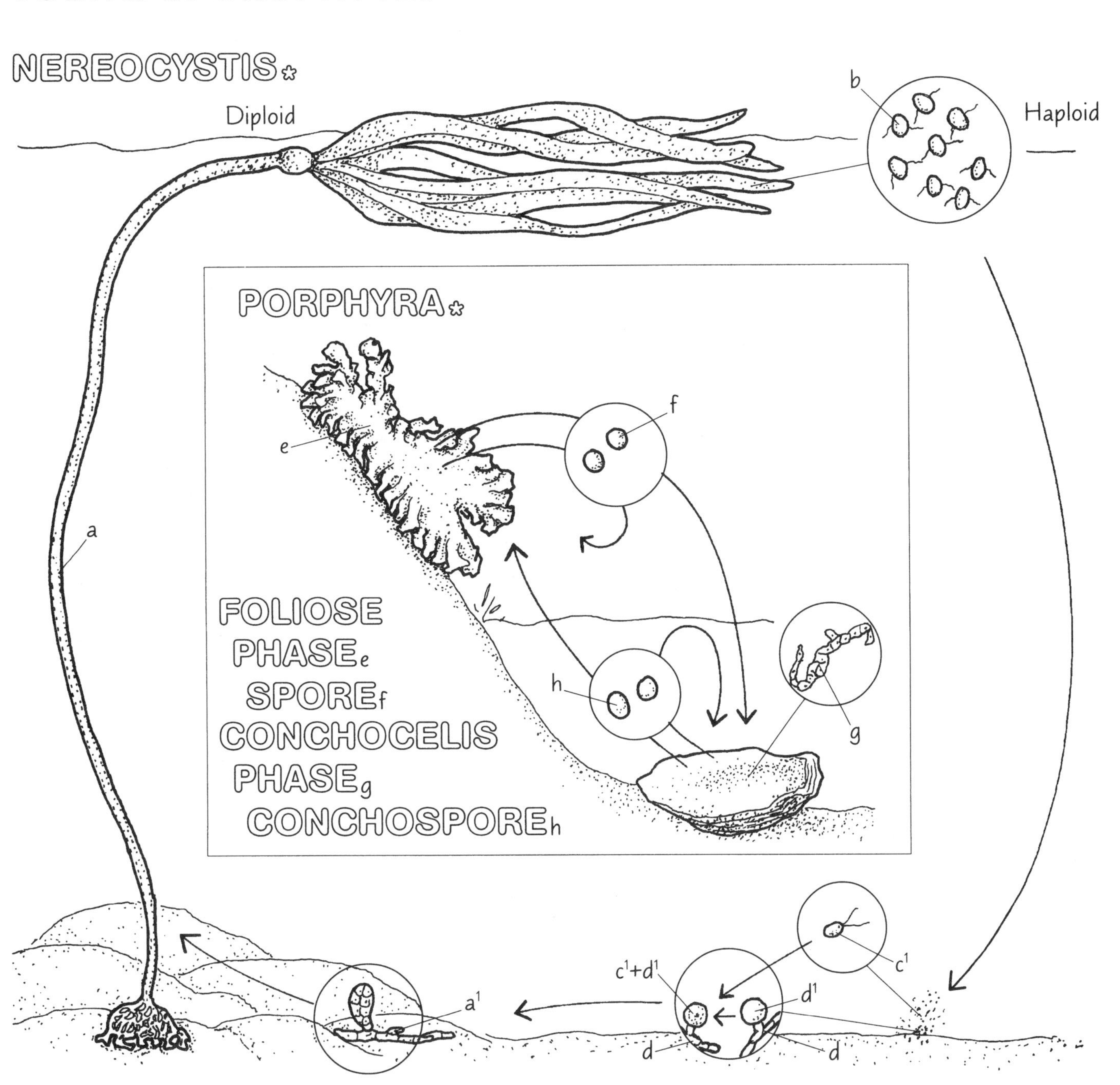

74
ASEXUAL REPRODUCTION

Sexual reproduction begins with the segregation of one-half of a cell's genetic material (chromosomes) into a sex cell, or gamete (sperm or egg). The union of sperm and egg (fertilization) restores the full complement of genetic material, creating a new individual with a combination of the genetic characteristics of the parents. Such genetic variation may result in organisms better adapted to their environments or that are able to adapt to new environmental conditions. This increased potential adaptability is one major survival advantage for species that utilize sexual reproduction. Some examples of these sexual processes are discussed in the following plates.

However, non-sexual (asexual) means of reproduction are used by many species to take advantage of favorable—and rapidly changing—environmental conditions. Great increases in numbers can be achieved more quickly by asexual than by sexual means. Five examples of asexual reproduction are presented in this plate.

Color each example of asexual reproduction as it is discussed. In the case of the sea star, both illustrated processes involve regeneration, however the example on the right also results in the asexual formation of two individuals from one (reproduction).

In the previous plate, it was explained how marine algae produce spores which asexually develop into new individuals. The single-celled diatoms and dinoflagellates (Plate 72) often reproduce asexually by simply dividing. The *turtle grass* shown here is an example of a marine flowering plant which produces new individuals by means of underground stems called *rhizomes* (Plate 18). Such *vegetative growth* also occurs as a type of asexual reproduction in many terrestrial plants.

Asexual reproduction is not always such a clearly definable phenomenon. For example, if a *sea star* loses all but one of its arms from the central disc it may gradually regrow the missing arms as shown. This process is called *regeneration*. However, if a sea star is cut evenly in two (a generally futile practice of some oystermen who hope to eliminate the predators from oyster beds), each half might regrow the missing portion. Obviously this process involves regeneration, but it must also be considered asexual reproduction since two individuals are produced from an original one without sexual activity.

The cnidarians are masters of asexual reproduction. The white-plumed anemone *Metridium* (Plate 23) exhibits a behavior known as *pedal laceration*. As the anemone creeps over the substratum on its pedal disc, portions of the disc are occasionally torn off and left behind where they grow into small anemones. Some species, such as the aggregating anemone, reproduce by literally pulling themselves in half from top to bottom in the process of longitudinal fission (Plates 5, 96).

The ability to reproduce asexually allows a single organism to take advantage of a favorable habitat. For example, if a planktonic larval form finds a suitable site for attachment and growth, chances are excellent that others of its species would also be successful in that area. If the organism can reproduce by asexual means, it can proliferate and maximize its use of the habitat in a relatively short period of time. Such opportunistic exploitation of habitat is exemplified by corals. The coral's planula larva (Plate 76) settles on an appropriate substratum and grows into a single *polyp*. If the polyp is successful, it does not simply continue to increase in size, but instead buds off a new polyp, as illustrated. This *budding* process continues and the polyps remain attached by a common sheet of tissue (not shown, see Plate 23). Budding can eventually produce large coral heads consisting of thousands of asexually produced individuals (Plate 13).

A number of marine polychaetes (*Autolytus* is shown here) utilize a variant of asexual reproduction as a preparation for sexual reproduction. For most of the year, *Autolytus* exists in a sexually unripe, but otherwise adult form called an atoke. Preparatory to breeding, the posterior portions of the atoke begin to develop as a series of sexually maturing *epitokes*, complete with head (see illustration). These remain connected until the time of breeding; there is a record of one individual having twenty-nine epitokes attached. Breeding involves the synchronous release and swarming of multitudes of epitokes of both sexes, and this occurs, in most species, on one or a few nights each year. The timing of the event is precisely linked to the lunar cycle in a manner not yet understood. Swarming takes place in surface waters, with tremendous predation by birds, fish, and humans—some epitokes are much prized in Samoa, for example (Plate 77). This asexual multiplication of reproducing individuals, together with the simultaneous shedding of gametes, seems to maximize the chance of successful reproduction.

ASEXUAL REPRODUCTION

VEGETATIVE GROWTH$_a$
TURTLE GRASS$_b$
RHIZOME$_{a^1}$

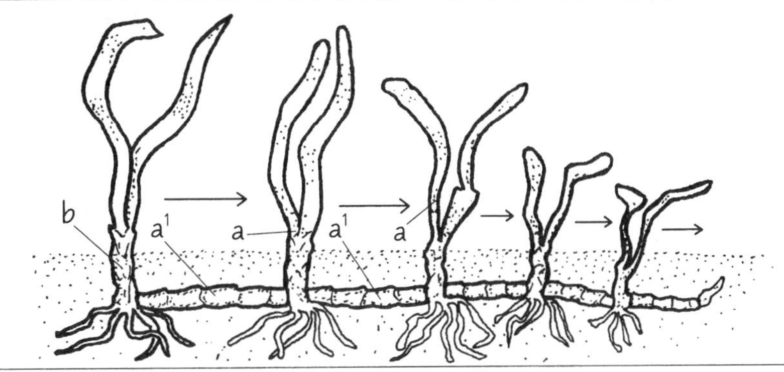

REGENERATION$_c$
ASEXUAL REPRODUCTION$_{a^2}$
SEA STAR$_d$

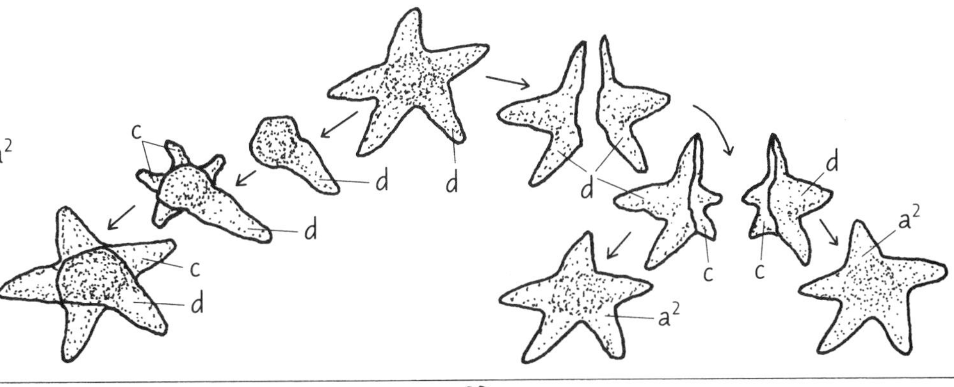

PEDAL LACERATION$_{a^3}$
METRIDIUM$_e$

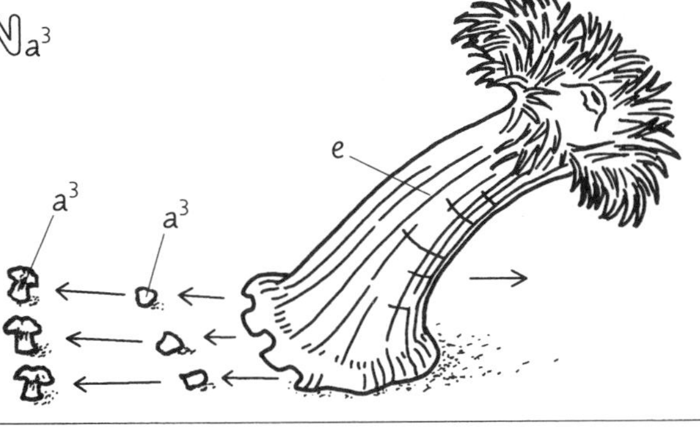

BUDDING$_{a^4}$
CORAL POLYP$_f$

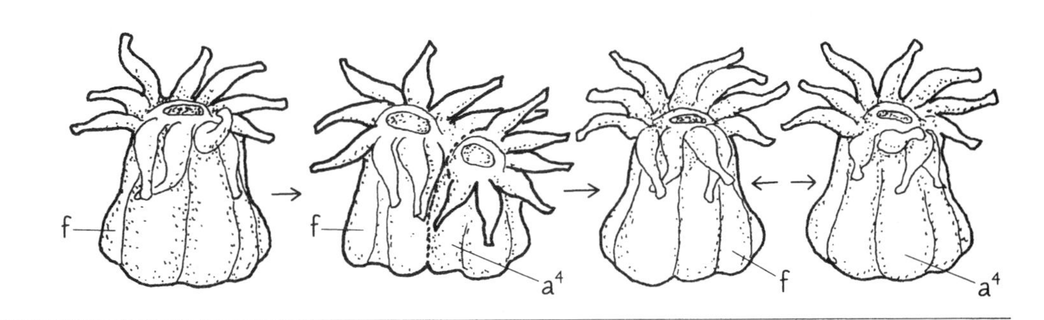

EPITOKE FORMATION$_{a^5}$
AUTOLYTUS$_g$

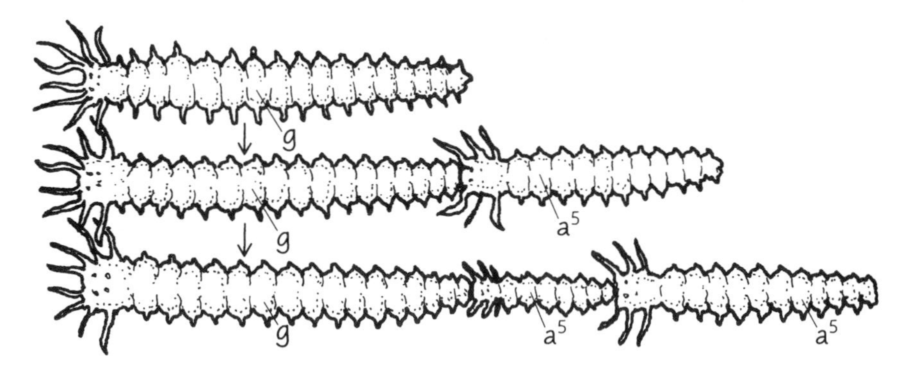

75
REPRODUCTION IN MARINE ORGANISMS: PLANKTONIC LARVAL FORMS

The embryonic development of many marine organisms includes a planktonic larval form. "Planktonic" indicates that the larvae float free of the bottom and are incapable of swimming strongly enough to avoid being carried by ocean currents. The occurrence of a planktonic larval stage in the life cycles of so many different animals indicates that it must have great advantages. First, the larva has an opportunity to feed on the richest of all marine food sources—the phytoplankton (Plate 19). Second, ocean currents may transport the larva over long distances and allow dispersal to new habitats as well as replenishment of areas already occupied. Third, competition between adult and larva is virtually eliminated, as each exploits a different environment.

The production of planktonic larvae is not without expense. Many larvae never find their way to habitats that are suitable for adult life. Planktonic life is precarious. Microscopic drifting larvae may be subjected to areas of unfavorable water conditions, predators, and lack of available food. To compensate for the usually very high mortality rates among these planktonic stages, the adults produce enormous numbers (tens or hundreds of thousands per female) of small larvae. Also, the release of larvae into the plankton is generally timed to occur during or just prior to periods of high food availability (spring and summer). Such high production of larvae can sometimes backfire. If larval success is extremely high, the result may be overcrowding of the adult habitat after settling (for example, as with barnacles; Plate 97).

As mentioned above, the habitats of larvae and adults are often totally different (for example, planktonic versus bottom-dwelling), and these two life stages usually consume different foods. Thus, it is not surprising that the larval body form is often entirely unlike that of the adult, as illustrated in this plate.

Color each adult/larva pair as it is referred to in the text. Begin with the brittle star. All the larvae illustrated here are shown greatly enlarged. The setae of the nectochaete larva are drawn with fine lines which may be lightly colored over. Only a simple outline and the eye of the adult sunfish are shown. Plate 45 includes a detailed drawing of this huge fish.

The larval form of the *brittle star*, or ophiuroids, (phylum Echinodermata, Plate 40) is the odd-looking *ophiopluteus*. It has eight *larval arms*, which are held somewhat rigidly by internal skeletal rods. The small (less than 0.5 mm, 0.02 in) ophiopluteus swims with the aid of tracks of cilia on the arms and body. It feeds on phytoplankton that are carried to the mouth by other ciliary tracks. The elongate arms help increase the surface area to volume ratio of the larval body and offer resistance to sinking. Notice how the ophiopluteus is organized bilaterally, while the adult brittle star is organized radially. This change in symmetry requires a wholesale modification of the larva when it metamorphoses to the adult form.

The *zoea larva* of the *porcelain crab* has elongated *larval spines* to help in flotation; the spines also serve as defense against predators. The adult porcelain crab bears little resemblance to its small (1.5 mm, 0.06 in) larval form. The larval spines are lost during metamorphosis, and the abdomen is tucked beneath the body—such extensions would hinder this much-flattened crab that hides in narrow places and under rocks.

The tiny (0.25 mm, 0.01 in) *nectochaete larva* of the polychaete *Nereis* (Plates 27, 77) shows the segmentation of the body so characteristic of the adult. Numerous spinelike *setae* on the larva's body offer resistance to sinking as this animal spends part of its life in the plankton. The setae also serve as paddles to propel the larva through the water.

Many fishes also have planktonic larval forms. The giant ocean *sunfish* has a tiny (3 mm, 0.1 in) larva that looks very little like the adult. The pronounced spines develop as the larva grows (at about 1.3 cm, 0.5 in) and are later lost as the adult form becomes apparent. These small larvae eventually become adults that can grow to more than 3 meters (10 ft) and 2000 kg (4400 lb)!

PLANKTONIC LARVAL FORMS

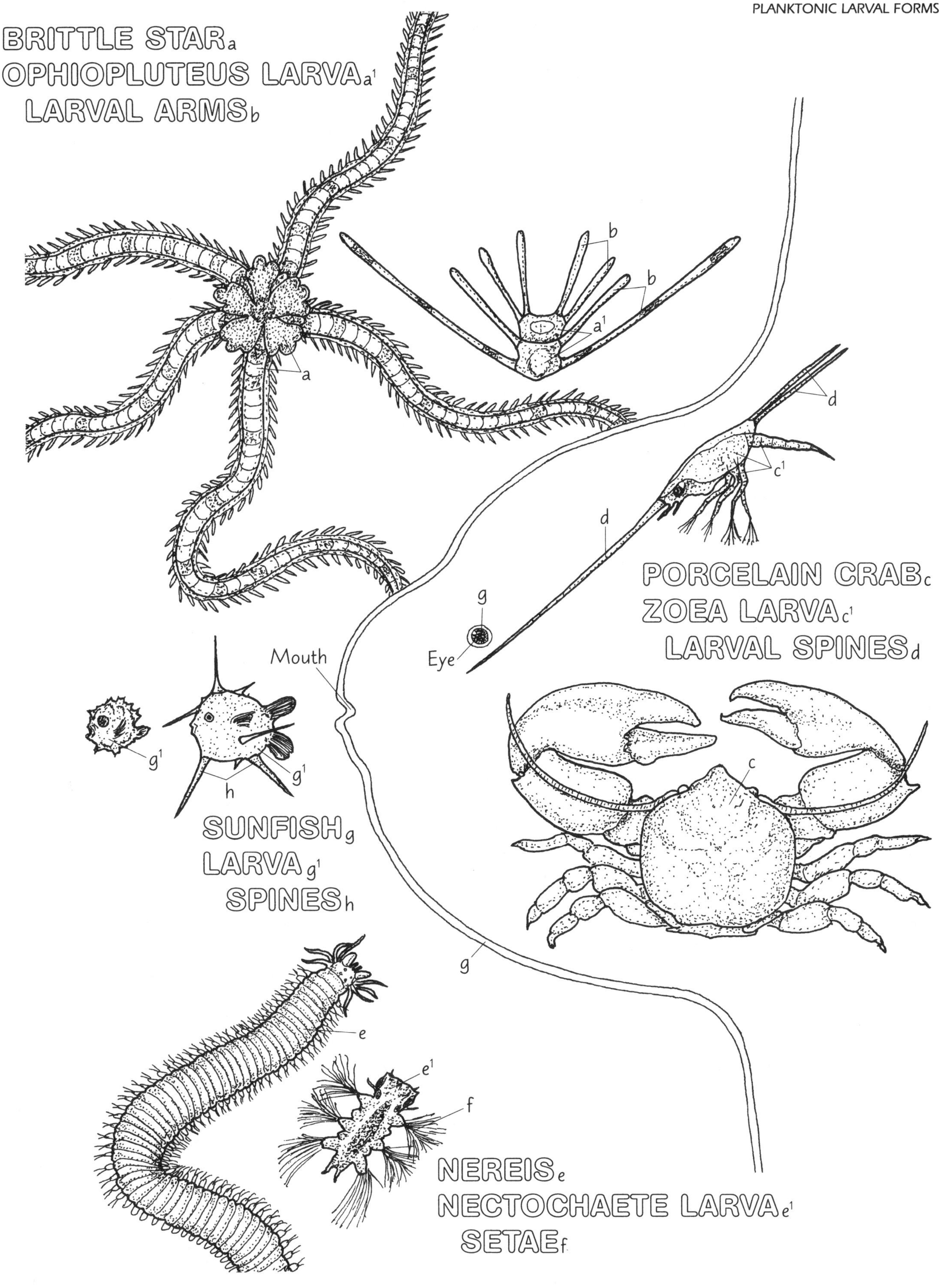

BRITTLE STAR$_a$
OPHIOPLUTEUS LARVA$_{a^1}$
LARVAL ARMS$_b$

b
b
a^1
a
d
c^1
d
g
Eye
Mouth

PORCELAIN CRAB$_c$
ZOEA LARVA$_{c^1}$
LARVAL SPINES$_d$

c
g^1
g^1
h

SUNFISH$_g$
LARVA$_{g^1}$
SPINES$_h$

g
e
e^1
f

NEREIS$_e$
NECTOCHAETE LARVA$_{e^1}$
SETAE$_f$

76
REPRODUCTION IN CNIDARIANS: CNIDARIAN LIFE CYCLES

Cnidarians (also known as coelenterates) exhibit two basic body forms, or morphs: the polyp and the medusa (Plates 23, 24). In general, polyps are sessile and reproduce by asexual methods (budding), while the medusae are free-floating and reproduce by sexual means. (There are many exceptions and modifications to this basic plan.) Both morphs occur in the life cycles of some cnidarians, while others lack one or the other form. In addition, most cnidarian life cycles involve both sexual and asexual reproduction. This plate discusses the life cycles of a representative of each of the classes of the phylum Cnidaria.

Begin with the life cycle of *Obelia*. First color the animal colony on the left, then follow the arrows through the cycle as it is discussed. The medusae are exaggerated in size, as are the sperm, eggs, and planula larva. Note that the entire colony is surrounded by a clear, non-living covering that is not colored.

The feeding polyp stage of the class Hydrozoa (here represented by *Obelia*) is called a *hydranth*. These hydranths occur singly in some species, but more commonly as colonies of asexually produced individuals connected by a branching *stalk* which contains their common gut tube or coelenteron (Plate 23). An *Obelia* colony resembles a small bushy alga to the naked eye, but microscopic examination reveals the tiny (0.2 mm, 0.008 in) hydranths along the colony's branches. At certain times of the year, one can find reproductive polyps or *gonozooids* interspersed among the hydranths. Visible in the illustration are numerous *medusa buds* forming on the gonozooids. These are produced asexually and released as tiny free-swimming *medusae*. Each medusa is either male or female, and as they mature, each produces either *sperm* or *eggs*, which unite to form a *zygote*. The zygote develops into a free-swimming *planula* larva, which eventually settles on a firm substratum and becomes a small hydranth that begins a new colony by asexual budding.

Color the *Aurelia* life cycle, beginning with the male (on the left) and female adult medusae. Continue as each stage is discussed.

In the common jellyfish *Aurelia* (class Scyphozoa), the medusa is the largest and dominant stage of the life cycle. Medusae 10 to 20 cm (4–8 in) in diameter often aggregate in very high numbers in the coastal waters of North America. The sexes are separate in *Aurelia,* and the female broods the zygotes on her *oral arms* until they reach a free-swimming planula stage. The planula swims for a short while, until settling on a solid substratum, whereupon it grows into the polyp stage called a *scyphistoma*. The scyphistoma feeds for some time in typical polyp fashion on small zooplankton and may produce other scyphistomas asexually. Eventually, each polyp begins to partition its body into a stack of tiny potential medusae. This process is called strobilation, and the polyp is now called a *strobila*. One at a time, these asexually produced (budded) young medusae are released, swim away, and mature to begin the cycle once again.

Color the life cycle of *Epiactis*. The arrow (e+f) indicates the transfer of the zygotes from the female's mouth to the base of her column.

Sea anemones and corals (class Anthozoa) have only the polyp stage in their life cycle, but most are still capable of both sexual and asexual reproduction. The small sea anemone illustrated here, *Epiactis*, is a common species on the west coast of North America, and has been subject to some past misunderstanding. The adults of this anemone are often found with many small *juveniles* attached to the base of their *columns*. Since anemones commonly reproduce asexually, it was assumed that these young individuals were the result of budding by the adult, hence the vernacular name "proliferating anemone." The faulty assumption was exposed some years ago by Dr. Daphne Fautin, who studied these anemones while at the University of California, Berkeley. Based on her work, it is now known that adult *Epiactis* retain their zygotes within their gut cavity where they are brooded for some time. The embryos eventually emerge from the anemone's mouth, crawl to the base of the column, and attach. Here juveniles grow to about 4 mm (0.16 in) in basal diameter, before crawling away to establish their own independent adult life.

CNIDARIAN LIFE CYCLES

OBELIA*
COLONY*
HYDRANTH a
STALK b
GONOZOOID c
MEDUSA BUD d
MEDUSA d[1]
SPERM e
EGG f
ZYGOTE e+f
PLANULA g

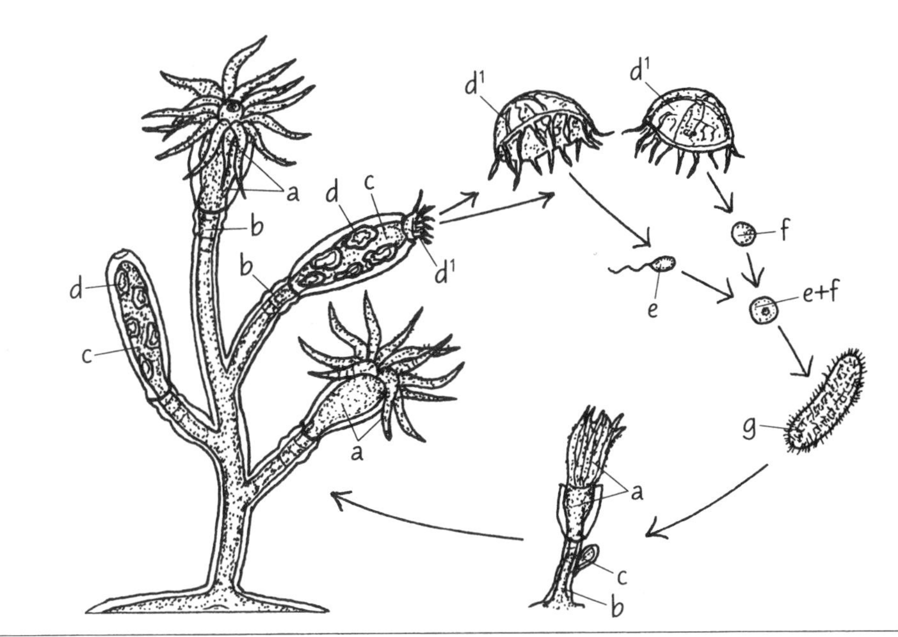

AURELIA*
MEDUSA d[1]
ORAL ARM h
SPERM e
ZYGOTE e+f
PLANULA g
SCYPHISTOMA i
STROBILA i[1]

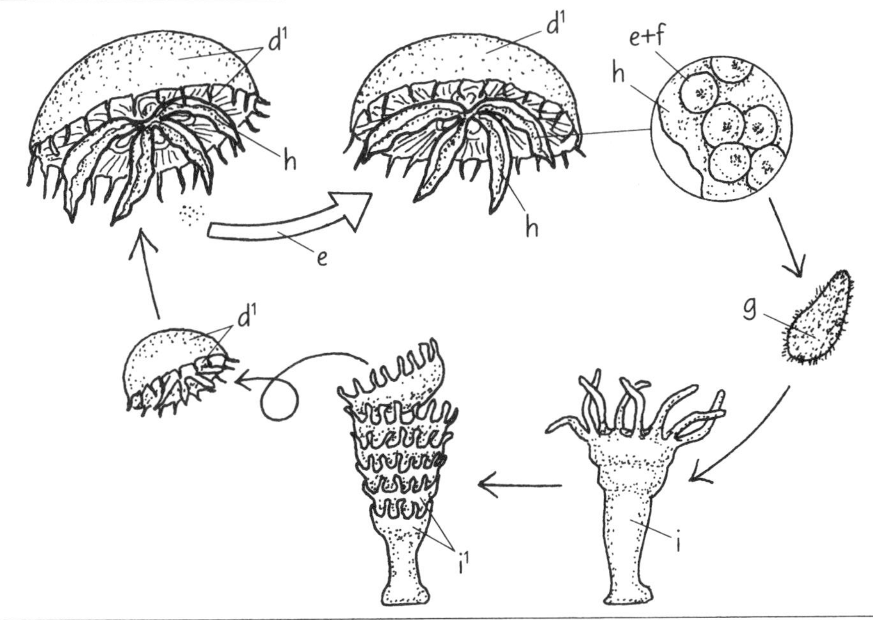

EPIACTIS*
ADULT*
ORAL DISC j
MOUTH k
TENTACLES l
COLUMN m
SPERM e
ZYGOTE e+f
JUVENILE n

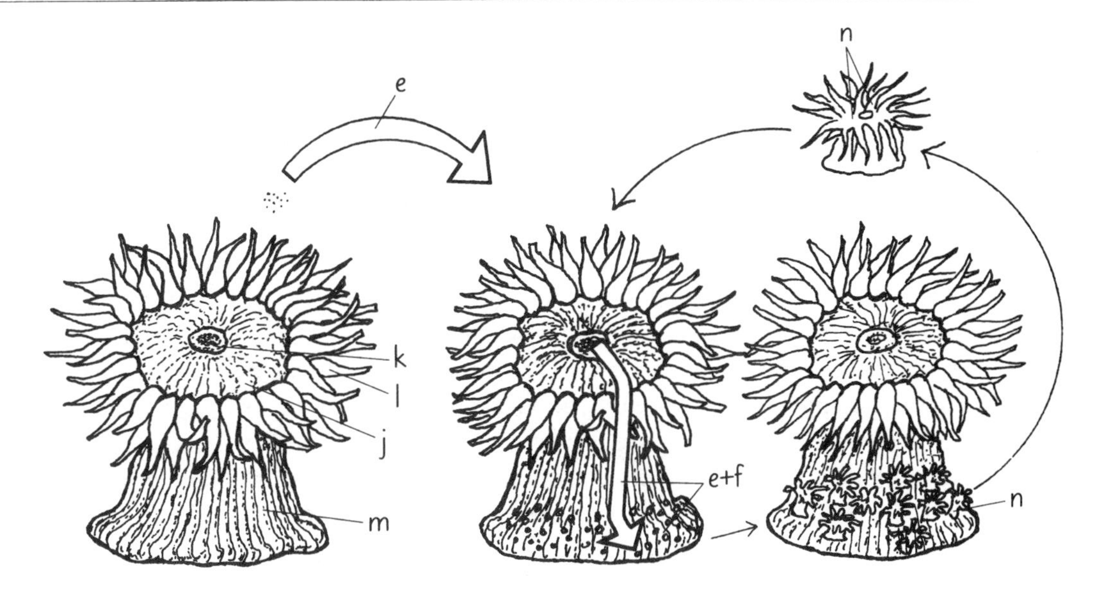

77
REPRODUCTION IN MARINE WORMS: POLYCHAETE LIFE CYCLES

The spawning of many species of polychaete worms and many other invertebrate animals, especially corals of the Great Barrier Reef, is tightly coupled to the phases of the moon. This results in local populations spawning simultaneously. Experiments done with polychaetes and sea stars suggest these invertebrates possess a mechanism that allows them to track changes in day length and synchronize their annual reproductive cycle accordingly. As spawning time comes closer, some combination of the monthly, moon-influenced change in tidal range along with changes in the amount of moonlight is thought to be the external cue that synchronizes the spawning of invertebrate populations with moon phase. This plate introduces the reproductive behavior of three polychaete worms and the role of the moon in each life cycle.

Begin with *Spirorbis* in the center of the page. The enlarged illustration shows the worm in its tube; portions in which the male and female segments lie are labeled. The newly settled larvae and adult worms are shown on a blade of oar weed.

The small, 1 to 2 mm (0.04–0.08 in) diameter, spiral, calcareous (lime) *tube* of the polychaete *Spirorbis* is a common sight on many marine substrata. These tiny filter-feeding worms settle on rocks, mollusc shells, and seaweeds. Unlike most polychaetes, in which the sexes are separate, *Spirorbis* is hermaphroditic. The anterior segments of the worm contain *female* sex organs, and the posterior segments contain *male* sex organs.

Many *Spirorbis* species follow a reproductive cycle that is closely tied to the moon's phases and occurs at the same lunar phase each month. Fertilized *eggs* of this worm are retained in the tube of the adult. This differs from most other polychaetes which release their eggs and sperm into the water. Each month free-swimming *larvae* are released. Settling behavior of a larva involves landing on a substratum to "feel" and "taste" it. If it is judged unsuitable, the larva swims off to try again. Different species of *Spirorbis* prefer different types of substrata. The tube-encrusted blade of the oar weed (Plate 21) shown here illustrates two aspects of settling behavior in *Spirorbis*. The individual worm tubes are evenly spaced. A settling larva crawls back and forth over the area and, presumably through some avoidance mechanism, will remain only if sufficient room is available to allow its growth to adult size. Note also that the larvae have settled on the innermost part of the blade, nearest the stipe. This is the point from which new blade growth arises. By settling on this area of new growth, *Spirorbis* is least likely to encounter other attached organisms, and, since the blade wears back from the tip, the innermost part offers the longest-lasting habitat.

Color the worm *Nereis* in the lower right corner. Then color the heteronereid which is shown swimming at the surface at the top of the page.

Many benthic polychaetes (Plate 27) spawn at the surface, rather than in their normal habitat. In *Nereis,* the bottom-dwelling adults undergo a radical transition in morphology (body appearance) into the sexual *heteronereid* form. The *eyes* become enlarged, and the appendages on the posterior segment change from a crawling to a swimming function. The pointed *crawling setae* are discarded and paddle-shaped *swimming setae* take their place. The parapodia also develop elaborate chemosensory organs that alert the worms to the presence of other heteronereids. The heteronereids mature and become packed with ripe sex cells (gametes). At precise phases of the moon, the heteronereids swim to the surface. The males release their sperm in controlled amounts, stimulating the females to rupture and release their eggs. Fertilization takes place in the water.

Color the palolo worm and its stolon in the lower left corner. Then follow the arrows to the stolons swarming at the ocean's surface. The swimming heteronereid is drawn in an exaggerated scale relative to the palolo stolon.

To ensure that individuals assemble for reproduction, a behavior called swarming occurs in many species. The best-known example of polychaete swarming involves the Samoan *palolo worm*. This polychaete lives in crevices of shallow coral reefs of Samoa, Fiji, and other South Pacific islands. The adult palolo worms grow an elongated posterior structure, called the *stolon*. The stolon segments are narrow and long and each has a light-sensitive ventral (bottom) *eye-spot*. On the eighth and ninth days after full moon in October and November, the adult palolo worm backs out of its crevice and releases the stolon. The stolons swim to the surface in vast swarms, rupture, and release their sex cells. Fertilized eggs may cover acres of the sea surface. Some islanders consider the palolo worm a delicacy and await the pre-dawn swarm with dip nets and hearty appetites.

POLYCHAETE LIFE CYCLES

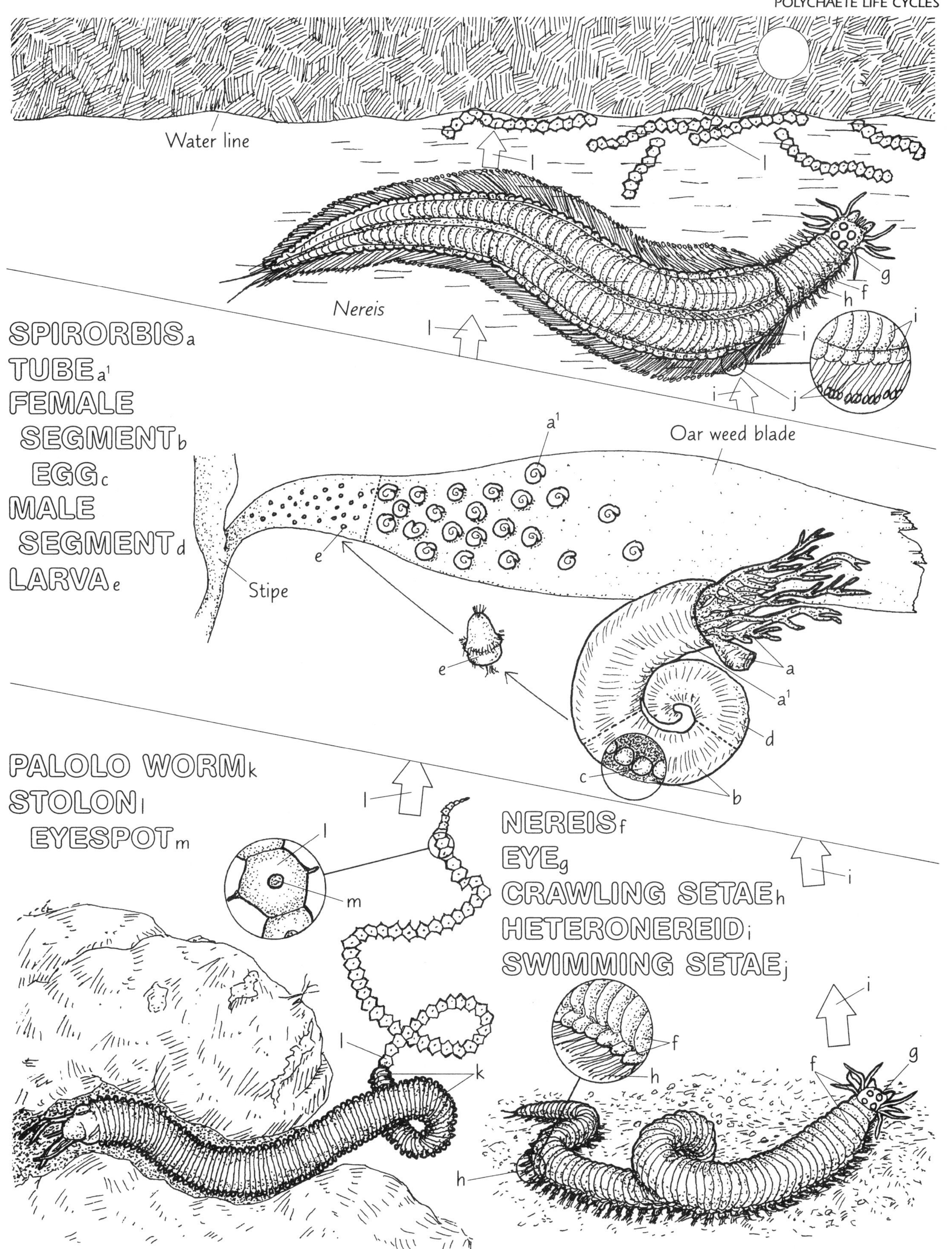

78
REPRODUCTION IN MOLLUSCS: BIVALVE REPRODUCTION: VIRGINIA OYSTER

Most species of bivalves are dioecious (sexes are separate) and reproduce by releasing their sex cells into the sea water. Fertilization occurs in the water, and a planktonic larva develops (Plate 75). This plate illustrates and explains the life cycle of the Virginia oyster and some aspects of the commercial importance of this animal.

Color the spawning male and female oysters and their sperm and eggs. Note that fertilization takes place in the overlying water. Follow the development of the fertilized eggs, and color each stage as it is mentioned in the text. Note that the veliger larva settles on an empty shell. New shell growth of the young oyster receives the same color as the settled spat.

Since oysters tend to settle in aggregations, the *males* and *females* are in close proximity and need only spawn into the water at the same time to ensure *fertilization*. The fertilized *eggs* undergo cell division and develop in the water column. As long as the developing zygotes remain encased in the egg membrane, they are considered *embryos*. Once they break free of the egg membrane they are considered larvae (or juveniles depending on the pattern of development). A ciliated *trochophore larva* hatches which then becomes a veliger larva with a rudimentary hinged bivalved *shell* and ciliated *velum* for swimming. The veliger larva feeds on *phytoplankton* (Plate 19) and continues to grow. After two to four weeks, a *foot* develops, and the veliger begins contacting the bottom with its extended foot. When a solid substratum is encountered, the veliger crawls about, testing the bottom. If the substratum is unsuitable (texture, crowding, or other factors), the veliger swims off to try again. When an appropriate substratum, such as an *empty shell*, is found, the veliger produces a cement (from pedal glands associated with the foot) and attaches its left valve to the substratum. Once secure, the velum is lost, the foot becomes very reduced, and the animal begins to grow. This newly attached oyster is called a *spat*. Its gills develop and it begins its life as an attached, filter-feeding organism (Plate 30). As the oyster's shell enlarges, it takes on the contours of the substratum and becomes irregularly shaped, as shown in the diagram of the two adults.

Virginia oysters are able to withstand reduced salinities and frequent exposure to air in the intertidal zone. Thus they can grow in estuaries, where rivers meet the sea and form quiet embayments and lagoons with broad mud and sand flats (Plate 8). Oyster larvae seek out and settle on hard substrata in these habitats, often in tremendous numbers. Their preference for settling on the shells of adult oysters (or other bivalves) may result in the development over time of thick oyster "beds." Humans recognized this tendency very early, and elaborate oyster cultivation practices date back to the Roman Empire. Today, oyster culture is big business, and it is an understanding of the natural life cycle and settling behavior that forms the basis for these endeavors.

Color the scallop shells to which the oyster spat are attached. After reading the text, complete the coloring of the oyster culture raft.

Oyster culturists maintain mature male and female oysters in aquaria and induce them to spawn by a sudden elevation of the water temperature. The fertilized eggs are collected and placed in tanks, where they are raised to veligers on a rich diet of phytoplankton. The maturing veligers are then transferred to other aquaria containing empty *scallop shells* to which the oyster spat attach. These scallop shells with their attached oysters are then moved out to the oyster beds. The oysters grow and can reach market size in two to five years, at which time they are harvested.

A more productive (more oysters per unit of available area) method of oyster culture is called raft culture. In this case, the scallop shells with spat attached are strung on *ropes* and suspended from a *floating raft*. By this arrangement the oysters are protected from benthic predators, such as sea stars, and spared the stress of exposure to air at low tide. In waters rich in phytoplankton, these oysters feed continuously and grow rapidly.

VIRGINIA OYSTER

MALE a
SPERM a1
FEMALE b
EGG b1
FERTILIZATION a1+b1
EMBRYO c
PHYTOPLANKTON d
TROCHOPHORE LARVA e
VELIGER LARVA *
SHELL f
VELUM g
FOOT h
SPAT i
EMPTY SHELL j

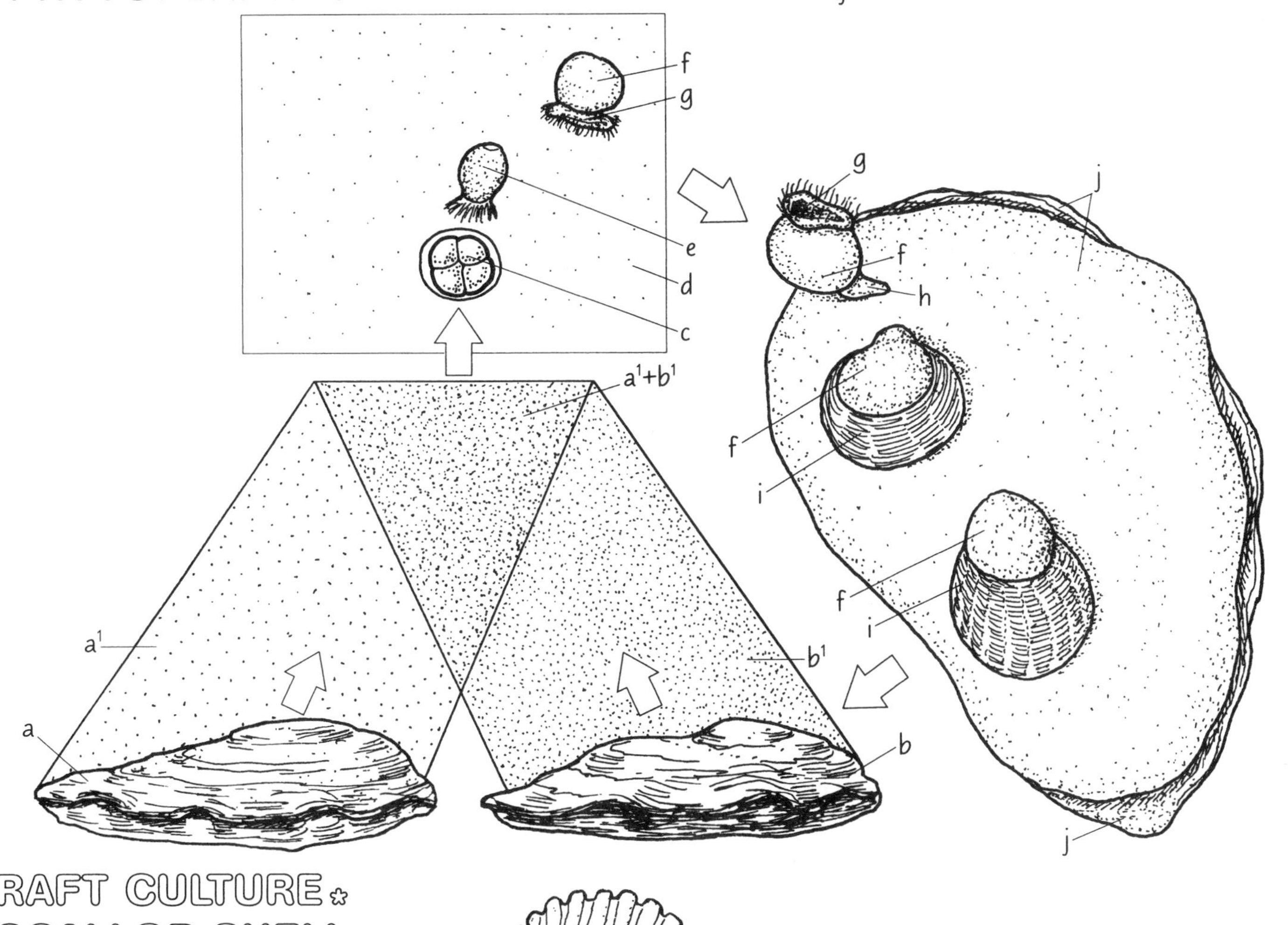

RAFT CULTURE *
SCALLOP SHELL j1
ROPE k
RAFT l

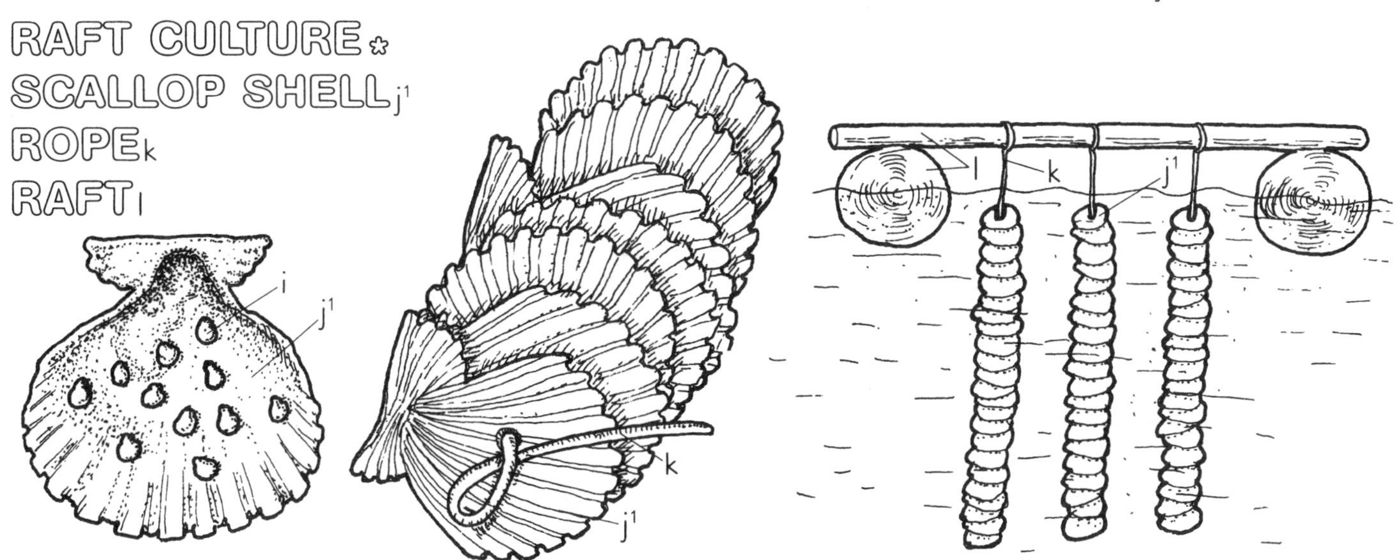

79
REPRODUCTION IN MOLLUSCS: GASTROPOD LIFE CYCLES

Gastropods are the most diverse group of molluscs. They include large herbivorous (plant-eating) abalones, shell-less nudibranchs, or sea slugs, the familiar garden snails and slugs, and many other forms. To illustrate this great variety, the life cycles of four representative gastropods are presented in this plate.

Color each life cycle as it is mentioned in the text. Choose light colors for (a) through (d). Beginning with the abalone, note that the excurrent openings toward the middle of each shell receive the color of the sperm or eggs that are released through them. The zygote receives the colors of the sperm and the eggs. The size of the trochophore and veliger larvae is greatly exaggerated here.

The life cycle of the abalone is fairly similar to that of the bivalve mollusc reviewed in the previous plate. The sexes are separate, and *sperm* and *eggs* are released into the water through excurrent openings in their shells. The *zygote* develops in the water, first into a ciliated *trochophore* larva, and later into a *veliger* larva. The gastropod veliger differs from the bivalve veliger, among other ways, in forming a snail-shaped, coiled shell. Abalone larvae differ from other gastropod veligers which are plankton-eating (planktotrophic). Instead, they rely on the yolk stored in the egg to nourish their development (referred to as lecithotrophic or yolk-eater). Abalone larvae mature rapidly and are competent to settle within a week's time. Settling is triggered by contact with a chemical produced by coralline algae, the food of the young abalone. Upon coming in contact with the algal chemical, the veliger detaches its ciliated *velum* and settles onto the rocky surface as a *juvenile*. The abalone becomes sexually mature in two to five years, depending on the species.

Color the moon snail and the whelk mating aggregation. The female moon snail is shown producing a sand collar after mating with the male.

The life cycle of the moon snail (Plate 31) is somewhat more complex than that of the abalone. The sexes are separate, but moon snails do not spawn into the water. The *male* possesses a penis with which he transfers sperm into the *female*, where the eggs are fertilized. The female lays the fertilized eggs in a case made of mucus and sand, called a *sand collar* because of its resemblance to an old-fashioned celluloid shirt collar. The embryos develop within the sand and mucus matrix. The sand collar eventually deteriorates, releasing the moon snail veliger larvae into the water. The veligers grow and seek out a fine sand bottom on which to settle and assume life as juveniles.

The whelk is a relatively advanced shelled gastropod, with separate sexes and internal fertilization. Reproduction is generally seasonal, and it is not unusual to find several dozen whelks clumped together in a mating aggregation in the rocky intertidal zone. The females lay their fertilized eggs in small (5 mm, 0.2 in), yellowish, vase-shaped *egg capsules* which are attached to rocks shielded from direct sunlight. The eggs pass through a modified development within the capsule; no free-swimming larval stages are formed. As the young snails develop within their capsules they become cannibalistic, eating one another until only a single (well-fed) juvenile emerges from each capsule. This direct, localized development and release of young ensures a proper habitat for the new snails, but relinquishes the potential dispersal and feeding benefits enjoyed by those animals with planktonic larvae (Plate 75).

Color the dorid nudibranchs. The adult nudibranchs are given both male and females colors since they have both reproductive systems and are pictured fertilizing each other.

The sea slugs, or nudibranchs (Plate 32), have a life cycle that combines some aspects of the patterns explained above with a few variations of their own. Most nudibranchs are predators, grazing on various hydroids, sponges, and so on, and are generally solitary in their habits. Nudibranchs are hermaphroditic (possess both male and female sex organs), and the meeting of two adults for reproduction involves the simultaneous exchange of sperm with the subsequent fertilization of each animal's eggs. Thus their solitary habits are somewhat countered by this double mating that results in two animals carrying fertilized eggs. Each nudibranch species lays its eggs in a characteristic egg cluster; shown here is the *egg spiral* of a dorid nudibranch. Other species lay their eggs in strings or flat sheets. After developing within the egg cluster, veliger larvae are released, each complete with a rudimentary coiled shell. This shell is lost when the larva settles and life as a juvenile begins.

GASTROPOD LIFE CYCLES

80
REPRODUCTION IN MOLLUSCS: CEPHALOPOD REPRODUCTION

The reproductive behavior of cephalopods is quite complex and involves some unique structures. Three species are discussed here to illustrate some of these features.

Begin by coloring the mating octopuses and proceed down the page as each species is discussed. Note that the male cephalopod has a modified, hectocotylized arm which receives a different color from the rest of the body. In the octopus, the small (2–3 cm, 0.8–1.2 in) spermatophore is exaggerated in size as are the juvenile octopus and squid. In the paper nautilus, note that the female's membrane which secretes the shell receives a separate color. The diagram at lower left illustrates a shell with the egg cluster attached.

The common shallow-water octopus of Europe and the east coast of the United States is a relatively hardy species. Because it does well in captivity, more is known of this cephalopod's reproductive processes than of most other species. This normally solitary animal displays little in the way of a courtship ritual. A *male* approaches a *female* and attempts to mate. He may mount the female or simply extend the third arm on his right side toward her, as shown. This arm is the *hectocotylized arm*; its tip is modified into a spoon-shaped palm called the *hectocotylus* (hecto: hundred; cotyl: cup) that is used to transfer sperm into the female's mantle cavity. The hectocotylized arm bears a groove along its posterior surface. The sperm are packaged in gelatinous sheaths known as *spermatophores* which are loaded into the sperm groove by the penis. Waves of muscular contraction along the arm carry the spermatophores along the groove to the hectocotylus, which is inserted into the female's mantle cavity, near her reproductive opening. (In some species, the hectocotylus is detached within the female's mantle cavity and remains there. Some early investigators mistook this detached arm as a parasitic worm and classified it as the sole species of the genus *Hectocotylus*. Although the genus has since been declassified, the name hectocotylus for the modified arm or its tip has been retained.) The spermatophores rupture inside the female's reproductive tract, assuring that the sperm reach their destination. A male may transfer as many as fifty spermatophores in an hour. After mating, the female leaves to lay her eggs, a process which may take a week as she delicately releases a few eggs at a time through her *funnel*. She positions thousands of eggs in grapelike *egg clusters* suspended from the rocky walls of her cave home. The female remains with her eggs for several weeks, constantly manipulating them and gently blowing water over them with her funnel to keep them clean and oxygenated. The *juvenile* octopuses hatch as miniature adults and spend weeks in the plankton before settling on the bottom. The female does not eat during her care of the eggs, and dies soon after the hatching of her young.

The common Pacific squid congregate and mate in large groups. As a male approaches a prospective female, his arms and head are ablaze with red and white markings which generally change to solid maroon when actual copulation ensues. The animals mate either head on (Plate 68) or with the male grasping the female from under the left side, as shown here. Using the lower left hectocotylized arm, the male picks up spermatophores delivered to his funnel by the penis and transfers them into the female's mantle cavity. He holds them within the mantle cavity until the sperm are released and then withdraws. The entire copulation process may take less than ten seconds. Following mating, the female produces long cases containing the fertilized eggs. The *egg cases* swell and become firm when exposed to sea water. They are attached to the bottom, often in large clusters or "mops," covering many square feet. After several bouts of mating and egg laying, the adult squids die. The young squids develop within the egg cases and eventually emerge as juveniles, complete with functional chromatophores and ink sacs (Plates 68, 104).

The female pelagic octopus, known as the paper nautilus, prepares an elaborate ark for her eggs. Each of the female's two dorsal arms bears a broad *membrane*. Each membrane secretes half of a *shell* in which the female attaches her egg cluster. The female remains in the shell with only her arms and funnel protruding. The smaller male often occupies the shell with the female and her egg cluster. This octopus is called the paper nautilus because the thin shell superficially resembles that of the nautilus (Plate 33), but the two cephalopods are not very closely related.

CEPHALOPOD REPRODUCTION

OCTOPUS*
MALE$_a$
HECTOCOTYLIZED ARM$_b$
HECTOCOTYLUS$_{b^1}$
SPERMATOPHORE$_c$
FEMALE$_d$
EGG CLUSTER$_e$
FUNNEL$_f$
JUVENILE$_g$

SQUID*
MALE$_a$
HECTOCOTYLIZED ARM$_b$
FEMALE$_d$
EGG CASE$_{e^1}$
JUVENILE$_g$

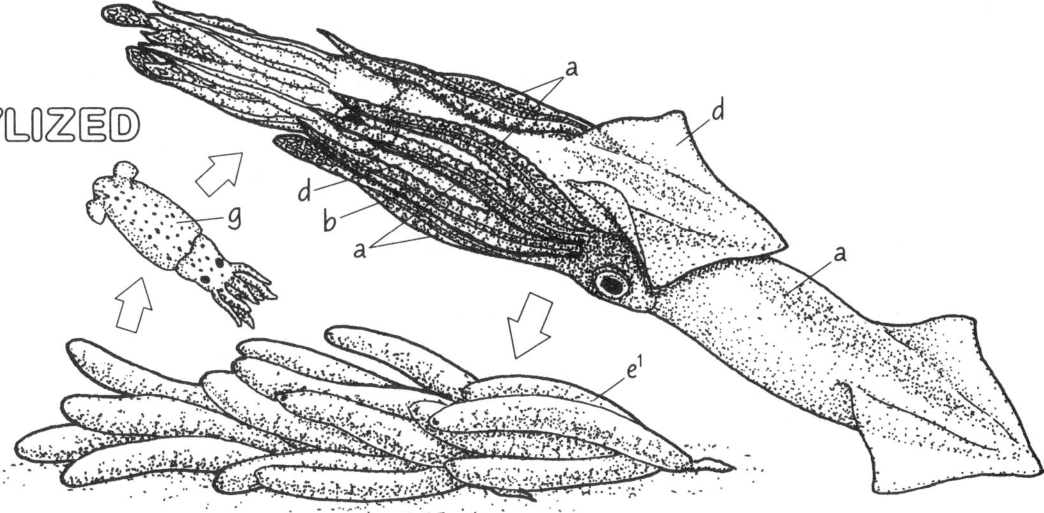

PAPER NAUTILUS*
MALE$_a$
HECTOCOTYLIZED ARM$_b$
FEMALE$_d$
MEMBRANE$_h$
SHELL$_i$

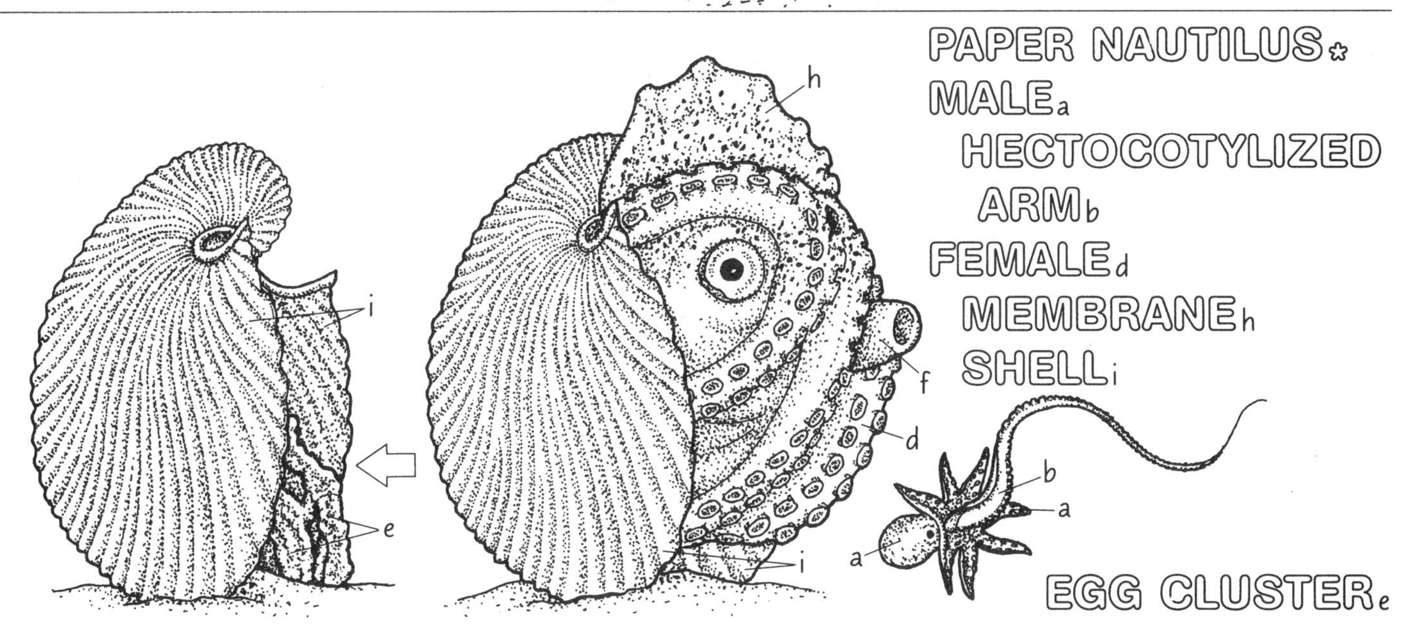

EGG CLUSTER$_e$

81
REPRODUCTION IN CRUSTACEANS: BARNACLE AND COPEPOD LIFE CYCLES

Reproduction presents a problem for attached organisms: How do they get together with others of the same species to mate and produce a new generation? Some species, such as oysters (Plate 78), simply release their gametes into the water, where fertilization occurs without contact between adults. However, this method is costly in the number of gametes wasted. In this plate, the reproductive methods and life cycles of barnacles and copepods are discussed.

Begin coloring the barnacle life cycle with the cross section of the adult at the top of the page. Continue with the larval forms; remember that these larvae are shown greatly exaggerated in size relative to the adults. Choose a light color for the carapace of the cyprid larva, as it is transparent and the body and appendages can be seen through it.

Barnacles do not release their gametes into the sea. When sexually mature, an adult barnacle uncoils its long, flexible, tubular *penis* and probes about for a nearby mate. Since barnacles are gregarious and larvae settle near their own kind, a receptive neighbor is generally available. The barnacle inserts the penis and transfers the sperm to its mate, as shown at the top of the plate. Most barnacles, like nudibranchs, are hermaphroditic, so any adult can serve as a mate to any other adult of the same species. Fertilized *eggs* are brooded within the shell of the adult until they develop into nauplius larvae. A single individual may brood and release as many as 13,000 larvae.

The nauplius is clearly a crustacean, with a single *naupliar eye*, *antennae,* jointed *appendages*, and a shield-shaped *body*. Like any crustacean, the barnacle nauplius molts (sheds its exoskeleton) as it grows and develops. After several molts as a nauplius, a cyprid larva emerges. The cyprid does not feed. It is equipped with large antennae and more appendages than the nauplius, and the body is enclosed in a hinged *carapace*. After a short time, the cyprid begins to crawl about the substratum, testing for an appropriate place to settle. Cyprids are attracted to rough surfaces and to adult barnacles of their own species. They recognize these features of their environment with tactile and chemical sense organs in their antennae. If the cyprid fails to find a suitable substratum, it can swim off and search further, often delaying settlement for several days (Plate 97).

When the cyprid does locate and select a substratum on which to settle, it attaches using special cement glands in its antennae. The cyprid then molts and rotates its body so the appendages are facing upward. The appendages, now called *cirripeds* (Plate 35), are long and feathery; they are used by the adult to filter food from the water and carry it to the *mouth*. The cyprid's carapace serves as a form around which the barnacle's plates or shells are secreted. The antennal cement glands serve to anchor the bottom or *basal plate* to the substratum. The *fixed plates* are attached to this basal plate and articulate with the *movable plates*, which swing open to allow protrusion of the cirripeds.

Color the copepod life cycle. The size of the larval forms is greatly exaggerated.

Copepods also pass through a nauplius larval stage. During reproduction, a male copepod grasps the female with its large antennae and transfers the sperm to her genital opening (not shown). Most copepods brood fertilized eggs in special pouches until the larvae are released. However, in some of the common planktonic copepods (calanoid copepods), the *zygotes* are released singly into the water, where they hatch as copepod nauplius larvae (I). The nauplius undergoes several molts (II, etc.) and develops into a copepodite stage. The copepodite looks somewhat like the adult, but it is smaller, has fewer appendages, and its abdomen is not yet clearly segmented (Plate 35). In most species, after five copepodite stages (Stage V is shown here), the adult form is reached. The entire cycle from egg to adult may take as little as a week or as long as a year, depending on the species involved.

BARNACLE AND COPEPOD LIFE CYCLES

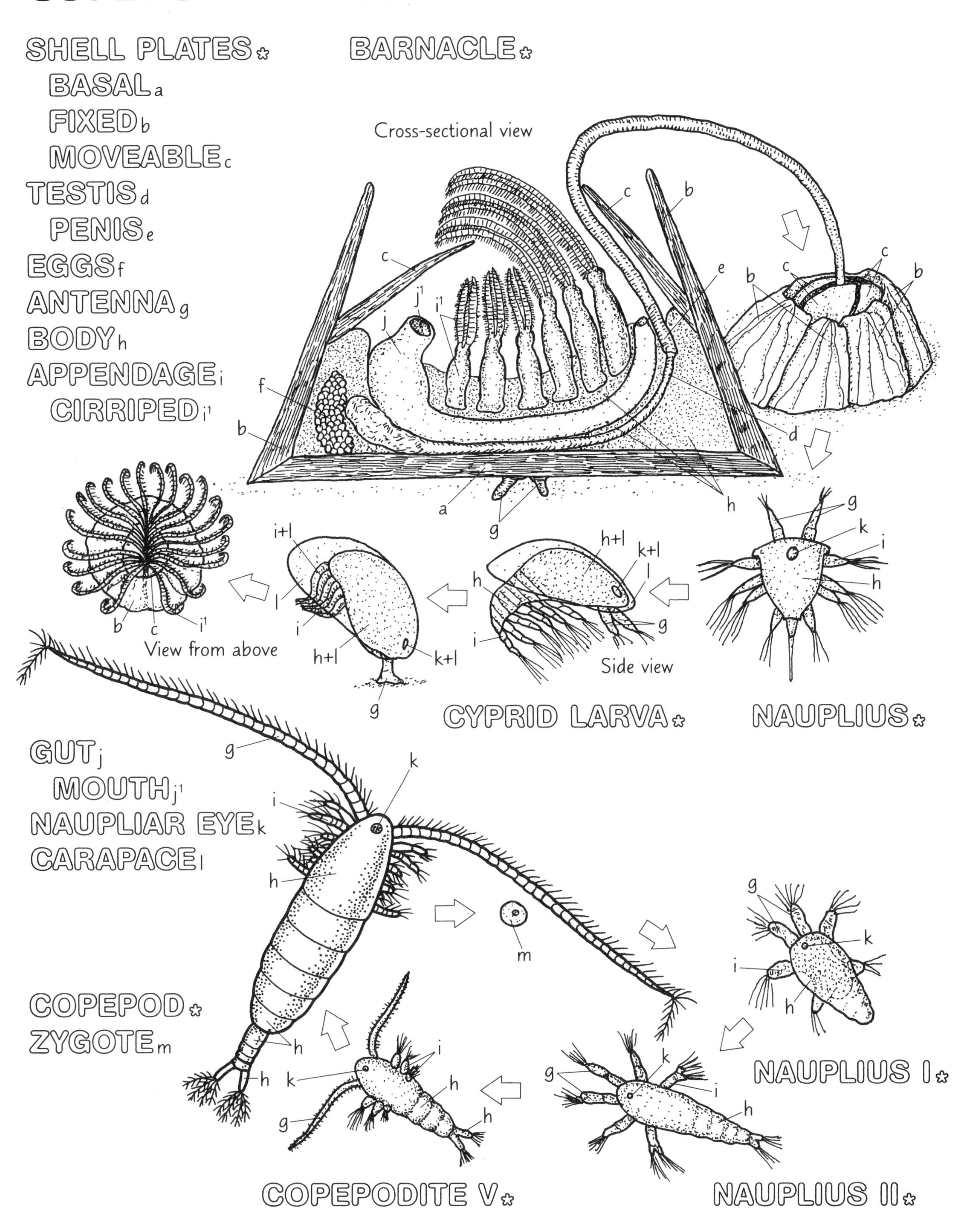

82 REPRODUCTION IN CRUSTACEANS: AMPHIPODS, STOMATOPODS, AND DECAPODS

The reproductive cycle of most crustaceans, as with barnacles and copepods (Plate 81), involves the transfer of sperm from the male to the female followed by brooding of the zygotes for varying lengths of time. The females may carry the embryos and release them as early nauplius larvae, as a later larval form, or as fully developed juveniles. Some of this variability is exemplified by crustaceans discussed in this plate.

Begin by coloring the amphipods. Note that the female gammarid amphipod has released a chemical scent which has attracted the male. The area around the gammarid amphipods can be lightly colored to indicate the presence of the pheromone in the water. Color the rest of the crustaceans as they are discussed in the text. Note that the larval forms are exaggerated in size relative to the adults.

Because the male must transfer sperm to the female at a time when she is sexually mature and ready to lay eggs, it is important that he be made aware of her condition. In many types of crustaceans (including the gammarid amphipod) the *female* releases a chemical (called a *pheromone*) into the water when she is approaching sexual readiness. This pheromone is sensed by the *male,* and he is attracted to the female. In many species of amphipods, the male holds on to the maturing female as shown in the illustration. The male releases his mate as she goes through a final molt, and then transfers sperm to her. Female amphipods have a *brood pouch* in which the male's sperm is collected; this pouch is easily seen in the drawing of the highly modified *caprellid amphipod*. The female releases her *eggs* into this pouch, where they are fertilized and retained until they hatch as juveniles. Thus, amphipods do not have planktonic larvae, and the young are released directly into the adult habitat.

Female mantis shrimps, or *stomatopods*, are also protective mothers. After the male deposits his sperm in a special pouch on the female's body, she releases her *eggs* (up to 50,000 of them!) and sticks them together in a mass about the size of a walnut. The egg mass is carried by the front legs and is carefully cleaned and rotated by the female. The female carries the developing embryos for several weeks and does not feed during this time. After this brooding period, the embryos hatch as stomatopod *zoea larvae,* which act as aggressive predators feeding on other zooplankters. After a few months, the zoea settles, metamorphoses, and takes up adult life on the bottom.

In the decapod crustaceans (crabs, shrimps, lobsters), the adults copulate, and sperm are transferred from the male to the female. Shrimp contact one another at the right angle, as illustrated. The male forms packets of sperm (*spermatophores*) that he attaches near the female's reproductive openings. In some shrimp species, the female bears a special pouch in which to store the sperm for later fertilization. In others, the eggs are extruded and fertilized soon after sperm transfer. With the exception of one large group of shrimp (the penaeids), female decapods brood their fertilized eggs by attaching them to their swimmerets (abdominal appendages). Brooding females are referred to as being "berried females," or "in berry." After the brooding period (duration varies from weeks to months according to species), the well-developed embryos are released as planktonic zoea larvae. The shrimp zoea illustrated here remains in the plankton for several weeks before metamorphosing to the adult form.

The larval forms of crustaceans vary considerably. The stomatopod and shrimp zoea larvae have recognizable features of the adult forms, but not so in the case of the spiny lobsters. Shown here is a berried female California *spiny lobster*. The larval stage of this animal is released after two months of brooding by the female. These larvae are called *phyllosoma* (leaf body) *larvae*; they are extremely flattened, almost paper-thin, and nearly transparent. The phyllosoma of the California spiny lobster remains planktonic for up to six months before settling. Only then does it assume its adult form.

AMPHIPODS, STOMATOPODS, AND DECAPODS

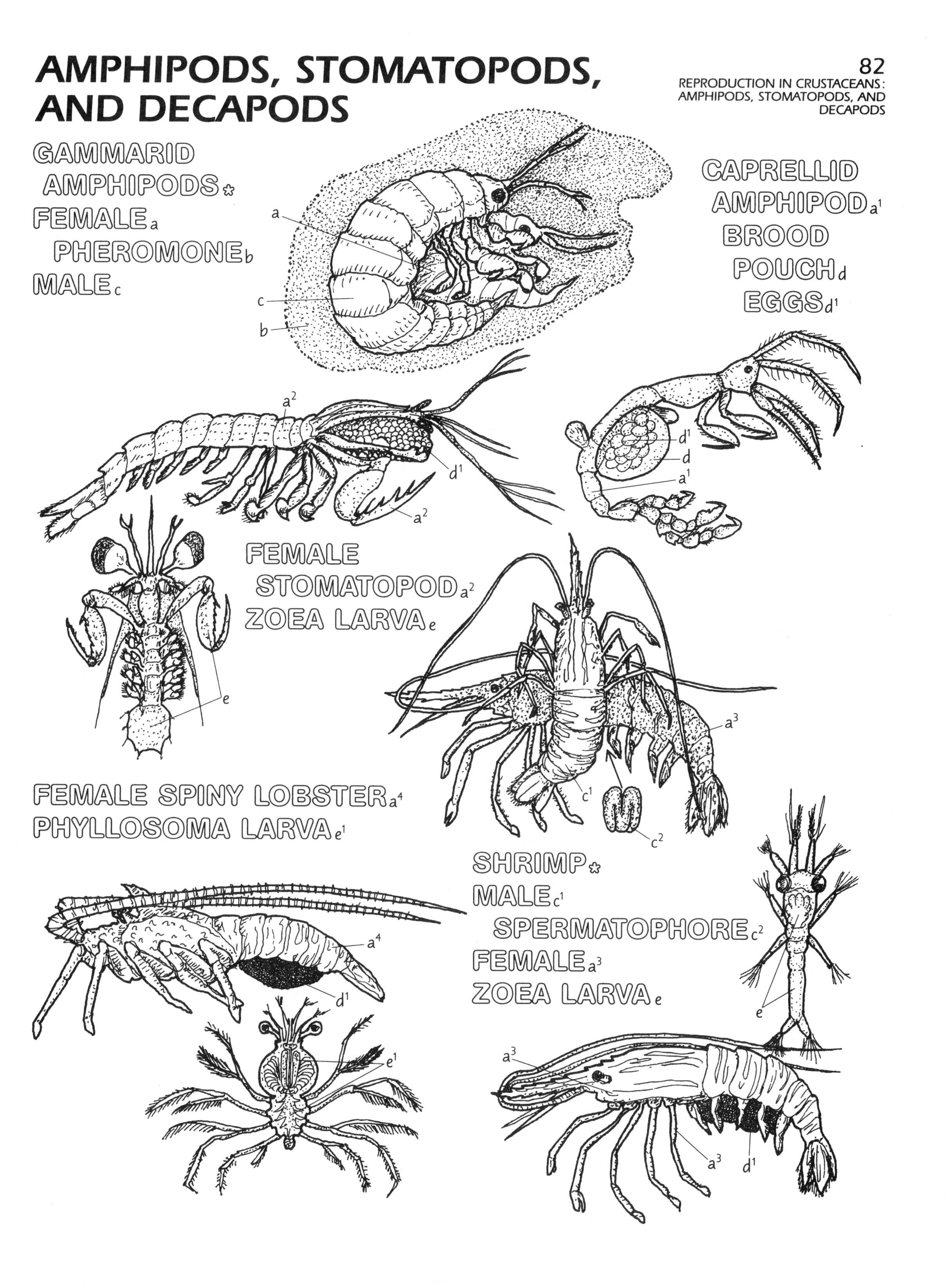

83
REPRODUCTION IN CRUSTACEANS: CRAB LIFE CYCLE

The group of decapods known as the "true crabs" (section Brachyura, Plate 37), are considered advanced crustaceans, and their reproductive biology is quite complex. The life cycle of the red rock crab is discussed here as an example of reproduction in this group.

Begin by coloring the male and female in ventral (underside) view at the top of the page. Locate and color each structure and body region as it is discussed in the text. Note that only a portion of the crab is colored here, and that the abdomen is shown bent away from its normal position against the thorax. The normal abdomen position is indicated by using the abdomen color in this location.

Adult *male* and *female* crabs vary in the form of their *abdomens*. Although both sexes have reduced abdomens (when compared with lobsters and shrimps, Plate 36) that are normally kept flexed beneath the *thorax*, the male's abdomen is much narrower than that of the female. The abdominal appendages (modified *swimmerets*) are visible in the male as two pairs of shaftlike processes, the smaller pair fitting into grooves on the larger pair. At the bases of the last pair of walking legs of the male are two small openings for the release of sperm from the reproductive tract. Sperm pass from the reproductive openings to the grooves on the larger pair of swimmerets. The smaller swimmerets push the sperm along the groove into the *genital openings* of the female as described below.

The female genital openings are located on the thorax, and the broad abdomen bears several pairs of feathery swimmerets on which the *egg cluster* is held during brooding.

Now color the illustrations of the mating behavior and life cycle of the crab. Color each stage as it is discussed in the text. The natural color of the early egg cluster is salmon/coral pinkish, and the late egg cluster is brownish purple. Note that the larval and juvenile stages are greatly exaggerated in size.

During the life cycle of the rock crab, the crabs must *molt* in order to grow. The crab accomplishes this by resorbing much of the calcium salts from its old skeleton and then swelling with water to burst the old skeleton. The new skeleton is soft, allowing the crab to crawl free from the old encasement. Once free, the crab continues to swell and increase in size, then gradually hardens the new skeleton by the redeposition of the conserved salts. Typically, the male molts before the mating season. The female, however, must be newly molted for successful mating to occur. It is critical that the male discover the female prior to her actual molting, so that he can be present to fertilize her eggs before her new exoskeleton hardens. The female probably releases a chemical attractant in her urine at the appropriate time. Once together, the mate grasps the female and holds her until she begins her molt. He releases her as she emerges from her old exoskeleton or "molt" (top right diagram of the life cycle). After her new exoskeleton becomes sufficiently firm, she lowers her abdomen and allows the male to insert his modified swimmerets into her genital openings for sperm transfer. Once this has been accomplished, the male holds the female in a protective embrace until her exoskeleton hardens.

The eggs are fertilized by the male's sperm as they are released from the female's genital openings. The sticky fertilized eggs adhere to one another, and the female attaches them to her "hairy" swimmerets. A female rock crab may carry several thousand eggs in her egg cluster. The mass of eggs is light in color when newly formed, but darkens substantially as the embryos develop. After several weeks, the developed crab embryos break free of their egg membranes and emerge as planktonic *zoea larvae*. The zoea larvae swim, feed on zooplankton, and molt several times until they reach a crablike *megalops larval* stage. The megalops finally sinks to the bottom, and at its next molt becomes a recognizable *newly settled rock crab*. *Juveniles* molt frequently, and the species discussed here may often be seen in the striped phase, with alternating black and red stripes on a light brown background. These crabs mature in three to four years and are then ready to participate in this complex adult reproductive behavior.

CRAB LIFE CYCLE

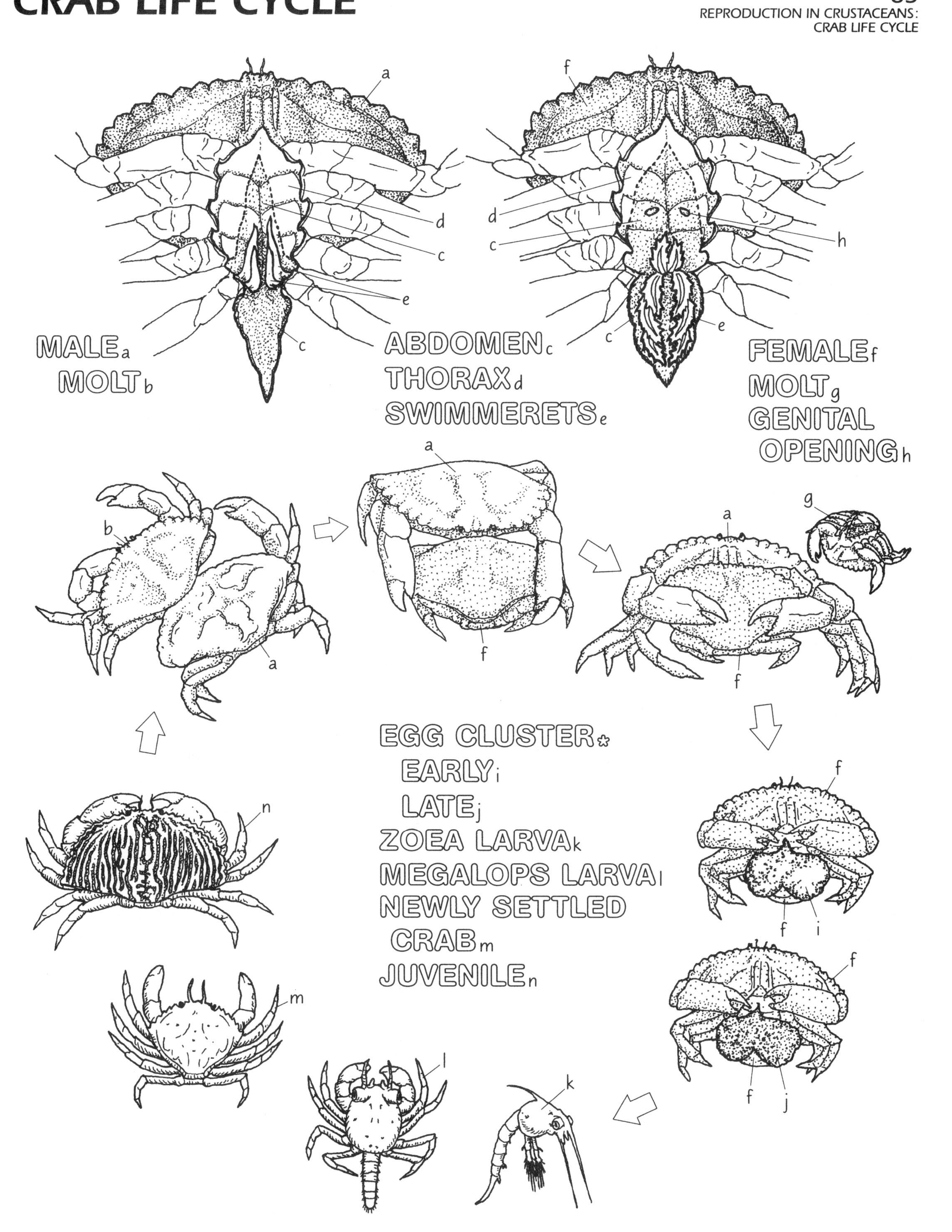

84
REPRODUCTION IN ECHINODERMS: ECHINODERM LIFE CYCLES

The echinoderms are almost all bottom-dwelling, slow-moving animals, and usually depend upon a planktonic larval stage for dispersal. However, there are some notable exceptions to this scheme. This plate discusses the typical life cycle of a sea urchin, and the rather unusual life cycle of a small sea star.

Color the sea urchin life cycle, coloring each stage as it is mentioned in the text.

Sea urchins usually occur in large aggregations (Plate 5). This gregarious habit allows urchins to "broadcast" their *sperm* and *eggs* into the water, in turn stimulating the spawning of other individuals.

After fertilization, the *zygote* is free in the plankton. It develops and hatches in a few days as an *early echinopluteus larva*. This larva has long, ciliated *arms* for locomotion and collecting phytoplankton, on which it feeds. This stage may persist for days or months, depending on the species of urchin, but it eventually begins to metamorphose. The *late echinopluteus larva* has reduced arms that are gradually resorbed by the body as the *juvenile* urchin develops. When the juvenile settles out of the plankton, it is only about 1 mm (0.04 in) across, with just a few spines and tube feet. It grows quickly, however, and soon looks like a miniature adult.

Color the typical sea star larval forms in the boxed area. These represent the planktonic larval forms of non-brooding sea stars.

Most sea stars undergo a life cycle that is similar to that of the sea urchin. The zygote (not shown) develops into a *bipinnaria larva* which has winglike folds that are ciliated for swimming. This stage changes into a more elaborate *brachiolaria larva,* with elongate arms and three *pre-oral arms* used for attachment when the brachiolaria settles and metamorphoses to a young sea star.

Now color the modified life cycle of the six-rayed sea star as it is discussed in the text. Note that the brooding female is shown with her body covering the juveniles, which receive the colors of both the mother and the juvenile.

In this sea star, the female retains and cares for her eggs while they develop. She attaches to the bottom or side of a rock by the tips of her arms and forms a cup-shaped brooding area with her oral surface. The female lays her eggs into this pocket, and the eggs are fertilized by sperm from nearby males. After fertilization, the female begins a two-month vigil during which she remains in this brooding position, moving only to manipulate the developing zygotes to keep them clean and well aerated. The brooded embryos emerge from their egg membranes as drastically modified brachiolaria larvae (the bipinnaria stage is bypassed completely). This brachiolaria lacks the long larval arms seen in the more typical planktonic form, and bears only the three adhesive pre-oral arms. At this stage in the larva's development, it must change from a bilateral (two sided) animal into one with radial (round) symmetry. As the larval body changes shape, five distinct bulges appear. These bulges develop into five of the sea star's arms, and a sixth soon forms. The *juvenile* sea star develops tube feet, and the larval body is resorbed. After two months of confinement, while developing in the brooding area, the juveniles begin to move about. The *brooding mother* reattaches her entire oral surface to the substratum, but remains in place with her young for a few more days. Soon the small (1 mm, 0.04 in) juveniles become more active and leave the "nest" forever. This type of life cycle insures a steady input of young sea stars into the adult habitat, but sacrifices the benefits of a planktonic larval stage (Plate 75). Also, because the female six-rayed sea star broods her eggs, she must provide each with a relatively large amount of nourishing yolk. This means she can produce proportionally fewer eggs compared to a sea star that broadcasts many tiny eggs into the water that will depend entirely on feeding in the plankton to grow and develop.

ECHINODERM LIFE CYCLES

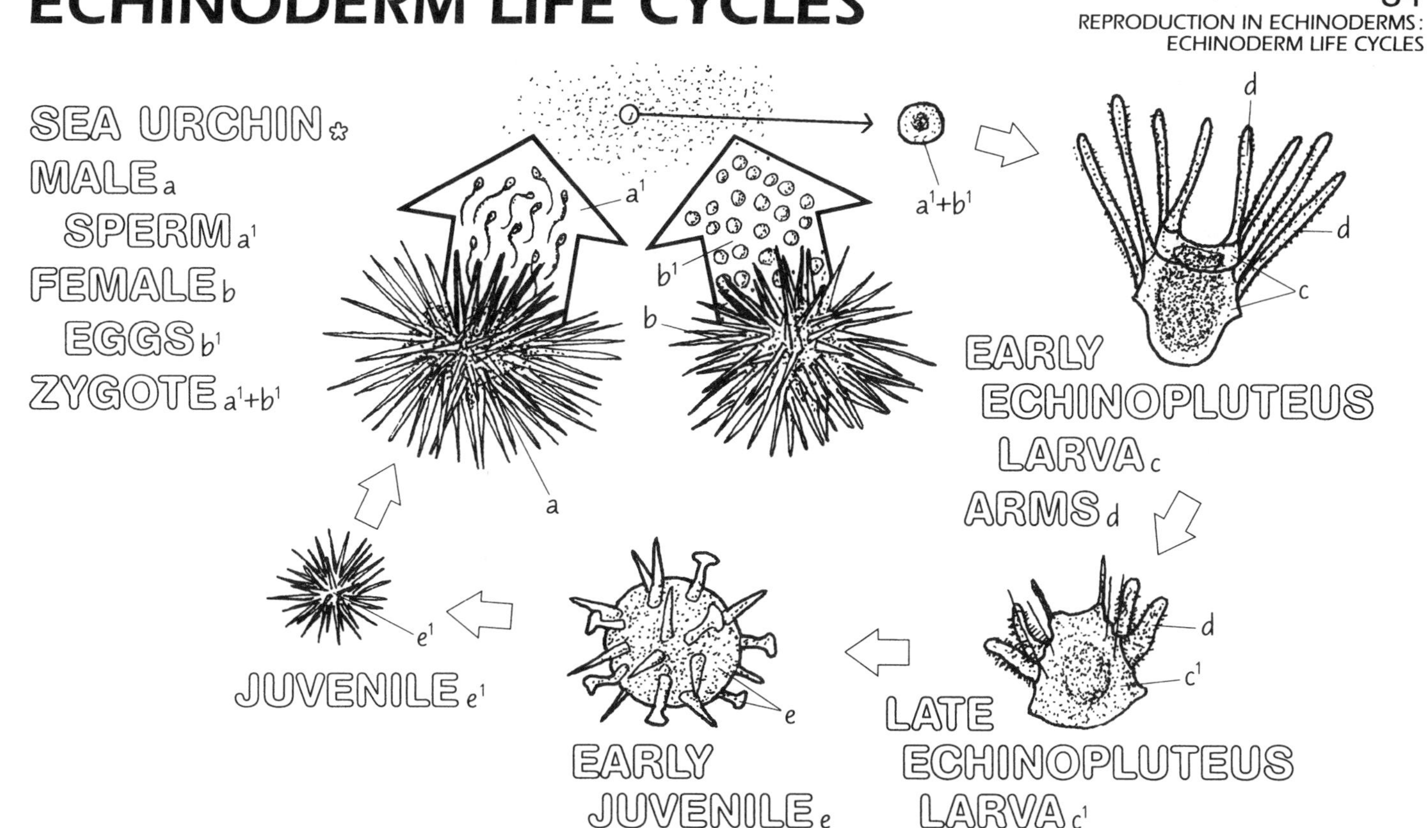

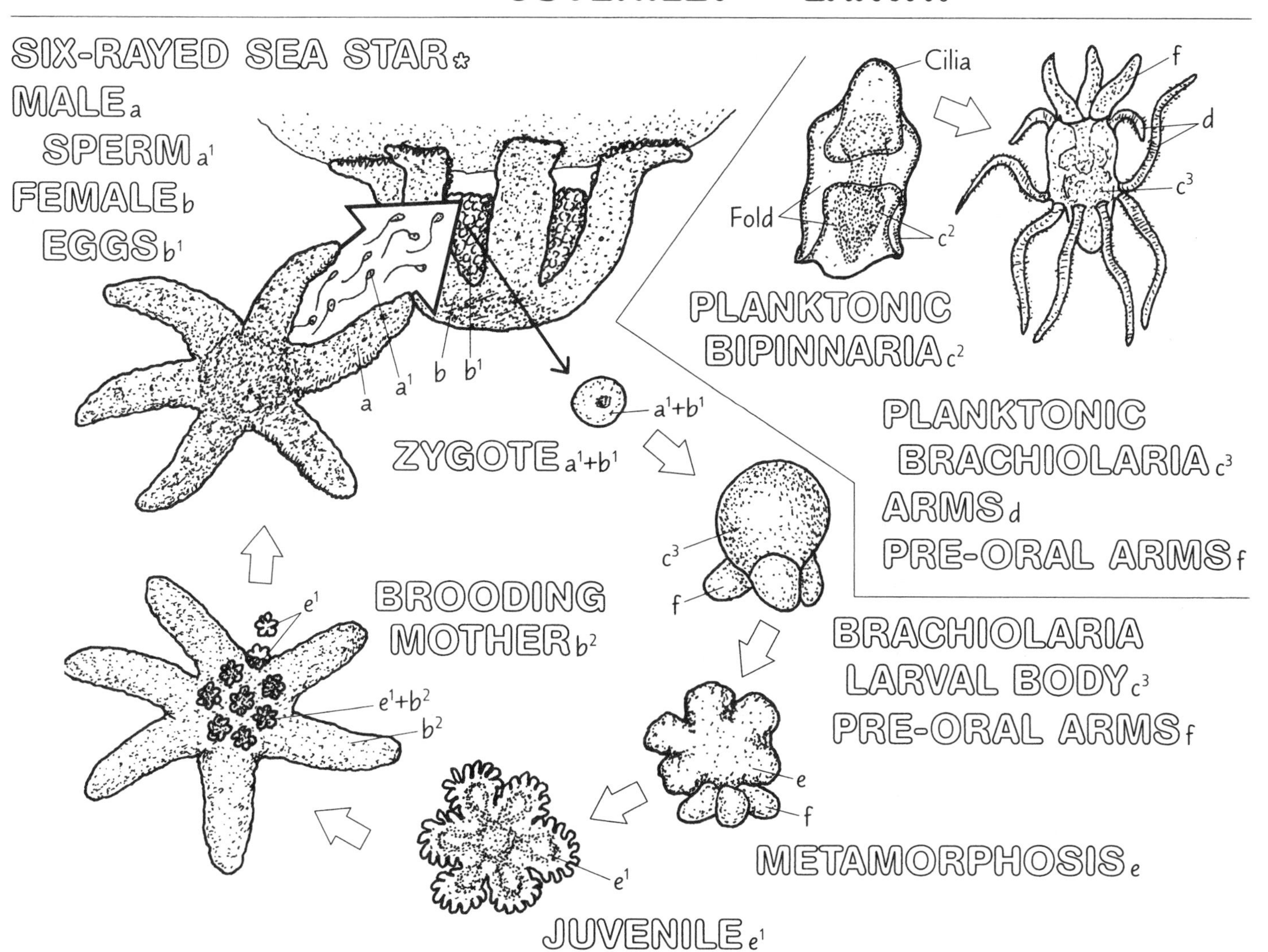

85
REPRODUCTION IN ELASMOBRANCHS: SHARKS AND SKATES

Begin by coloring the shark's pelvic fin and clasper; use the male color (c) for the body. Next, color the copulating (mating) spotted dogfish sharks.

Sharks, skates, and rays practice internal fertilization, which is a more efficient and less wasteful means of fertilization than releasing gametes into the environment. To accomplish the transfer of sperm, *male* elasmobranch fishes have special intromittent organs (transfer organs) called *claspers*. The upper drawing shows the clasper of a horn shark; the clasper is derived from the shark's *pelvic fin*. During copulation, the clasper is inserted into the *female's* genital opening, and the *spur* is erected to insure that the sharks stay coupled long enough for sperm transfer to occur. In spotted dogfish sharks, the male wraps his body around the female, inserts a clasper and transfers his sperm. In sharks that are less supple than the spotted dogfish, the male and female lie side by side. The male holds onto the female's pectoral fin, erects the clasper at a right angle to his body, and inserts it into the female.

Color the female skate and its egg case, embryo, and yolk sac. Note that in the drawing on the right, the embryo is large and well developed. Note also that only the intact egg cases are to be colored. Then color the female horn shark and its egg case.

Once the eggs have been fertilized within the female, one of three possible patterns of development occurs, depending on the species of fish. In one pattern, the fertilized eggs are packaged in a special *egg case* and released by the female to develop outside her body. In this situation, each egg is provided with a large amount of yolk to nourish the developing *embryo* during its growth. Skates employ this reproductive strategy that frees the female from prolonged maternal care. The illustration shows a female *skate* swimming away from her large (23 cm, 9 in) egg case, sometimes called a "mermaid's purse." Inside this egg case are one or several developing embryos. The two drawings of opened egg cases show the embryos increasing in size and their *yolk sacs* dwindling as the yolk is used for nourishment. Under normal conditions, the egg case gradually deteriorates and begins to fall apart as the young skate is ready to emerge and begin a life of its own.

Some sharks also produce an egg case, exemplified by the spiral egg case of the *horn shark* shown here. The young sharks emerge from this case when they are about 12 cm (4.7 in) long and appear as miniature adults.

A second developmental pattern seen in elasmobranchs is the retention of developing fertilized eggs inside the female's reproductive tract (not shown). Upon completing their development, the young are released as fully formed juveniles. This type of development is known as ovoviviparity (ovo: egg, viviparity: live birth). These eggs are also provided with a yolk sac that serves as their source of nourishment; the female is simply protecting the embryos within her body.

Now color the embryonic shark within the mother's uterus. Note that the "placenta" and "umbilical cord" receive shades of the same color as the yolk sac from which they derived.

Some sharks undergo a third type of development known as viviparity. In this situation, the mother provides the developing embryos with nourishment in addition to the yolk within the egg. Supplying this nourishment may occur in a number of different ways. The drawing shows a method that resembles the mammalian placental connection between female and embryo. A smoothhound shark embryo is shown folded in an *embryo sac* within its mother's *uterus*. An *"umbilical cord"* and *"placenta"* containing embryonic blood vessels connect the embryo to the wall of the uterus. The "placenta" receives nourishment from the mother's circulatory system, and it is transferred to the developing embryo through the "umbilical cord." Actually the "placenta" here is a modified yolk sac, and the "umbilical cord" is the elongated connection between the yolk sac and the embryo. The result of this arrangement is the direct provision of nutrients by the mother to the embryo.

SHARKS AND SKATES

PELVIC FIN$_a$
CLASPER$_b$
SPUR$_{b^1}$

COPULATION*
MALE$_c$
FEMALE$_d$

SKATE$_{d^1}$
EGG CASE$_e$

EMBRYO$_f$
YOLK SAC$_g$

HORN SHARK$_{d^2}$
EGG CASE$_{e^1}$

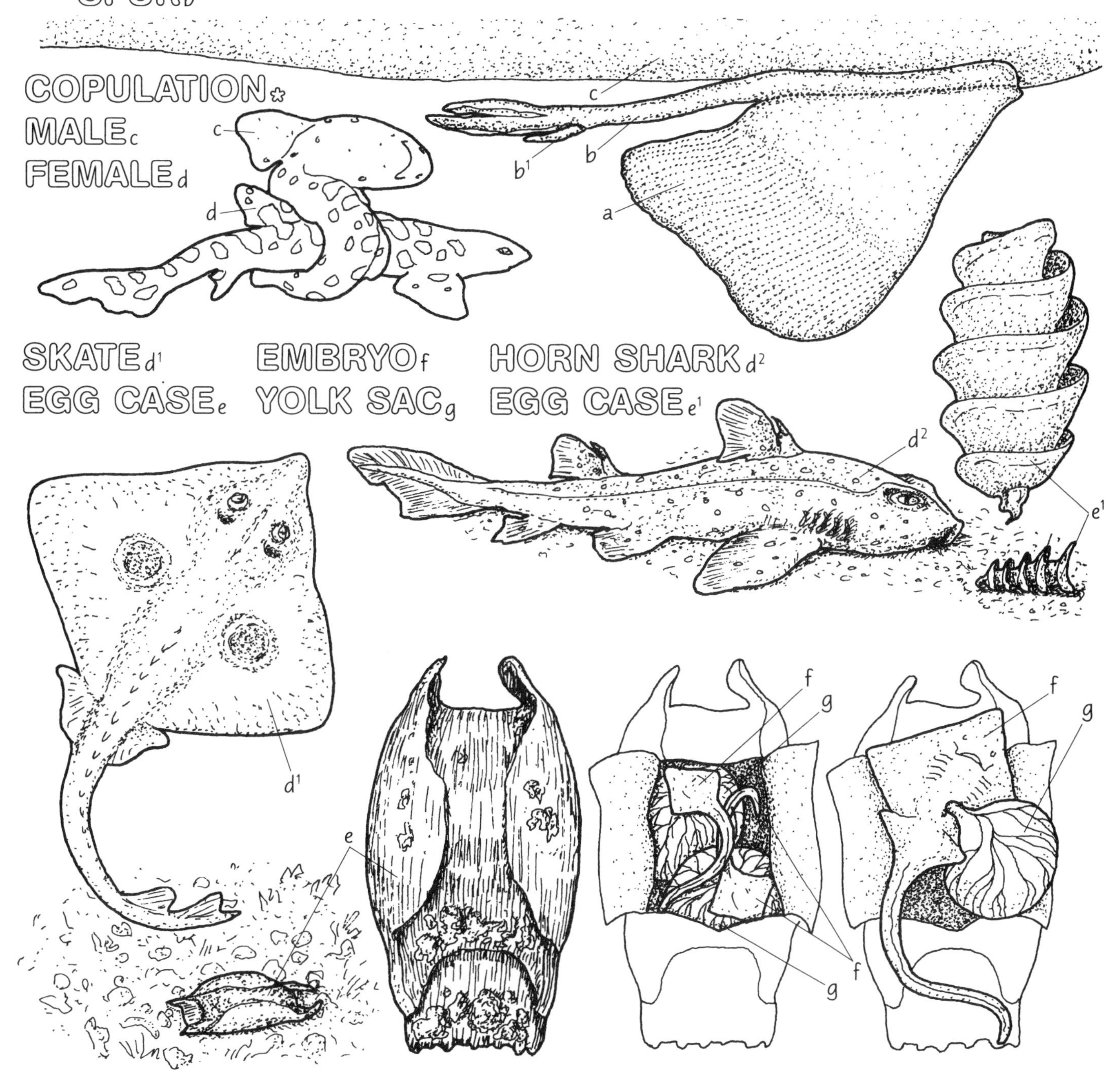

UTERUS$_h$
EMBRYO SAC$_i$
EMBRYO$_{f^1}$
PLACENTA$_{g^1}$
UMBILICAL CORD$_{g^2}$

"PLACENTAL" ARRANGEMENT*

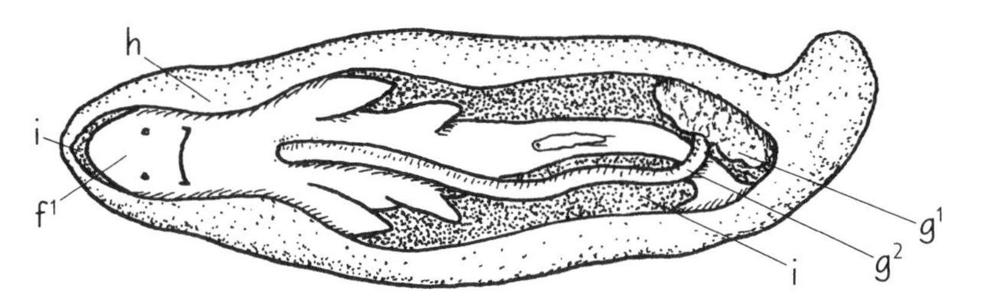

86
REPRODUCTION IN BONY FISHES: LIVE BEARERS AND BROODERS

Bony fishes employ a variety of reproductive strategies to ensure continuation of their species. Some of these methods of reproduction are discussed in this and the next two plates.

Parental care of the young is expensive in terms of the parents' time and energy, but it does tend to ensure the survival of a high percentage of offspring. The strategy is to produce relatively few young—that is, a number that can be cared for—and protect them until they are able to take care of themselves (a very different situation from the free spawning of various invertebrates, as discussed earlier). This approach also ensures that juveniles will be "turned loose" in the proper habitat.

Color the various fishes gray, except for the areas located within circles and the seahorse's pouch. Within the large circles below each fish, color only those structures outlined with a heavy line. Begin with the live-bearing surfperch, and then color each fish separately as it is discussed in the text.

As seen in the previous plate, all of the elasmobranch fishes (sharks, rays, skates) practice internal fertilization followed by some protection of the developing young, either within the female or in an elaborate egg case. The retention of developing young within the parent's body (or in some special brooding area on the parent) is the most effective way to secure the safe development of the offspring. As mentioned above, however, it is costly to the adults engaged in such activities. This type of reproduction is practiced by several families of bony fishes, most of which are called *live bearers*.

The surfperches, common along the Pacific coast of North America, are classic examples of live bearers. The female surfperch is inseminated by the male following rather complicated mating behaviors. The fertilized eggs are retained in the female's ovary where they develop into miniature adults during a five-month gestation period. During development, nourishment is provided by the mother in the form of a nutrient-rich secretion from the highly vascularized (richly supplied with blood vessels) ovarian wall. The secretion is absorbed by the young through their hindgut and their enlarged fins which are also richly supplied with blood vessels for this function.

Young surfperch are quite large (about 50 mm, 2 in) when released compared to the young of other groups of live bearers. For example, rockfishes of the very large genus *Sebastes* are also live bearers, but the individual juveniles are tiny in comparison to surfperch (about 5 mm, 0.2 in) and are still referred to as larvae by many because of their small size. The female surfperch produces a small number of eggs because the individual embryos grow to such a large size. A two-year-old female barred surfperch may produce a clutch of 20 embryos while a four year old might produce 70 embryos. This is compared to a clutch of several thousand embryos produced by a female rockfish.

A number of other strategies involve protecting the developing embryos either in some sort of nest (Plate 87) or by means of a special brooding area on the adult fish. The gafftopsail catfish of the Gulf of Mexico is an example of a *mouth brooder*. As the name indicates, the fertilized eggs are brooded within the parent's mouth, in this instance the male's. The male carries 40 to 60 marble-sized eggs in his mouth for about nine weeks while the young develop to the hatching stage. After hatching, the juveniles may stay with the father for an additional month, during which time they enter and leave the mouth at will, thus remaining protected until finally venturing off on their own. This exceptional gesture of parenthood is especially impressive in that the male catfish does not eat during the entire brooding period.

There are a variety of other brooding methods among the bony fishes. The *Kurtus* of the South Pacific uses its head. Again, it is the males which carry the embryos; they are called *forehead brooders*. The male *Kurtus* has a special hook on its forehead to which he attaches a mass of fertilized eggs. In this case, the adult can continue to feed while he protects the embryos until they hatch.

A very effective brooding technique is seen in the seahorses and pipefishes. In these small, slender fishes, the female lays her eggs into the special *pouch* on the ventral surface of the male, who then fertilizes and broods the eggs, out of harm's way. The young remain within this pouch, nourished by their father's blood supply, until they are fully formed young juveniles. The male then "gives birth" by writhing to and fro and flexing the pouch muscles, forcing the young fish out on their own.

LIVE BEARERS AND BROODERS

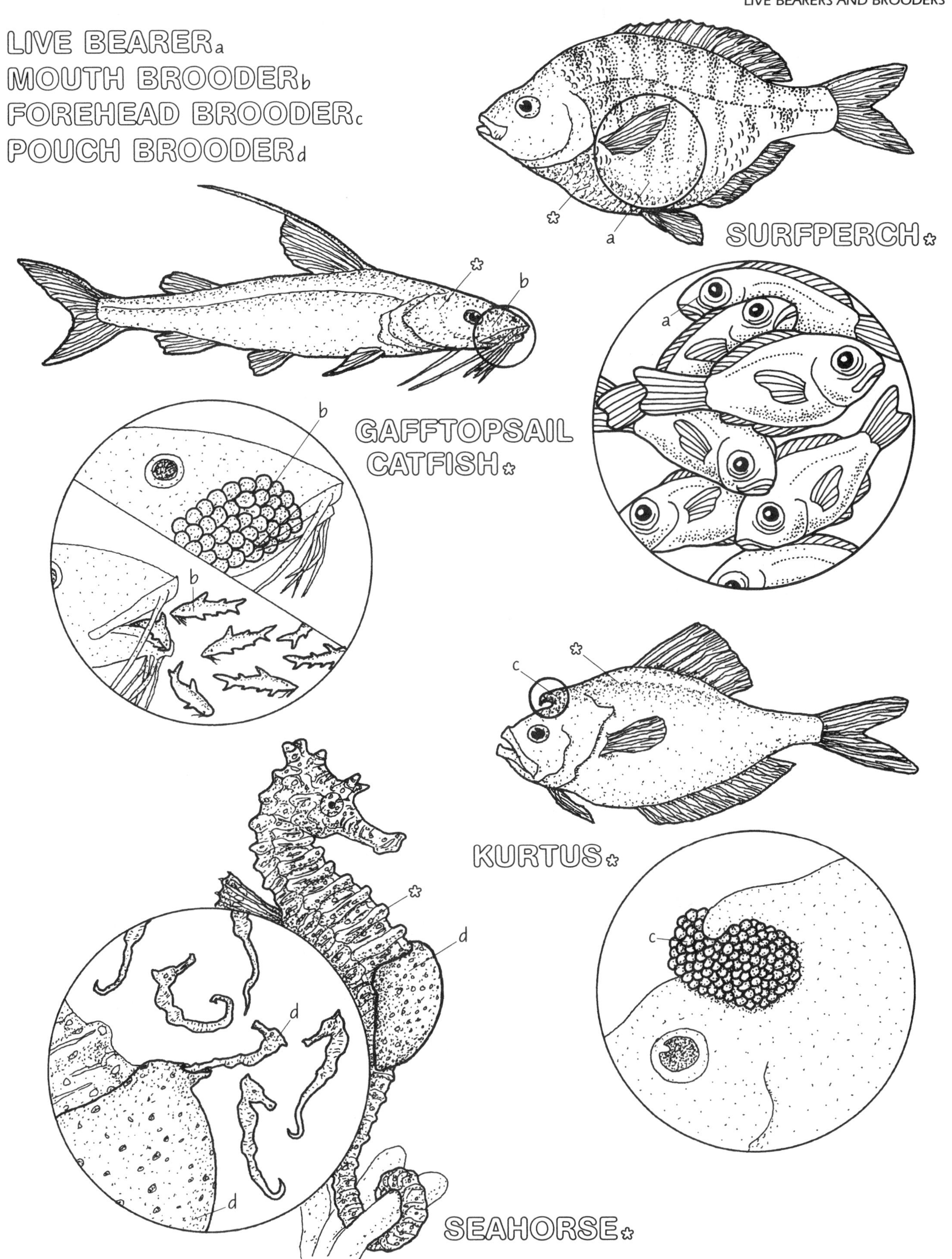

87
REPRODUCTION IN BONY FISHES: NEST BUILDERS

Many species of bony fish prepare some type of nest or protected area in which to deposit their fertilized eggs. Three very different types of nesting behavior are discussed in this plate.

Reserve bright blue for the male and dull blue for the female damselfish. Begin by coloring the nesting female and spawning male salmon using bright red for the body (a) and grayish green for the head (b). The jaw, pelvic fins, pectoral fins, ventral surface, and the edge of the caudal fin in the adult salmon are very light. You may leave them uncolored. Color the other stages in the life cycle. The alevin, the parr, and the smolt are light gray in life.

Nest building by the sockeye salmon is only a small part of this fish's remarkable life cycle. After spending two to four years at sea in the North Pacific, the adult sockeye return to their home stream in the late summer to reproduce (spawn). This journey involves a migration from the ocean, through estuarine areas, into freshwater rivers and streams. Some of these fish travel as far as 2400 km (1500 miles) to eventually return to the small tributaries where they themselves were hatched years earlier. Scientists believe the salmon employ a number of cues to guide them in their journey, including the earth's magnetic field and the unique chemical make up of the home stream which they are able to discern with their acute sense of smell.

As the adult sockeye salmon move upstream, their silver-blue sea-run colors change to bright red mating colors. The *male's* jaws become grotesquely hooked, and he develops a pronounced hump on his back just ahead of the dorsal fin. When the *female* reaches her home stream, she uses her tail to prepare a shallow trough or nest in the gravel bottom. The male joins her over the nest, and they simultaneously release sperm and *eggs*, which drop into the depression. After a short time, the female moves upstream and digs another nest; the materials from which are carried downstream by the current and cover the eggs retained in the first nest. The female may prepare a half-dozen nests until she has released 3000 or more eggs. The combined series of nests with their fertilized eggs is called a "redd." After migrating and spawning, the formerly sleek sockeyes are emaciated and very weak; their mission completed, they soon die.

The fertilized salmon eggs remain buried under several centimeters of sand and gravel through the winter. By spring, each embryo has developed into a 2.5 cm (1 in) *alevin* that still carries the remainder of the egg's yolk in the attached *yolk sac*. The alevin grows into a *parr*, which remains in fresh water for about two years. Toward the end of this period, it matures to a 15 cm (6 in) *smolt*, which moves downstream and out to sea where it feeds and grows to adulthood.

Color the grunion spawning cycle. These figures represent a bird's eye view of the edge of a sandy beach.

The California *grunion* lay their eggs high up on sandy beaches during the spring and summer and let the warm southern California sun incubate the developing embryos. For this strategy to be successful, the activity must coincide with the season's highest (spring) tides. For three or four nights following the highest tide (the highest tides are always at night during the spring and summer in southern California), male and female grunion ride inshore on *waves* at the peak of the *high tide*. As the wave recedes, the female wriggles tail first into the sand, and one or more males wrap themselves around her. She releases her eggs into the tunnel "nest" she has made, and the males release their sperm (spawning). The adults swim off at the next high wave, and the nest is buried by the waves' action. It is important that the grunion accomplish their egg laying immediately following the highest tide of a cycle so that no more waves will reach the nests for at least 10 to 12 days. When the waves do finally wash over the nests, the *juveniles* burst out of their protective membranes, wriggle to the sand's surface, and swim to sea with the receding wave. They will return to breed the next year as 12.5–15 cm (5–6 in) adults.

Color the damselfish cycle. Leave the edge of the male's caudal fin uncolored.

As the season for reproduction approaches, the *male damselfish* takes on bright breeding colors and busies himself preparing a patch of *red algae* on the coral reef. He then attracts a prospective mate by special movements of his white-edged caudal fin. The *female damselfish* is enticed into his nest of red algae and releases her eggs, which the male fertilizes by releasing sperm over them. The female then leaves the nesting site (the male may even chase her away), and the male begins a several-week-long vigil, guarding the eggs and fanning them with his fins to keep them clean and well oxygenated during development.

NEST BUILDERS

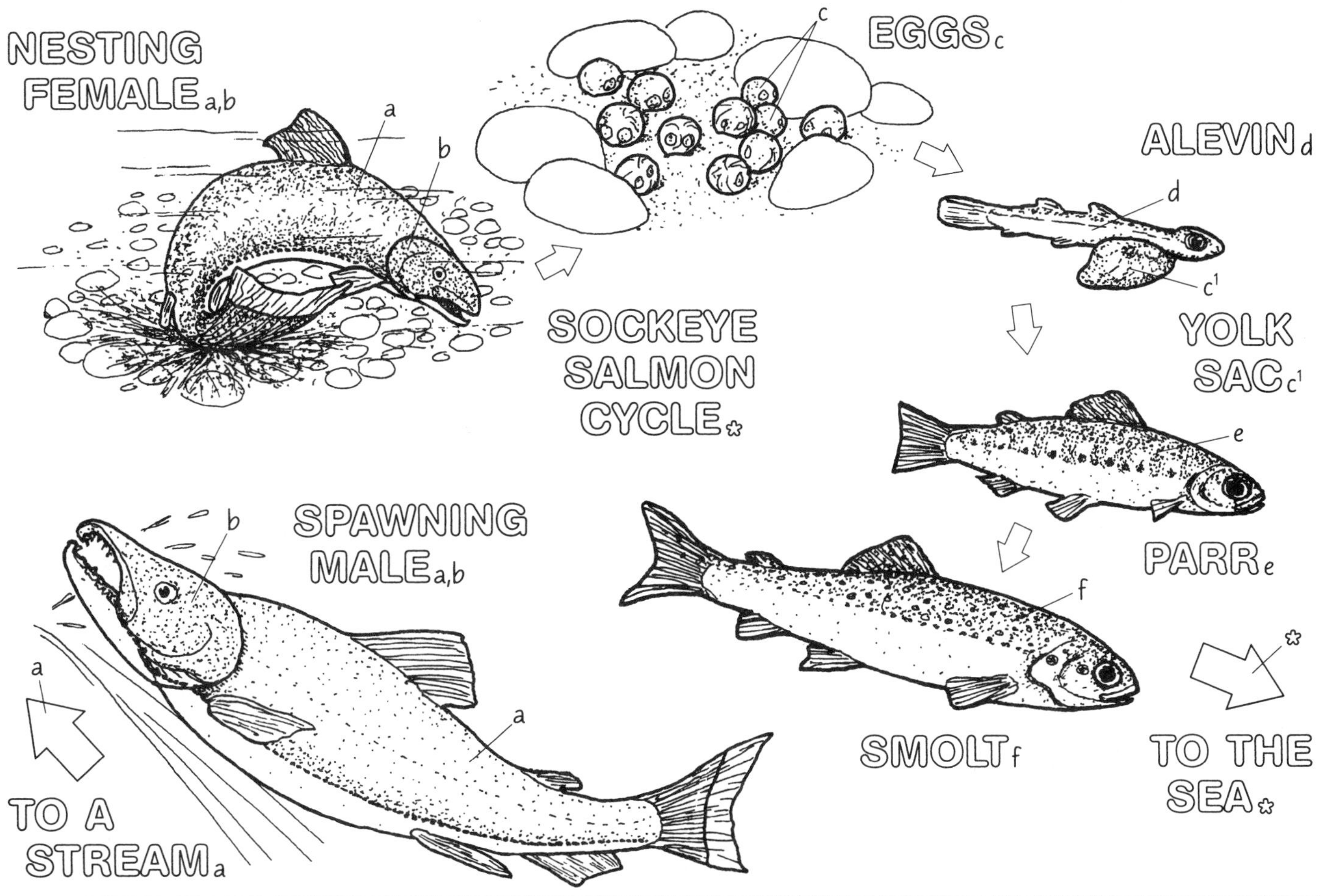

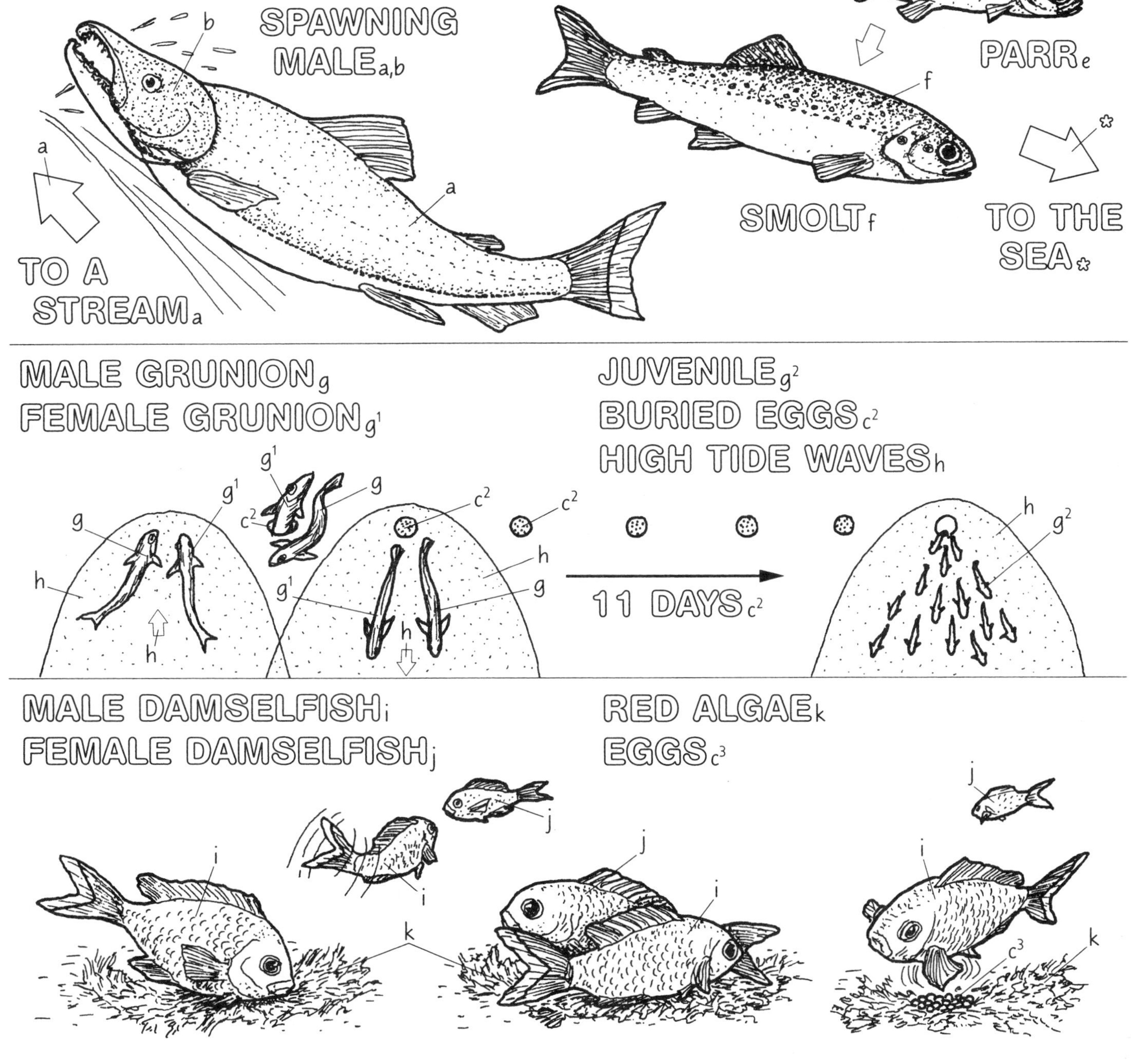

88
REPRODUCTION IN BONY FISHES: BROADCASTERS

Many marine fishes neither carry their young nor build nests for their protection. Instead, these fishes release their gametes into the water, where fertilization and development take place without further parental involvement. Such fishes are commonly called "broadcasters," as their gametes are distributed, or broadcast, by the waves and currents.

A number of factors contribute to the success or necessity of broadcasting as a reproductive strategy. Pelagic fishes (Plate 45) are generally not associated with a suitable substratum for nest building and cannot stop swimming to brood their young. The release of gametes directly into the water takes advantage of the ocean's free dispersal service and the rich supply of plankton potentially available to young fishes (Plate 75). The reproductive habits of two species of broadcasting fishes are discussed in this plate.

Color the male and female herring and their sperm and eggs. The larval herring are not drawn to scale with the adults.

The northern herring (Plate 45) is found in the Atlantic Ocean, and a closely related subspecies occurs in the Pacific. Herring range widely in both oceans, often in shoals of millions of fish. The herring is a pelagic fish that feeds on zooplankton and usually lives over deep water. It is a broadcaster and usually moves into shallow water to spawn. Along the Pacific coast of the United States, herring come into estuaries and shallow bays. *Females* release their *eggs* near *males*, which then release *sperm*. The fertilized eggs are heavier than water (termed demersal eggs) and sink to the bottom, where they adhere to *algae* and eel grass. Within 10 to 15 days, the *attached embryos* hatch as *yolk-sac larvae* at a length of 6 to 8 mm (0.24–0.32 in). The yolk-sac larvae remain near their hatching site for a few days, feeding on reserves of yolk. As they grow into larger *larvae* (29 mm, 1.2 in), they swim upward into the water, where currents carry them away. The herring larvae continue to grow into *prejuveniles* (about 40 mm, 1.6 in) that soon join other herring in the open water. These young fish will mature to spawn in two to seven years.

Color the migration routes and life history stages of the European eel. The small eel on the map receives the separate colors of the three advanced stages of the eel: glass, yellow, and silver, which occur in and around the coastal streams of Europe. The map includes the coastal waters and streams inhabited by the American eel as well.

The common or European eel is a catadromous fish; that is, it lives in fresh water and migrates to the ocean to spawn (the reverse of salmon, which are termed anadromous; Plate 87). The migration of these eels is truly amazing. In European fresh water *streams*, the adults change from the common *"yellow eel"* to the silver-white *"silver eel*," with enlarged eyes and reduced mouths. These silver eels migrate across the Atlantic Ocean to an area known as the *Sargasso Sea*, a floating "island" of the seaweed *Sargassum* (a brown alga), located in a calm area of the western Atlantic near Burmuda. Details of the journey are not known, but once at their destination, the silver eels spawn in deep water (500 meters, 1640 ft) and die. The *eggs* slowly float to the water's surface, and an exotic-looking larva called the *leptocephalus* (thin head) emerges in the spring. This transparent, leaf-shaped larva then sets off back toward Europe, a migration of more than 4000 km (2500 miles) that takes two years or more to complete. As it swims and drifts eastward in the Gulf Stream current, it grows much larger, becoming a 7.5 cm (3 in) *"glass eel,"* or elver, by the time it reaches the European coast. The glass eel enters coastal streams and acquires the yellow adult pigmentation. Young eels migrate upstream by the thousands, sometimes over a distance of many kilometers. The urge to migrate comes again when the eels are anywhere from four to twenty years old, and the adults begin their journey back to the Sargasso Sea to breed.

The life cycle of the European eel is still imperfectly understood. Few adult eels have been caught offshore in the Atlantic along the proposed migration route. The entire migration has been inferred from the distribution of the leptocephalus larvae; small ones are found in the Sargasso Sea, and progressively larger ones are found as one moves eastward towards Europe. *American eels* spawn in the same area of the Sargasso Sea as the European eel, but their larvae reach the coast of North America in only one year. There has been some question whether the European eel and American eel are separate species. Recent studies of molecular variation in DNA support the contention the two eels are genetically isolated and should be considered two species.

BROADCASTERS

HERRING*
MALE$_a$ SPERM$_{a^1}$
FEMALE$_b$ EGGS$_{b^1}$
ALGAL SUBSTRATE$_c$

ATTACHED EMBRYO$_d$
YOLK-SAC LARVA$_e$
LARVA$_f$
PREJUVENILE$_g$

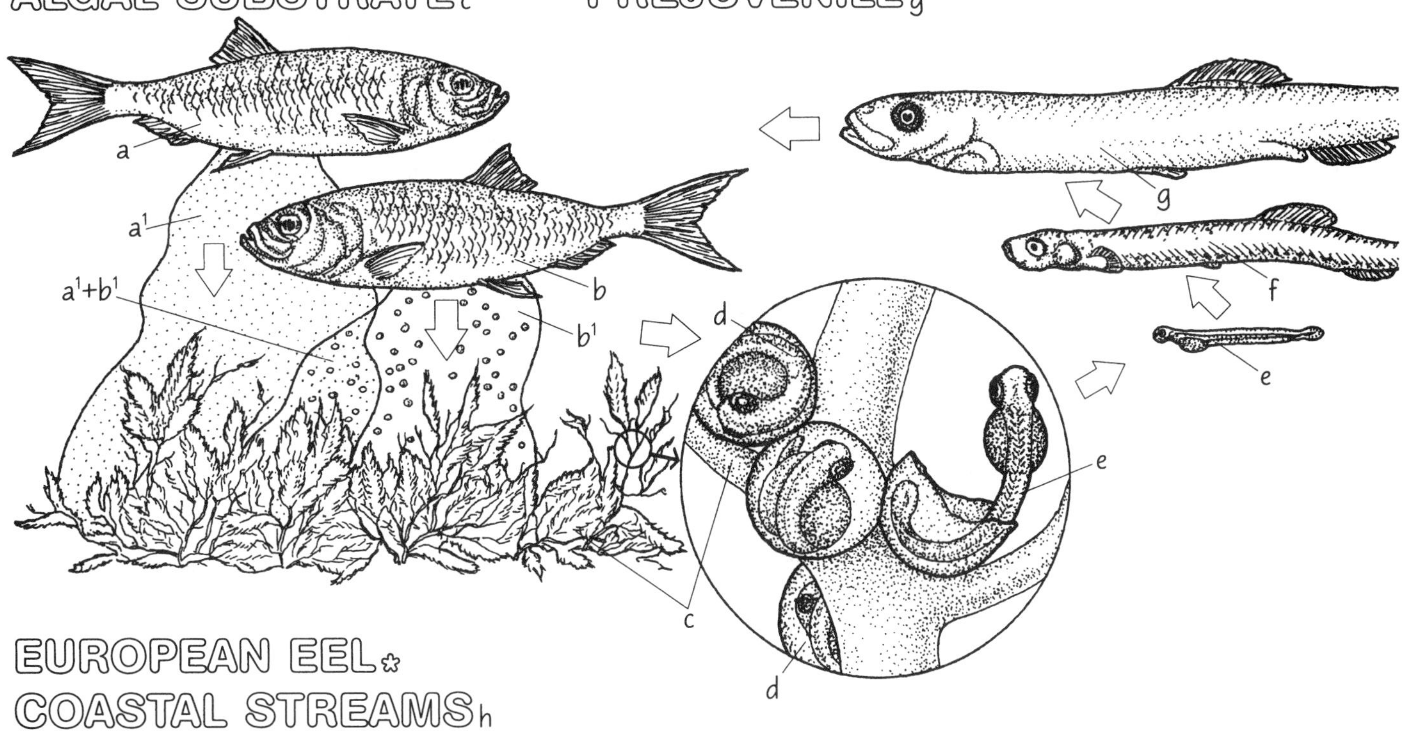

EUROPEAN EEL*
COASTAL STREAMS$_h$
YELLOW EEL$_i$
SILVER EEL$_j$
SARGASSO SEA$_k$
EGGS$_l$

LEPTOCEPHALUS LARVA$_m$
GLASS EEL$_n$
AMERICAN EEL$_o$

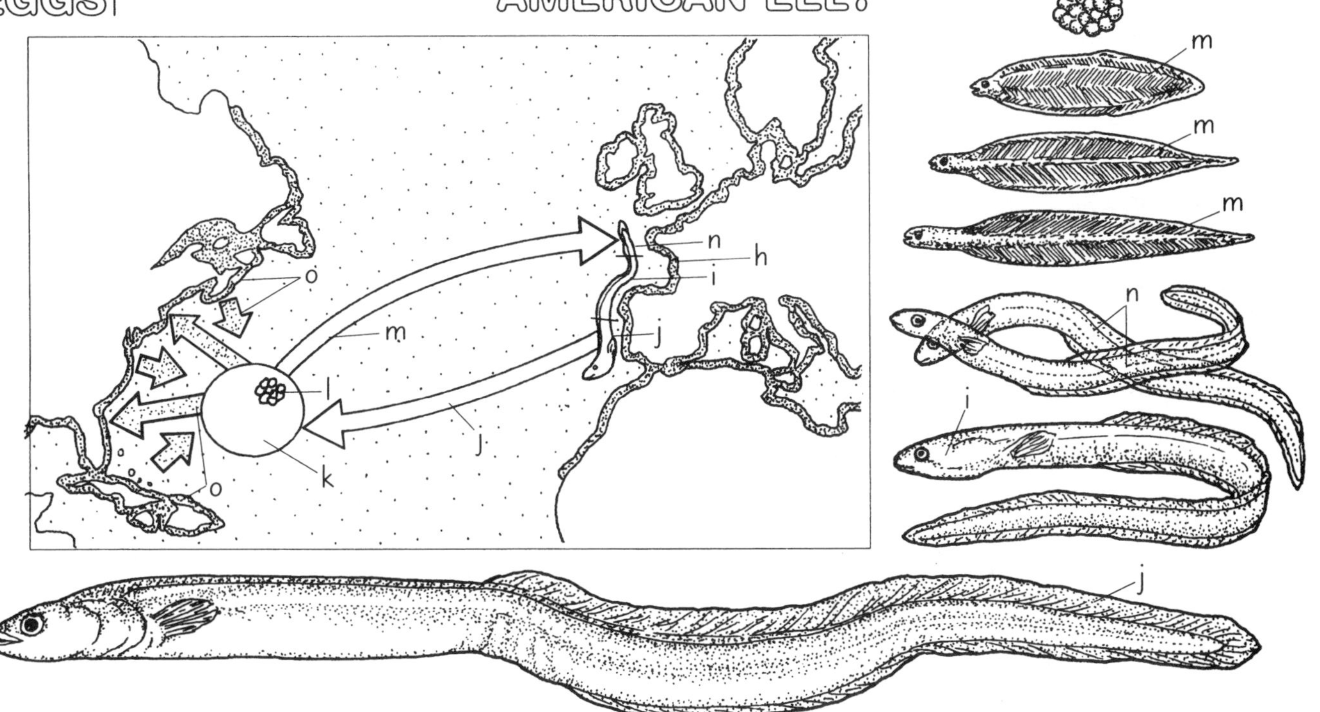

89
CALIFORNIA GRAY WHALE ANNUAL MIGRATION

Many species of whales undertake annual migrations. These animals require an exceptional amount of food to fuel their warm-blooded metabolisms and huge bodies. They travel to feed in rich polar feeding grounds in the spring and summer, and retreat to warmer waters during the winter to breed. The timing of the reproductive cycle of the whales is intimately associated with this annual pattern. This plate examines the annual migration of the California gray whale. This migration is well known because the gray whales often travel close to shore and spend the winter in shallow coastal lagoons where they are easily observed. Most other whale species migrate on the high seas and are far more difficult to study. Gray whales were once hunted to very low numbers, and their extinction was feared. However, since the whaling moratorium of 1946, the gray whale's recovery has been so spectacular that it was removed from the endangered species list in 1993.

Color the migration route of the gray whale; then color the feeding grounds and the mating and nursing sites. Color the illustrations of gray whale behavior as each is mentioned in the text.

In late spring, the gray whales arrive on their northern *feeding grounds* in the Bering Sea and Arctic Ocean, north to the edge of the polar ice pack. Gray whales feed on benthic invertebrates (mostly small crustaceans called amphipods; Plates 35, 62), and are limited to rather shallow water for feeding. During the long days of the Arctic summer, the gray whales feed almost continually, consuming up to a ton of food each day as they grow and replenish their blubber supplies. During this time the females wean their calves, which will make the southward fall migration on their own. In September, as the pack ice spreads over the feeding grounds, the whales begin their 6500 km (4000 mile) trek southward (the longest known migration of any mammal). The *route* follows the west coast of North America, and in December and January whales are frequently sighted along Oregon and California, often very close to shore. They are sometimes observed holding their heads out of water, cocked at an angle, in a behavior called "*spyhopping*." It is speculated that they may be looking for, literally, landmarks on shore. Migrating gray whales are also observed to *breach*, leaping completely out of the water. The reason for this behavior is also unclear; it may involve visual searching by the whales, or serve to dislodge parasites from the skin, or it might be termed play behavior.

Migrating whales swim steadily all day and perhaps through the night. After three months, they reach their destination in one of several warm-water lagoons on the western shore of Baja California, Mexico or along the west side of mainland Mexico. In the relatively calm lagoons, the pregnant females give birth to a single *calf* (in rare instances twins). A gray whale calf may be 5 meters (16 ft) long and weigh nearly a ton at birth. For the next two months the *mother* suckles her calf on her rich (50 percent fat content) milk. The teats of the mammary glands are located in slits on the female's underside. They are equipped with special muscles to actually force the milk into the calf's mouth. While nursing, the female generally lies on her side, with one fin out of the water. She is often assisted in the nursing process by another adult female called an "*aunt*." An aunt is a female between pregnancies (adult females give birth every two years).

Females begin to ovulate (produce fertilizable eggs) in November and may mate during the fall migration or during the winter stay in the lagoons. Mating sometimes involves a prolonged courtship during which two or more *males* pursue a *female*. When she becomes sexually receptive, the female turns on her back and a male lies alongside, also on his back. The male's penis is positioned near the female's genital opening, and they roll together to complete the coupling. The second male may stand by and assist the mating couple in maintaining their position in the water. The threesome may remain together for a day or more after mating. Pregnancy in gray whales lasts thirteen months (gestation period); nursing and weaning occupy another six to eight months.

By early spring, the new calves have grown to over 6 to 8 meters (20–26 ft) in length, gaining up to 100 kg (220 lb) each day, and are ready to begin the migration to the Arctic feeding grounds with their mothers. In three months, the gray whales reach the feeding grounds where the females with new calves replenish their taxed energy reserves, and the pregnant females feed to nourish their unborn young.

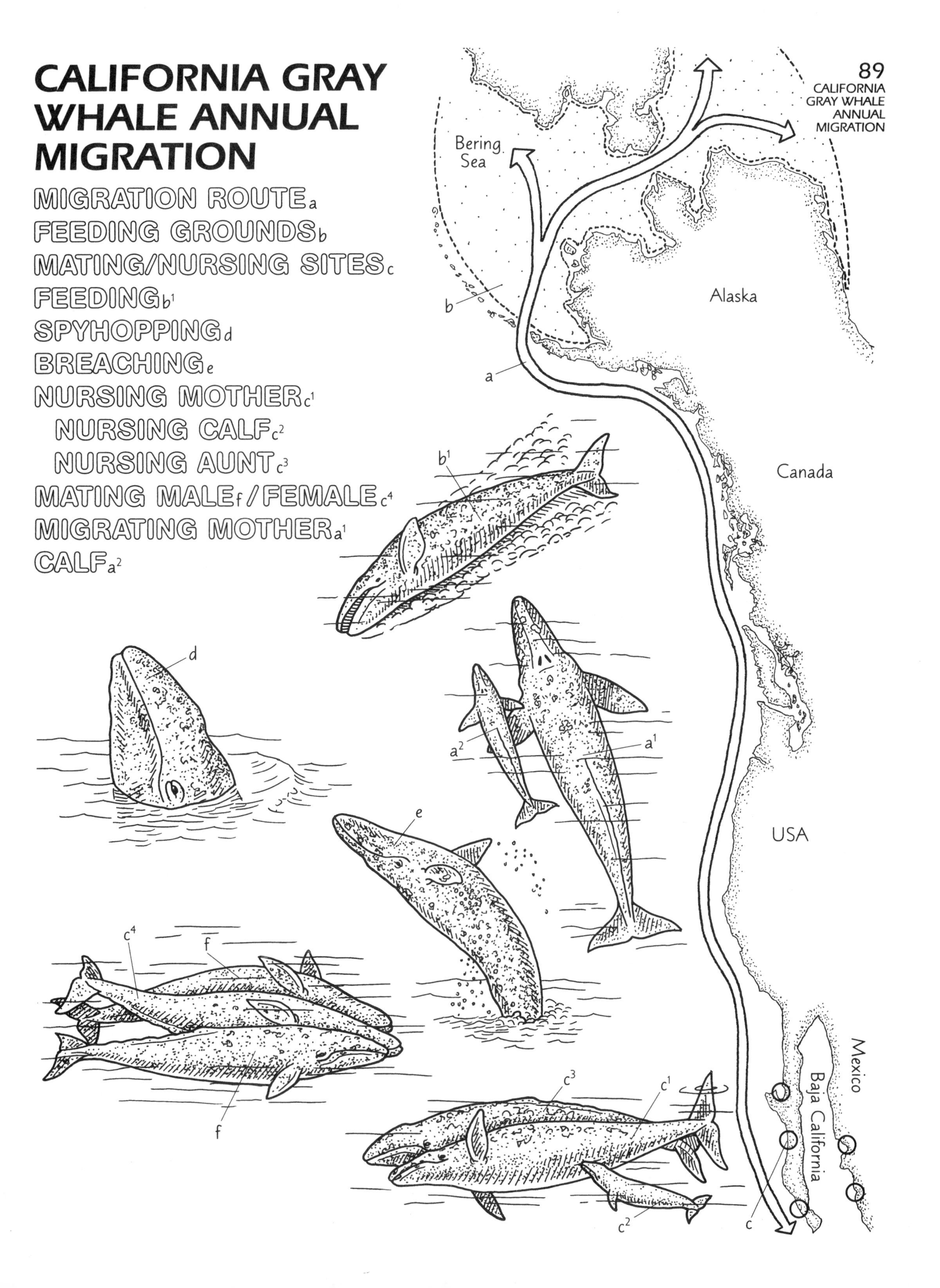
CALIFORNIA GRAY WHALE ANNUAL MIGRATION
MIGRATION ROUTE a
FEEDING GROUNDS b
MATING/NURSING SITES c
FEEDING b^{1}
SPYHOPPING d
BREACHING e
NURSING MOTHER c^{1}
NURSING CALF c^{2}
NURSING AUNT c^{3}
MATING MALE f / FEMALE c^{4}
MIGRATING MOTHER a^{1}
CALF a^{2}
Bering Sea
Alaska
Canada
USA
Mexico
Baja California
a
b
c
b^{1}
d
e
f
a^{1}
a^{2}
c^{1}
c^{2}
c^{3}
c^{4}

90
ELEPHANT SEAL ROOKERY

Adult elephant seals spend much of the year alone at sea. During this time, the population is widely dispersed as individuals search for food (Plate 115). However, every year the adults congregate for breeding in certain areas called rookeries. As with most pinnipeds (seals, sea lions, and walruses; Plate 60), elephant seal pups are born on land. The rookeries are usually located on remote islands, safe from mainland intruders and predators. An elephant seal rookery typically occupies a sandy beach, and at the peak of the winter breeding season it is crowded with seals. The breeding behavior of these animals is quite complex and apparently acts to ensure that the strongest possible offspring are produced each year.

Color the top panel, which illustrates the struggle for social position among the bull elephant seals. The most aggressive bull is designated beachmaster. Color the second panel showing the cows beginning to arrive. Note that the beachmaster is in the center of the rookery.

The males, or *bulls,* arrive at the rookery in early winter, several weeks before the arrival of the females. During this waiting period, the bulls fight repeatedly to establish a pecking order (hierarchy) among themselves. The strongest bulls remain on the beach as *beachmasters* and the rest are chased out to sea. The number of beachmasters depends on the overall size of the rookery. When the adult females, or *cows*, begin to arrive, the dominant beachmaster establishes his territory in the portion of the rookery most favored by the cows. The lower-ranking beachmasters partition the remaining rookery area among themselves. The cows do not arrive at the rookery all at once, but over a two to three month period.

Cows give birth to a single *pup* soon after arriving at the rookery (usually beginning in January). The pups are nursed for about four weeks and grow to a weight of 140 to 180 kg (300–400 lb). The cows rarely feed during this nursing period and tend to remain close to their pups. If a cow is separated from her pup, she is able to locate and distinguish it from others by its unique smell and cry. After the pups are weaned, they congregate at the edges of the rookery, away from the mating adults. After weaning her pup, the cow enters a short period of sexual receptiveness (called estrus) when she will allow a beachmaster to mate with her. After mating, the cows leave the rookery and return to the sea; little is known about their activities during the non-breeding season.

Color the third panel which depicts the rookery at the peak of the breeding season. The beachmaster is confronting a challenging bull, while another bull waits at the edge of the rookery.

During the two or three month period that cows are arriving at the rookery, the males do not feed or return to the water. At the peak of breeding activity, new cows are arriving to give birth while others nurse, or come into estrus and mate. The cows are highly gregarious, and their various activities appear almost as ripples in a sea of plump seal bodies. Adding to this confusion are constant challenges by bulls attempting to "dethrone" the beachmasters. Other bulls sneak into the rookery area and try to mate with receptive cows. The beachmaster must remain aware of cows in estrus and forcefully drive away intruding bulls to maintain his dominant position. Many nursing pups are crushed in this chaos. The cost to the beachmasters is also high, as they lose a great deal of weight during the breeding season. Studies indicate that bull elephant seals only survive one or two seasons as a beachmaster, and tend to die much younger than non-dominant bulls. They do, however, leave their marks on the new generation, as three or four beachmasters account for 90 percent of the mating that occurs each season. Thus, the strongest, most aggressive bulls father the majority of each new generation.

Color the lowest panel showing the rookery area being utilized as a haul-out.

When the breeding season ends in early spring, the bulls leave the rookery. Only the weaned pups remain for a while, before going to sea to feed. During the non-breeding season, the rookery is used as a haul-out or resting area for molting. The seals shed their fur and outer layer of skin in a process that takes several weeks. The females come in to molt in March, the *juvenile males* in May and June, and the adult males from July to September. During this molting period, bulls tolerate or ignore one another, and it is not unusual to see a "pile" of elephant seals hauled out on the beach. However, as winter approaches, the bulls return to the rookery and resume the struggle for a position of dominance.

ELEPHANT SEAL ROOKERY

BULL$_a$
BEACHMASTER$_b$
COW$_c$
PUP$_d$
JUVENILE MALE$_e$

91
SYMBIOSIS: MUTUALISM: INVERTEBRATE FARMERS

Photosynthetic plants are the basis for most of the food chains of the sea. The exceptions are food chains that are dependent on chemoautotrophic bacteria. These occur around deep-sea hydrothermal vents and cold seeps (Plate 17). The complicated chemical pathways involved in photosynthesis require sunlight for energy, the light-trapping chlorophyll molecule, carbon dioxide (CO_2), water (H_2O), and various inorganic nutrients (chiefly those containing nitrogen and phosphorus). Through these chemical processes, the sun's energy is utilized to convert water and carbon dioxide to oxygen and simple sugars. These products are, of course, used by the plants, but virtually all animals are ultimately dependent on them as well, either directly as herbivores or through the food chain. This plate discusses several marine invertebrates that take shortcuts in the food chain and literally "farm" plants within their own tissues, benefiting directly from the photosynthetic products.

Color each invertebrate farmer as it is discussed. The sea slug *Elysia* and the alga *Codium* receive the same bright green color. Do not color the green sea anemone located under the rocky ledge.

The bright green sea slug *Elysia viridis* (a gastropod mollusc) is the same color as *Codium,* the green alga on which it feeds. This situation appears at first glance to be a case of substratum mimicry (Plate 67), but in fact, the sea slug's color is acquired directly from the alga. *Elysia* feeds on *Codium* by slitting open the plant with its radula and sucking in the plant's tissues. Somehow, the green chloroplasts (the photosynthetic organelles of the plant's cells) are transferred intact to the gut lining of the sea slug. These chlorophyll-containing chloroplasts give *Elysia* its green color, and a few additional benefits. The chloroplasts continue to photosynthesize, and the sugars and oxygen they produce are utilized by *Elysia*.

Another mollusc that farms algae is the giant clam (182 kg, 400 lb), *Tridacna*, found on shallow coral reefs throughout the eastern tropical Pacific. Like the sea anemones and corals discussed below, *Tridacna* harbors single-celled plants called *zooxanthellae* (modified dinoflagellates, Plate 19) within special cells in its *mantle*. *Tridacna* is normally found lodged in coral rock, with its hinge directed downward, and its fluted, gaping *shells* facing upward. The thick mantle, containing zooxanthellae, projects over the edges of the gaping dull white shells in a dazzling display of blue, green, violet, and brown pigments. Besides being beautiful, these pigments serve to shield the zooxanthellae from the bright tropical mid-day sun; too much light can inhibit the photosynthetic process. Each night, *Tridacna* harvests a portion of its ever-growing plant crop by actually consuming the zooxanthellae with special amoebocytes (motile, feeding cells). *Tridacna* still feeds as a normal bivalve filter feeder, but a substantial amount of its nutrition is derived from its crop of zooxanthellae. An interesting development is that the "farmer clams" themselves are being farmed on South Pacific islands in recently initiated aquaculture projects.

Zooxanthellae play a significant role in the nutrition of many cnidarians. The giant *green sea anemone* was one of the first cnidarians in which the presence and importance of zooxanthellae were discovered. When this anemone occupies a habitat in which it is shielded from sunlight, such as in a cave or under a ledge, it is not bright green but very pale, nearly white. In sunlight, the animal is distinctly green, a color imparted by resident zooxanthellae. When a green anemone is experimentally transferred to a dark place for several days, the tissues pale dramatically.

The reef-building corals are very much dependent on their partnership with zooxanthellae. The corals receive a significant portion of their nutritional energy from the zooxanthellae, much of which the corals utilize to secrete their calcium carbonate skeletons (Plates 12, 23). The zooxanthellae are compensated for their contribution by having a safe place to live, and access to the waste products of the corals' cellular metabolism — nitrogenous and phosphate compounds — as a source of plant nutrients. This type of symbiotic relationship, in which both parties benefit, is called mutualism or mutualistic symbiosis.

Some corals derive more of their requirements from the zooxanthellae than others. The difference appears to be related to the size of the individual *polyps*. Corals with small polyps, such as the *eklhorn coral*, have a high ratio of surface area to volume; this facilitates absorption of sunlight by the zooxanthellae. Corals with larger polyps, such as the *brain coral*, have a lower surface-to-volume ratio, but the bigger polyps are more efficient in trapping zooplankton for food. They still depend upon the zooxanthellae for energy to use in the production of their skeletons.

INVERTEBRATE FARMERS

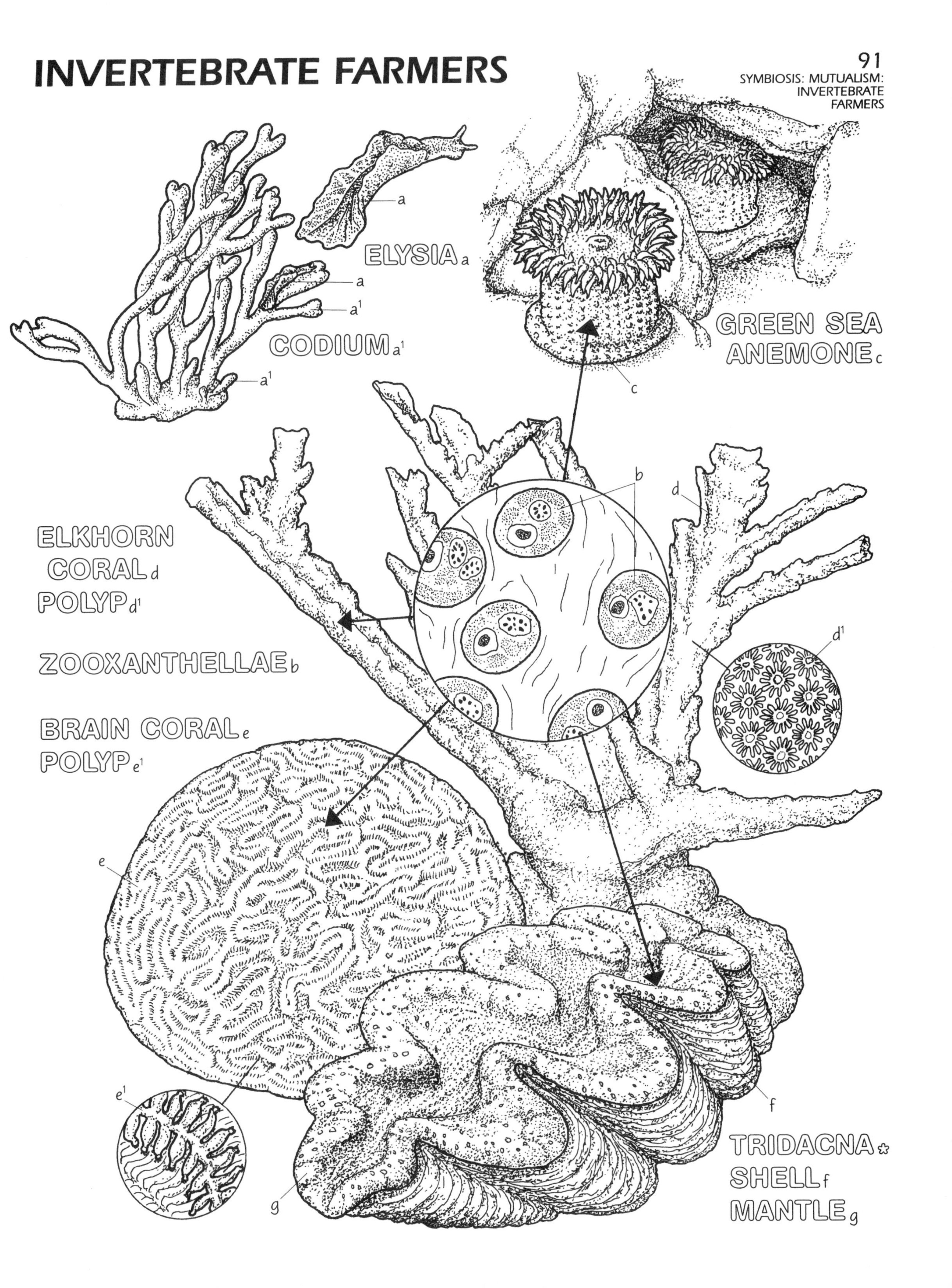

92
SYMBIOSIS: MUTUALISM: CLEANER SHRIMP AND FISHES

Many marine fishes are plagued with parasites and diseases, just like other organisms. Small crustaceans (isopods and copepods) and certain worms attach themselves to the outside of the fishes' bodies, onto the gills, or in their mouths. These parasites live at the expense of their hosts and do varying degrees of damage. Bacterial infections also occur on the fins, body surface, and gills of fishes. So common are these parasites and bacteria that some animals specialize in cleaning the affected hosts while feeding on the organisms they pick from the fishes' bodies. Several species of fishes and shrimps engage in cleaning behavior, some examples of which are explained in this plate.

Begin by coloring the illustration of the cleaner shrimp on the anemone at the upper left; color only the body and antennae. The anemone is white with purple-tipped tentacles. When coloring the enlarged view of the shrimp in the inset, include the claw. In the cleaning scene to the right, color only the operculum and mouth of the largest customer fish, leaving the body blank. Color over the entire body of each of the other fish.

Pederson's cleaner shrimp is a common inhabitant of Caribbean coral reefs. This small transparent, violet-spotted shrimp (4 cm, 1.5 in) occurs singly or in pairs, and is always associated with a particular species of *sea anemone*. The shrimp either clings to the anemone or lives in the same crevice with it; the shrimp derives protection from the anemone and is apparently immune to its sting. Many researchers believe that each shrimp occupies a *"cleaning station"* in the form of some conspicuous landmark, like a particular coral head. They may station themselves at such points for extended periods of time. The reef fishes apparently recognize and seek out these cleaning stations and their tenants when in need of care. When a fish approaches a cleaning station, the shrimp rocks back and forth and whips its long, white *antennae* to signal that it is ready to clean. The fish then swims up and stops within a few inches of the cleaner shrimp, which leaves the protection of its anemone and begins to scour the fish's *body* for parasites. The shrimp uses its *claws* to dislodge parasites, and may even make incisions to remove invaders from beneath the fish's skin. When the shrimp approaches the customer's head, the fish opens each *operculum* (gill cover) and allows the shrimp to enter the gill chamber to clean. The fish then opens its *mouth* for cleaning by the shrimp.

During the actual cleaning process, the shrimp is usually safe from predation by its customer fish, even though the fish may be a carnivore that normally feeds on small crustaceans. The reef fishes depend on the cleaners, and, if heavily parasitized or infected, may visit a cleaning station several times a day.

Now color the cleaning scene at the bottom of the plate. Color the horizontal bars on both the cleaner wrasse and the imposter blenny black. Leave the space between the bars uncolored as their bodies are white. As in the diagrams above, color only the mouth and operculum of the customer being cleaned, and the entire bodies of the waiting fishes.

Many different types of fishes also exhibit cleaning behavior. In some cases, the young of a species (e.g., butterflyfish and angelfish) are cleaners before they acquire adult feeding habits. These cleaner fishes service a variety of marine animals in addition to bony fishes. Stingrays, mantas, green sea turtles, and even crocodiles have been observed being attended by cleaner fishes.

One conspicuous full-time cleaner is the small (8 cm, 3 in) *cleaner wrasse* of eastern Pacific coral reefs. This fish is marked with bold black, horizontal stripes that apparently designate it as a "cleaner" to other fishes. Cleaner wrasses often live in pairs near a prominent cleaning station. Specific body movements and head nodding by the wrasse serve as an invitation to potential customers. The customer fish then advances and presents itself to the cleaner, often with its opercula and mouth open. Like the shrimp's clientele, lines of customers may be observed waiting for the services of the wrasses.

The relative immunity to predation enjoyed by the cleaner wrasse is capitalized on by the *false cleaner blenny*. This small fish has nearly identical markings to the cleaner wrasse. However, instead of picking parasites, this fish grabs mouthfuls of healthy tissue from other fishes that have presented themselves to the imposter for cleaning!

CLEANER SHRIMP AND FISHES

SEA ANEMONE$_a$
CLEANER SHRIMP*
BODY$_b$
ANTENNA$_c$
CLAW$_d$

CLEANING STATION$_e$
CUSTOMERS*
OPERCULUM$_f$
MOUTH$_g$
BODY$_h$

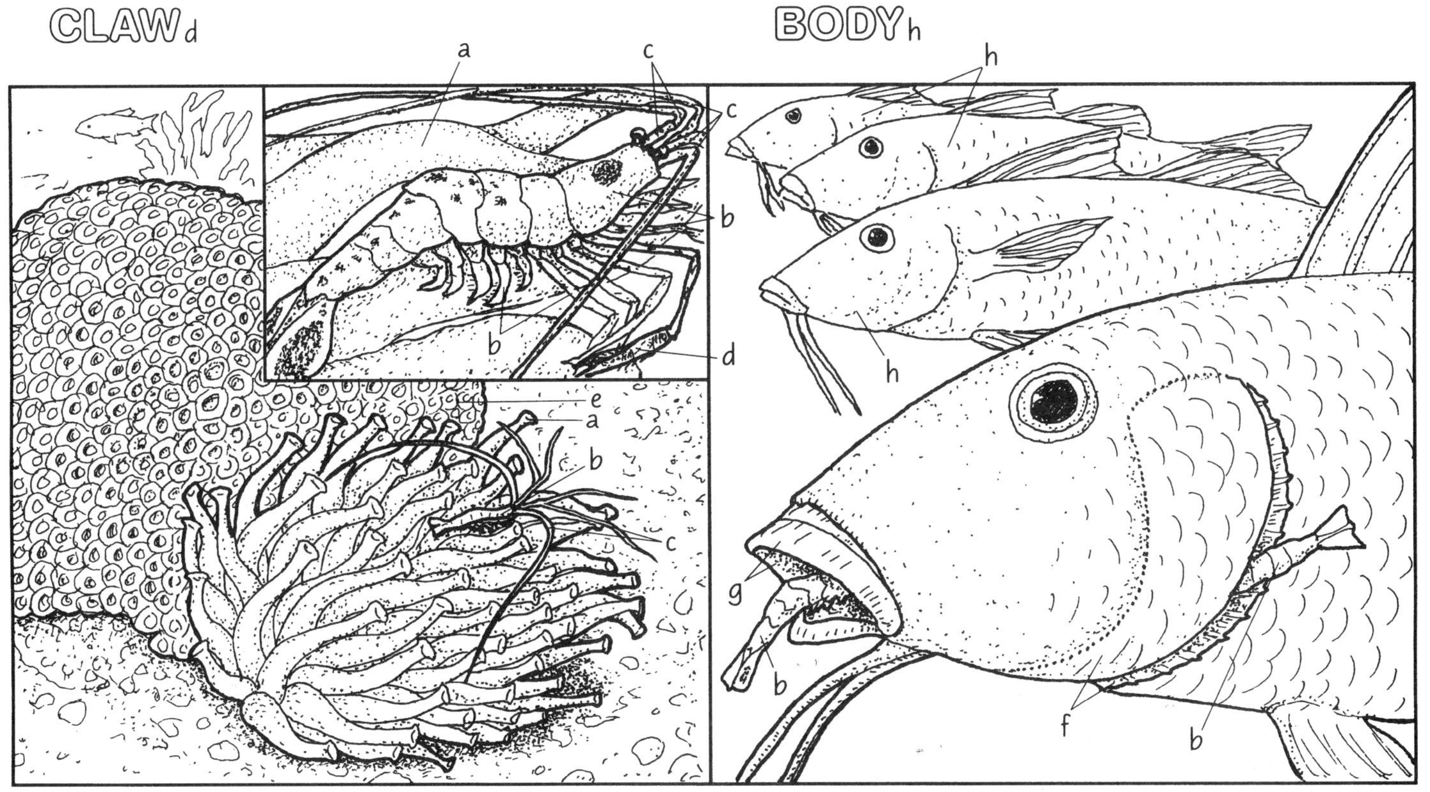

CLEANER WRASSE$_i$

FALSE CLEANER BLENNY$_j$

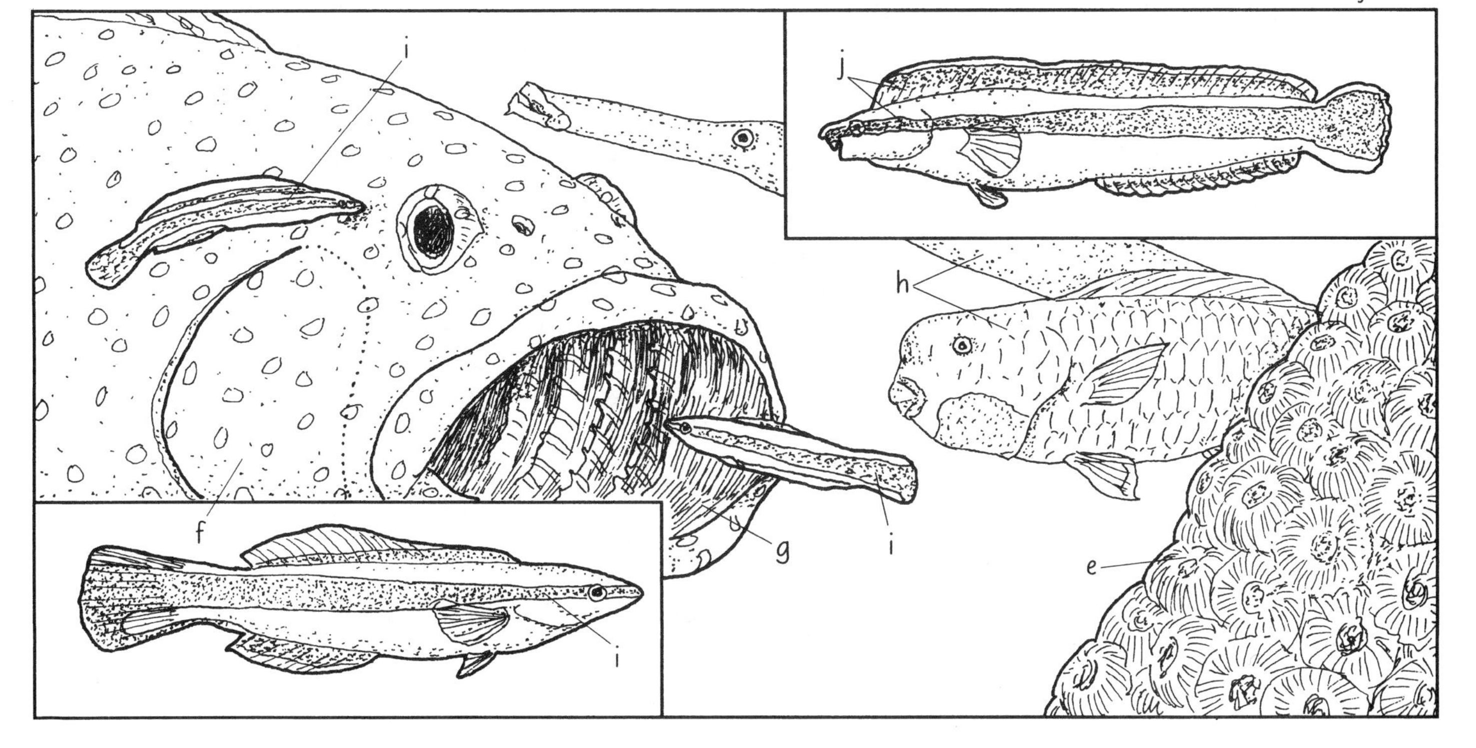

93
SYMBIOSIS: MUTUALISM: ANEMONEFISH AND SEA ANEMONE

Anemonefish (also called clownfish) are small (6–14 cm, 2–6 in), brightly colored fish abundant on the coral reefs of the western Pacific and Indian oceans. They are closely related to damselfish (Plates 13, 87) and the garibaldi of California's kelp forests (Plate 63). Anemonefish are dependent on the protection offered by their host sea anemone, and are found only in the anemone's company.

Begin with the top part of the plate. The anemone has green tentacles and a coral-pink column. The four boxed drawings represent the process of acclimatization of the anemonefish to the anemone. The first two show the anemonefish's quick withdrawal from the anemone's stinging tentacles. Leave the white stripes of the anemonefish uncolored and use orange for the rest of the body.

Sea anemones are carnivores that subdue their prey by stinging them with the venomous nematocysts on their *tentacles* (Plate 23). The *anemonefish*, through a process of acclimatization, is able to live among, and be protected by, the anemone without being stung. Acclimatization may take as little time as a few minutes, or many hours, depending on the species of anemonefish and anemone. The fish approaches the anemone and gingerly brushes against its tentacles with its tail or ventral surface. It quickly pulls away upon being stung. The anemonefish returns and gradually brings more of its body into contact with the tentacles, until it is able to be engulfed in the tentacles with total impunity.

There are several theories regarding the anemonefish's acquired immunity to the anemone. One theory suggests that during the process of acclimatization, there is a change in the quality of the mucus coating on the outside of the fish. This change entails removal of chemicals from the mucus that trigger the nematocysts, so that when contact is made between fish and anemone, the deadly discharge of the nematocysts is prevented. Another theory suggests that the fish gradually becomes coated with a combination of the anemone's mucus and its own, until the anemone does not distinguish the fish from itself, and therefore does not fire its nematocysts. A third theory suggests that each mechanism may operate in different situations, depending on the anemone-anemonefish species pair involved in the symbiosis.

Many species of the anemonefish occur in single, tightly structured family groups, living with a single anemone. There is an adult *female* and a *male*, which establish a *nest* site on the substratum adjacent to the *column* of the anemone, within the protective overhang of the anemone's tentacles. The two fish clear away all algae and other benthic organisms, and the female attaches her *eggs* to the bare substratum. The male keeps the eggs clean by fanning them with his fins while the female continues with the normal feeding behavior of foraging near the anemone for algae, or darting out to grasp zooplankton. The female may also exhibit territorial behavior in protecting the host anemone from the anemone-eating *butterflyfish*. This protective behavior suggests there is a mutualistic benefit to the anemone for harboring the anemonefish.

The anemonefish constantly re-establishes contact with the anemone. During feeding forays or nest-tending, the fish will frequently return to brush against the anemone's tentacles.

Anemonefish eggs hatch in six to eight days and the larvae live several weeks in the plankton, feeding and growing. Upon settling out of the plankton, the small fish seek shelter among the tentacles of a host anemone, feeding directly on the tentacles or on material egested by the anemone. When a small fish ventures away from the anemone, it is noticed and chased by the resident adults and other *juveniles*. Many juveniles are chased away and must find a different anemone with which to live, and go through the acclimatization process with this new host.

Among those that remain in the original group, a complicated hierarchy is established. The adult female is usually the largest and most dominant fish in the group. If she is removed, the adult male of the pair changes his sex and becomes the dominant female. The next fish down in the hierarchy becomes the adult male of the pair. The ability to change sex is not unusual for fishes that live in closely structured social groups; many groups such as wrasses, sea basses, and parrotfishes show this behavior. However, the pattern of sex reversal is normally from female to male, not male to female as is seen in the anemonefish.

In a commensal (from Latin co-mensa, to share a table) symbiosis, one party benefits and the other is not harmed. The arrangement between the anemonefish and the sea anemone considered in this plate is classified as a mutualism by some and a commensal relationship by others.

ANEMONEFISH AND SEA ANEMONE

ACCLIMATIZATION*
SEA ANEMONE*
TENTACLE a
COLUMN b
MALE ANEMONEFISH c

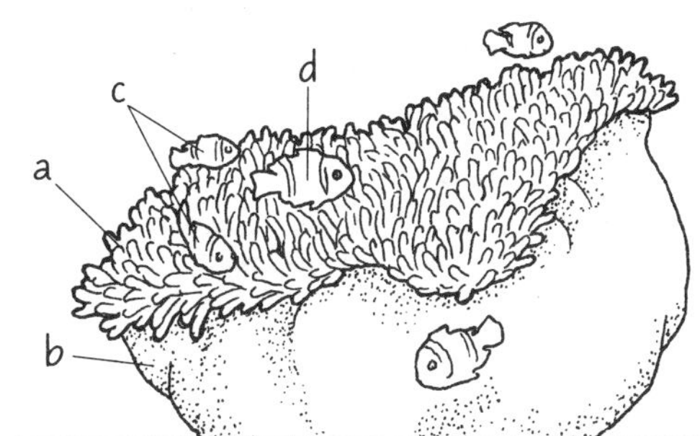

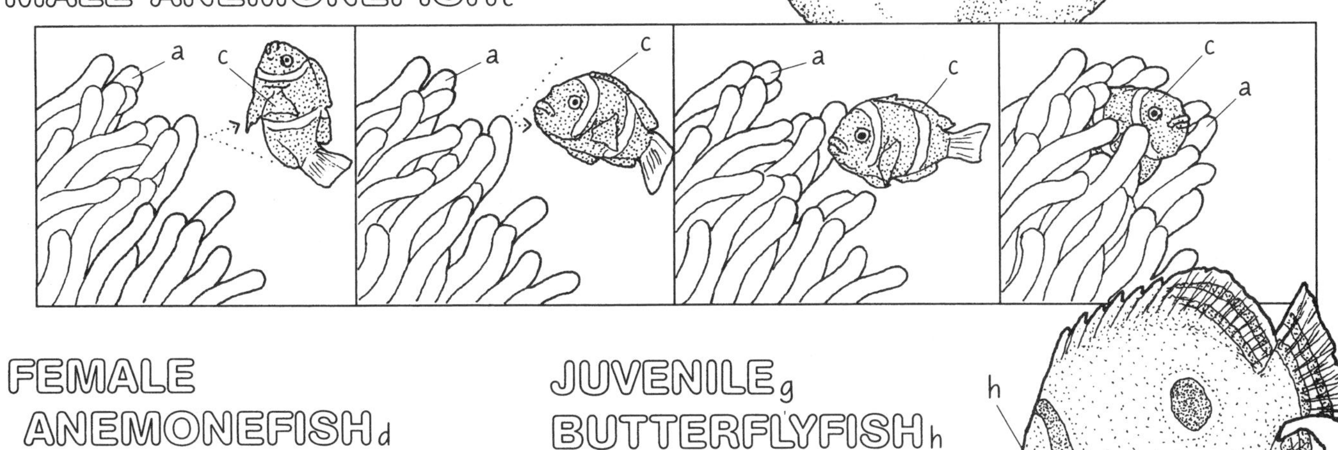

FEMALE ANEMONEFISH d
NEST e
EGG f

JUVENILE g
BUTTERFLYFISH h

94 SYMBIOSIS: MUTUALISMS AND COMMENSALISMS

Color the hermit crab pale brown. The sea anemone is light gray with red stripes.

Many species of *hermit crabs* form symbiotic relationships with *sea anemones*. The mutualistic nature of the relationship seems obvious. The sea anemones are attached to the hermit's shell and serve as a deterrent to predators. The sea anemone gets a solid place to attach and a free ride to potential feeding opportunities. Hermits (Plate 36) have been observed transferring sea anemones to new shells when their increasing size requires them to switch shells, clearly suggesting the anemones are beneficial to the hermits. What evidence is available to suggest there is a mutual benefit to the anemone? In some relationships, the anemone initiates the move itself, using its tentacles to attach to the hermit's new shell and releasing its base from the old one. One especially remarkable symbiosis is found in the deep sea off Hawaii. Here the anemone *Stylobates aeneus* attaches to a small hermit crab *Parapagurus dofleini* in a small shell. As the hermit grows and would normally require a bigger shell, the anemone enlarges the old shell with a chitinous secretion from its base. The anemone's base wraps around the entire shell, and its secretions follow the corkscrew-shape of the hermit's abdomen which is specially modified to fit into a snail shell. The resulting golden-colored *false shell* so resembles a snail's shell that a famous malacologist identified the shell as a mollusc and gave it a scientific name. This mutualistic relationship is especially appropriate as new, larger shells are not readily available to hermit crabs on the deep-sea floor, and sea anemones would be precluded from the soft deep-sea bottom without the solid attachment site provided by the hermit's shell.

Color the stingray a soft brown. The bar jack is silver with a black bar that extends from under the dorsal fin onto the caudal fin and is underlain by a blue bar.

A clearly commensal relationship is seen in the Caribbean between the *Southern stingray* and the fish known as the *bar jack*. The swift-swimming jacks are excellent hunters (Plate 109), however Southern stingrays are frequently seen with a single bar jack "riding shotgun" just above them as they swim along the bottom. When the ray stops to feed by digging into the soft bottom with its large pectoral fins to uncover buried prey, tasty food morsels are thrown into the water with the sediment and the jack swoops in quickly to gobble them up. Similar commensal pairings of jacks and stingrays are also found in other tropical seas.

Color the sea bat and its commensal dark brown polychaete. The sea bat varies considerably in color, but is commonly red-orange.

The *sea bat* is an omnivorous scavenging sea star common on shallow rocky bottoms along the west coast of North America. It frequently harbors one to several thin *polychaetes* in the ambulacral grooves among its tube feet (Plate 39). When the sea bat everts its stomach on food, the polychaetes move in to help themselves to part of the meal. The polychaete is referred to as a "facultative" commensal because it is not obligated to this commensal lifestyle; it is also found living free in the environment. A commensal animal that cannot live without its host is called an "obligate" commensal.

Color the coral, pistol shrimp, crab, and sea star. The coral is pale brown to off-white, the crab has a reticulate (netlike) pattern of orange and gray outlined in red, and the shrimp's color varies from red to brown. The body of the sea star is gray and the spines are brown.

Many small invertebrates live commensally in the burrows or on the shells or bodies of larger invertebrates. Examples are the small pea crab and scale worm that share the burrow and food of the innkeeper worm (Plate 26). Large colonial invertebrates are especially targeted by commensal dwellers, and may resemble apartment houses given the number of hangers-on. One such relationship is found on the Pacific coast of Panama between small *pistol shrimps* and *crabs* that live among colonies of the branching coral *Pocillopora*. However, when the coral-eating *crown-of-thorns sea star* approaches the *Pocillopora* colony, what appeared to be a commensal relationship quickly reveals itself to be a mutualistic one. The small crustaceans rush to the tips of the *Pocillopora* branches, the pistol shrimp bangs away repeatedly on the star using the concussive snap of its large claw (Plate 71), and the crab nips the star's tube feet. The star hesitates and then retreats, forced to eat other, less preferred prey. If the shrimps and crabs are removed, the coral colony is soon eaten.

MUTUALISMS AND COMMENSALISMS

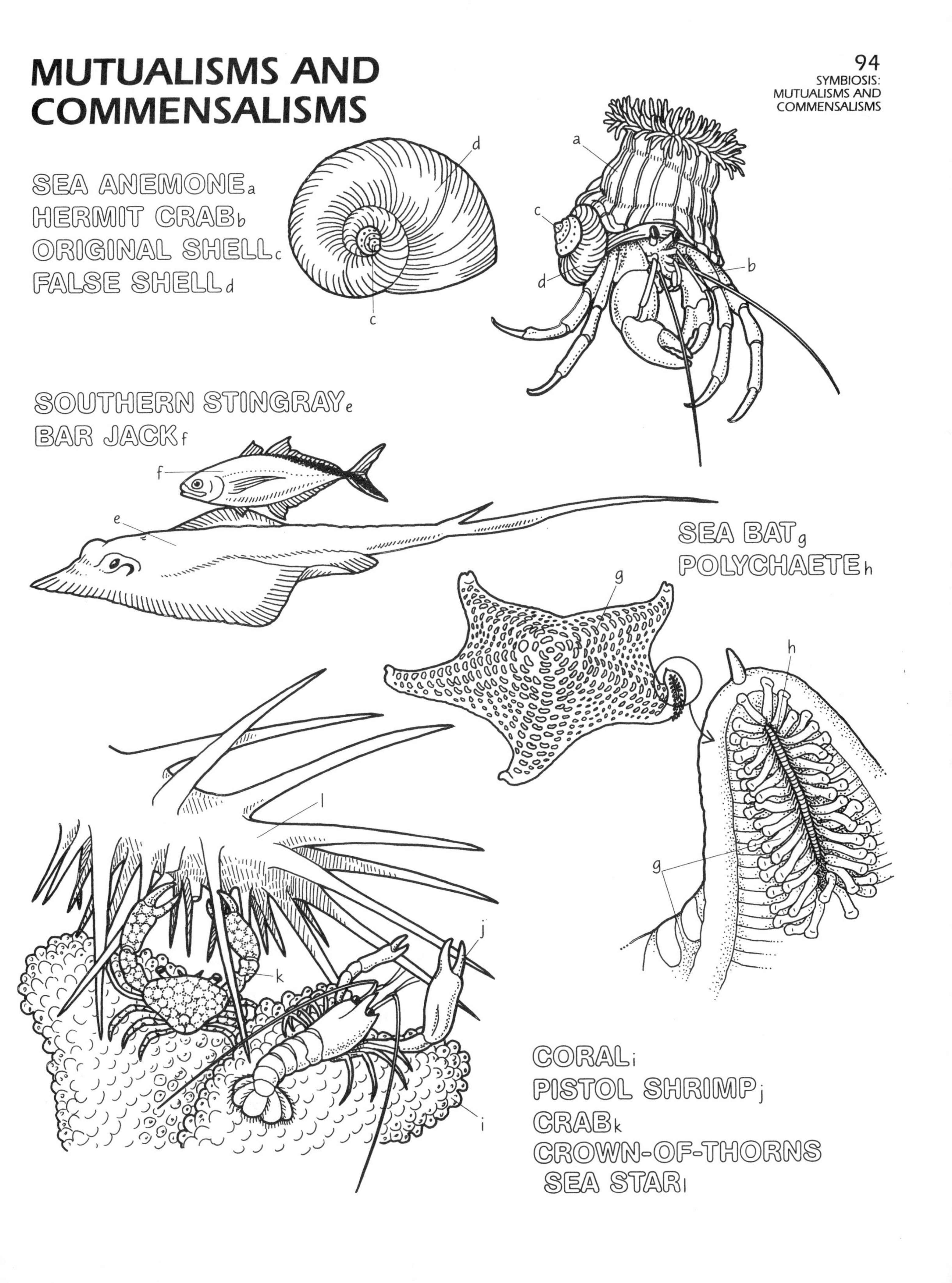

95 SYMBIOSIS: PARASITISM

Parasitism involves an interaction between two species in which one, the parasite, benefits and the other, the host, is harmed. The parasites that bedevil humans — tapeworms, leeches, pin worms, round worms, flukes and the like, are also represented by species that utilize marine vertebrates as their hosts. In this plate, some uniquely marine parasites are introduced. Ectoparasites are external, living on the surface of their host, and endoparasites are internal, typically living in the circulatory system or within an organ system. Ectoparasites usually bear a somewhat close resemblance to their non-parasitic relatives, while endoparasites, protected by the host's body, may be quite modified.

Color the flying fish and its ectoparasitic copepods. The fish is silver gray and the copepods are brown. The commensal barnacles are white with purple stripes.

Copepods are widely distributed in plankton and bottom communities, and are recognized as the most important herbivores in the sea (Plates 14, 35). However, over 1000 copepod species (one-quarter of the total number of species) are parasites. Shown here are parasitic *copepods* living on the body of a *flying fish* (Plate 45). The copepod's head and mouth parts (not shown) are modified for piercing flesh and attaching to the fish. It feeds on the fish's body fluids. The long *neck* connects to a bulbous *body* and trailing behind are two long *egg sacs*. Like most parasites, finding the next host is always of paramount importance, and a large portion of the parasite's energy is utilized for reproduction. The parasite in this case doesn't get away unencumbered; *commensal barnacles* attach to the copepod's body, enjoying a free ride while they filter feed.

Now color the pearl fish and its host sea cucumber. The pearl fish is pale silver and blue. The cucumber is mottled with orange, brown, and yellow.

Many different invertebrates parasitize fish. It is rare when the tables are turned. *Pearlfishes* live inside the bodies of *sea cucumbers* and bivalves. The name "pearl" comes from the presence of these pale, translucent fish inside the shells of pearl oysters. Pearl fish species that utilize sea cucumber hosts live in the cucumber's cloaca which they enter through the anus. The sea cucumber is vulnerable to such an intrusion because in order to breathe it inhales water through its anus into a pair of respiratory trees that branch off its cloaca (Plate 41). Pearl fish emerge at night to feed; later they track down their sea cucumber hosts by following a chemical scent emanating from the cucumber's anus. The fish approaches the anus head first then quickly turns and backs in. The pearl fish is highly modified for its tight quarters. It is very thin and has lost its scales and pelvic fins; its anus has moved well forward, an apparent adaptation to avoid defecating inside its host. One large Caribbean cucumber (35 cm, 14 in) often harbors pairs of pearl fish, and as many as ten of these slender, 10–15 cm (4–6 in) fish have been found in a single host. On first inspection, the pearl fish might appear to be a commensal freeloader, however it feeds on the cucumber's gonads and respiratory tissues, making it a true parasite.

Color the sacculinid barnacle life cycle as you read the text. Use a light color for the crab host, shades of one color for the female stages, and shades of a contrasting color for the male stages. Begin with the female cyrid larva. Color the interna the female color, and color the externa a combination of the male and female colors.

One of the most insidious marine parasites is the sacculinid barnacle which parasitizes crabs. In the nauplius and cyprid larval stages, the sacculinid looks like any other free-swimming barnacle larva (Plate 81). The *female cyprid* finds the body of a *crab* and changes into a larval form called a *kentrogon*. The kentrogon pierces a thin part of the crab's exoskeleton and injects a portion of its cells into the crab. The parasite cell mass settles in the crab's gonads and over a several-month period, grows into an invasive rooted process call the *interna*, that spreads throughout the host's body. The parasite takes over the crab's metabolism, preventing it from reproducing or molting. When the parasite becomes reproductively mature, it grows a large brood chamber, the *externa*, on the underside of the crab's abdomen. *Male cyprid* larvae are attracted to the externa by a water-borne pheromone released by the female. A few males enter the structure and deposit cells that eventually produce sperm. The female's eggs are fertilized in the brood chamber and released as *nauplius larvae* which later molt to the cyprid stage. Female cyprid larvae search for a crab host to begin the cycle over again, while male cyprids seek out a female's pheromonal signal. When the female parasite completes its reproduction, the host crab can molt and shed the externa. However, the crab has been rendered sterile by the parasite.

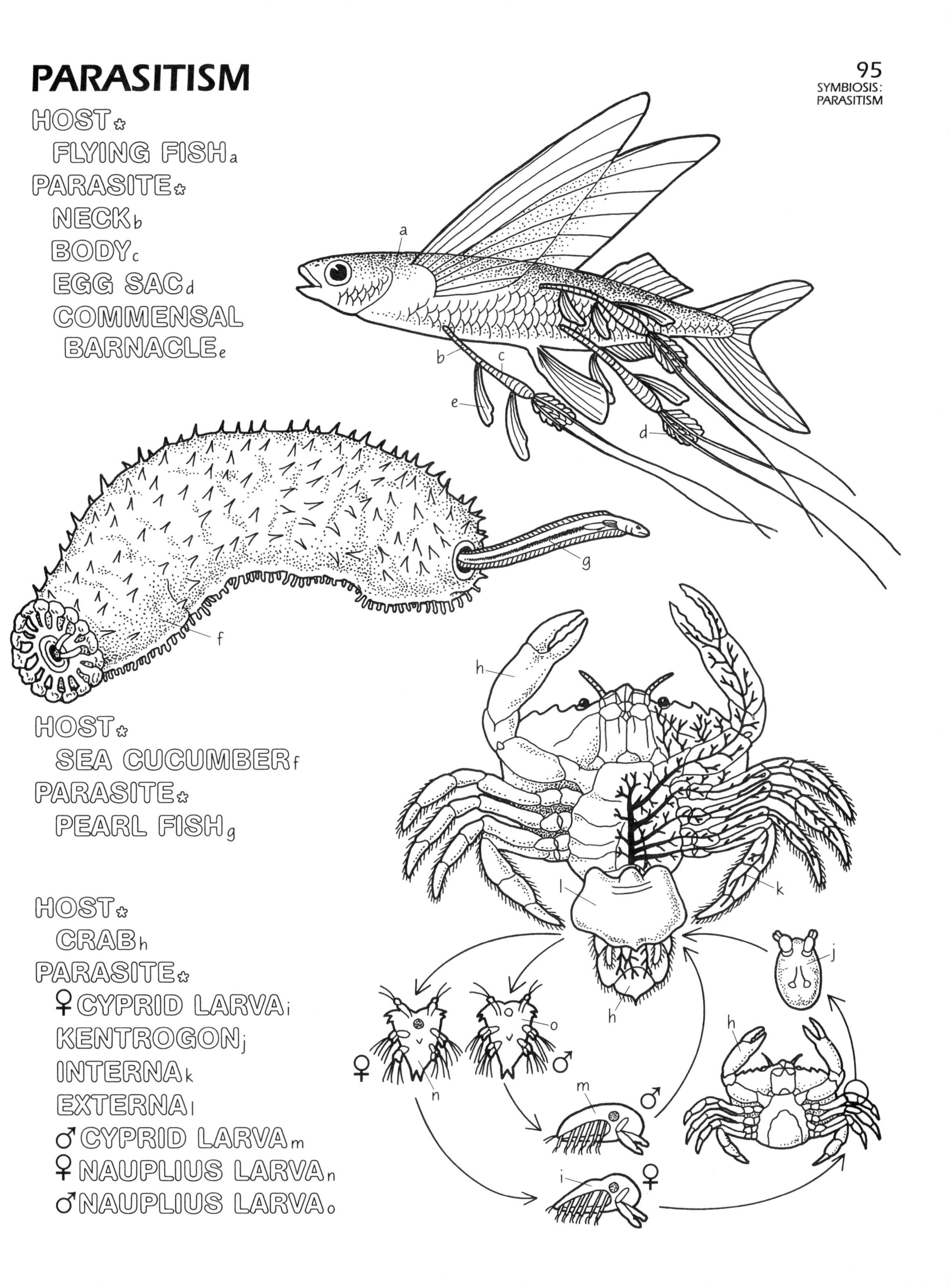
PARASITISM
95
SYMBIOSIS: PARASITISM
HOST
FLYING FISH a
PARASITE
NECK b
BODY c
EGG SAC d
COMMENSAL BARNACLE e
HOST
SEA CUCUMBER f
PARASITE
PEARL FISH g
HOST
CRAB h
PARASITE
♀ CYPRID LARVA i
KENTROGON j
INTERNA k
EXTERNA l
♂ CYPRID LARVA m
♀ NAUPLIUS LARVA n
♂ NAUPLIUS LARVA o

96 INTRASPECIFIC AGGRESSION IN SEA ANEMONES: CLONE WARS

Color the uppermost sea anemone. The anemone is undergoing asexual reproduction. Next, color the two sea anemone clones in the rectangle, and note that clone 1 is derived from the continued asexual reproduction of the anemone at the top of the plate, and receives the same body color. The anemone-free area between clones should be colored gray.

The small gray-green aggregating sea anemone, commonly found in the middle rocky intertidal zone of the Pacific coast of North America, reproduces asexually by binary fission. The *pedal disc* moves in two directions at once, stretching the sea anemone and ultimately pulling it in half. Each half becomes a new anemone, which can grow and split again. If the sea anemone is in a favorable spot with abundant food and room to spread, it will continue to reproduce asexually, and soon a group of identical sea anemones is established. Since all the anemones originated from the same individual, each anemone possesses exactly the same genetic material; the group as a whole is known as a genetic *clone*.

Different clones in the same locale can be recognized by distinct color patterns on the *tentacles* and *oral discs*, which are specific to each clone. When two different clones are found in a given area, the interface is often marked by a patch of bare rock, called an *anemone-free area*. Sometimes barnacles or other animals inhabit these spaces, but both anemone clones avoid it. Laboratory experiments conducted by Dr. Lisbeth Francis, while at the University of California at Santa Barbara, provide our best understanding of the anemone-free area and the clone wars.

Color the four stages of aggressive encounters. The two anemones are members of different clones. The acrorhagi and the ectodermal tips on clone 1 are each colored a different color than the body. Note that in the illustration of withdrawal, the ectodermal tips end up on the body of the victim clone.

When the tentacles of two members of the same clone come into contact, both anemones initially contract. Then they gradually extend their tentacles so that they interlace, and the two anemones have no further interaction.

If two anemones from different clones have tentacle contact, they also initially retract. This expansion and retraction of the tentacles may occur several times. Eventually, one of the anemones begins to inflate special structures called *acrorhagi*, which are located just beneath the oral disc. Water is forced into the hollow acrorhagi, and they become distended and cone shaped. The anemone may also rise up by contracting the circular muscles in its column or body and become elongated. The towering aggressor then sweeps down with extended acrorhagi and attacks the alien anemone. The acrorhagi have large, penetrating nematocysts in their *ectodermal* tips that discharge into the victim.

This attack may be repeated by the aggressor anemone, or the victim may retaliate with similar behavior. More often, however, the victim pulls its body away from the aggressor in an attempt to avoid further contact. If there is no room to move back, the victim may release its hold on the substratum and be washed away by the water. If there is absolutely no avenue of escape, and the victim is left in contact with the nonclone-mates, it may be killed in a matter of days.

This interaction between clones explains the existence of anemone-free areas between adjacent clones in the natural habitat. Those anemones on the leading edges of interacting clones pull back to avoid the mutually aggressive encounters, and an open area remains.

Subsequent study of aggregating anemone clones have revealed their intraspecific aggressive behavior to be quite complex. There is considerable difference in aggressiveness among clones. Some attack nonclone mates immediately, some wait and counterattack, still others are nonaggressive. There is also a suggestion that the level of aggressive behavior demonstrated by a clone differs depending on which clone it encounters. The discovery of intraspecific aggression in the aggregating anemone has spurred research on other cnidarian species that form clones or colonies. To date, a number of sea anemone species have been found to show intraspecific aggression between clones. Very strong intra- and interspecific aggressive behaviors have also been discovered between coral reef corals.

CLONE WARS

SEA ANEMONE*
TENTACLES a
ORAL DISC b
BODY c
PEDAL DISC d

ASEXUAL REPRODUCTION*

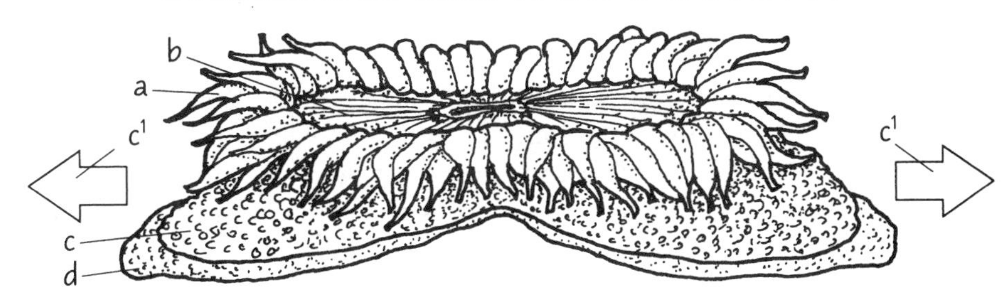

CLONE 1 c1
ACRORHAGI e
ECTODERM f
CLONE 2 g

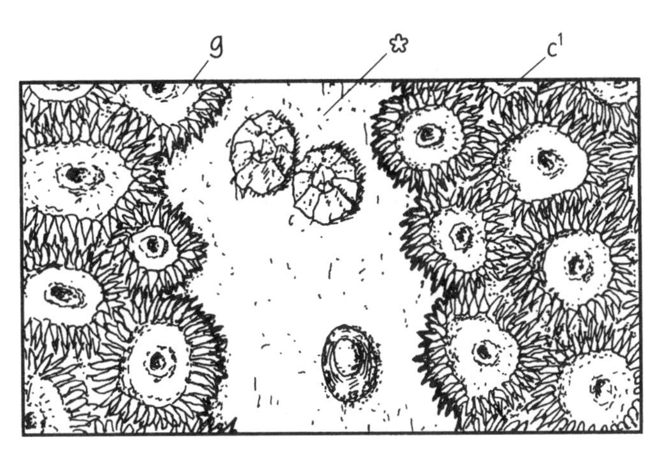

ANEMONE-FREE AREA*

CONTACT*

INFLATION*

ATTACK*

WITHDRAWAL*

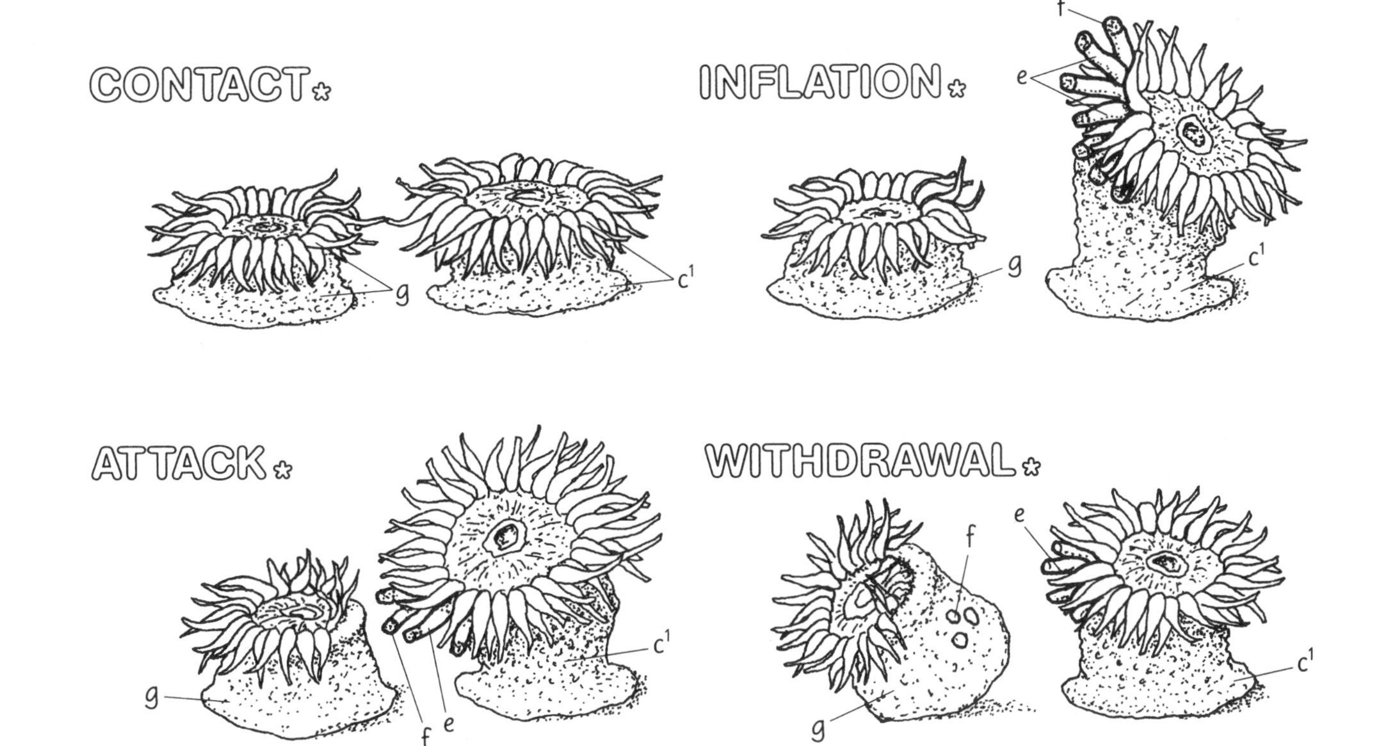

97 COMPETITION: BARNACLE INTERACTION

The acorn barnacle is a familiar marine organism. Although sessile as an adult, it is first a swimmer and then a crawler in the larval stage. Thus, the choice of attachment site by a settling cyprid larva is literally a life or death decision for the individual barnacle—it could settle in an area where no food is available; and for the species—reproduction requires close proximity of members of the same species (Plate 81). Barnacle species, therefore, have evolved complex mechanisms to ensure successful and neighborly settlement. In this plate, how such mechanisms operate in a single species and how settlement affects interactions between barnacles competing for space will be explored.

Begin by coloring the illustration of settling behavior at the top of the plate.

The settling mechanism of the common Atlantic intertidal barnacle *Balanus balanoides* was studied by Dr. Dennis Crisp and others. They have demonstrated that the *cyprid larva* of this barnacle can recognize a proteinaceous substance that is found on the *adult barnacle's* body as well as the *basal plate*, which is left behind if the adult barnacle is removed. When the crawling cyprid larva contacts this protein substance with its antennae, it begins crawling in a circular pattern. As long as it intercepts adult barnacles of its own species, it continues to crawl. Once it encounters a clear area, it tightens up its circular pattern to make sure the area is large enough for adequate growth. If the site is acceptable, the cyprid attaches (Plate 81). This settling behavior ensures that the selected area is suitable for feeding and reproduction (indicated by the presence of other adults).

Color the two rectangular drawings; note the size of the young barnacles relative to the thumb in the drawing. The pencil-form adults and previously occupied basal plates are illustrated on the right.

In *Balanus glandula*, a U.S. Pacific coast intertidal species, this gregarious settling behavior may result in overcrowding. A collection of young barnacles (*spat*) may settle out very close to one another, suggesting a capacity for recognition quite similar to *Balanus balanoides*. However, the spacing between individuals in areas of particularly dense settlement is insufficient for lateral growth. The barnacles have to grow upward to reach full size. This results in a *pencil-form adult* instead of the typical volcano-shaped barnacle. Because the attachment surface is small relative to overall size, the pencil form is more easily dislodged by predators or floating debris, leaving its vacant basal plate behind.

Color the two examples of interspecific settlement. Note that the arrows receive the *Balanus balanoides* adult color in the bottom example.

Settling behavior by *Balanus* barnacles does not include recognition of species other than its own. If there are no adults of its own species present, the cyprid selects a coarse or rough-textured substratum for settlement. In the example shown here, spat of *Balanus glandula* have chosen the rough shell of the much larger thatched barnacle, *Tetraclita squamosa*.

Another, perhaps more dramatic example of interspecific indifference was discovered by Dr. Joseph Connell. The Atlantic species *Balanus balanoides* and *Chthamalus stellatus* both grow successfully in the high rocky intertidal zone in Scotland. Although they both settle in groups of mixed species, the more rapidly growing and larger *Balanus* spaces itself as described above but does not recognize the presence of the smaller *Chthamalus*. This results in *Chthamalus* being: 1) overgrown and smothered; 2) uprooted by the growing edge of a *Balanus*; or 3) literally squeezed to death by the rapid growth of several adjacent *Balanus*. In this manner, *Balanus* effectively eliminates *Chthmalus* from the substratum on which they both reside. *Chthamalus*, as a species, is able to persevere in the face of this competition because it has a greater tolerance to exposure and can live higher in the intertidal zone than *Balanus*.

BARNACLE INTERACTION

BALANUS BALANOIDES SETTLING BEHAVIOR*
CYPRID LARVA$_a$
MOVEMENT$_{a^1}$
ATTACHMENT SITE$_{a^2}$
ADULT$_b$
BASAL PLATE$_{b^1}$

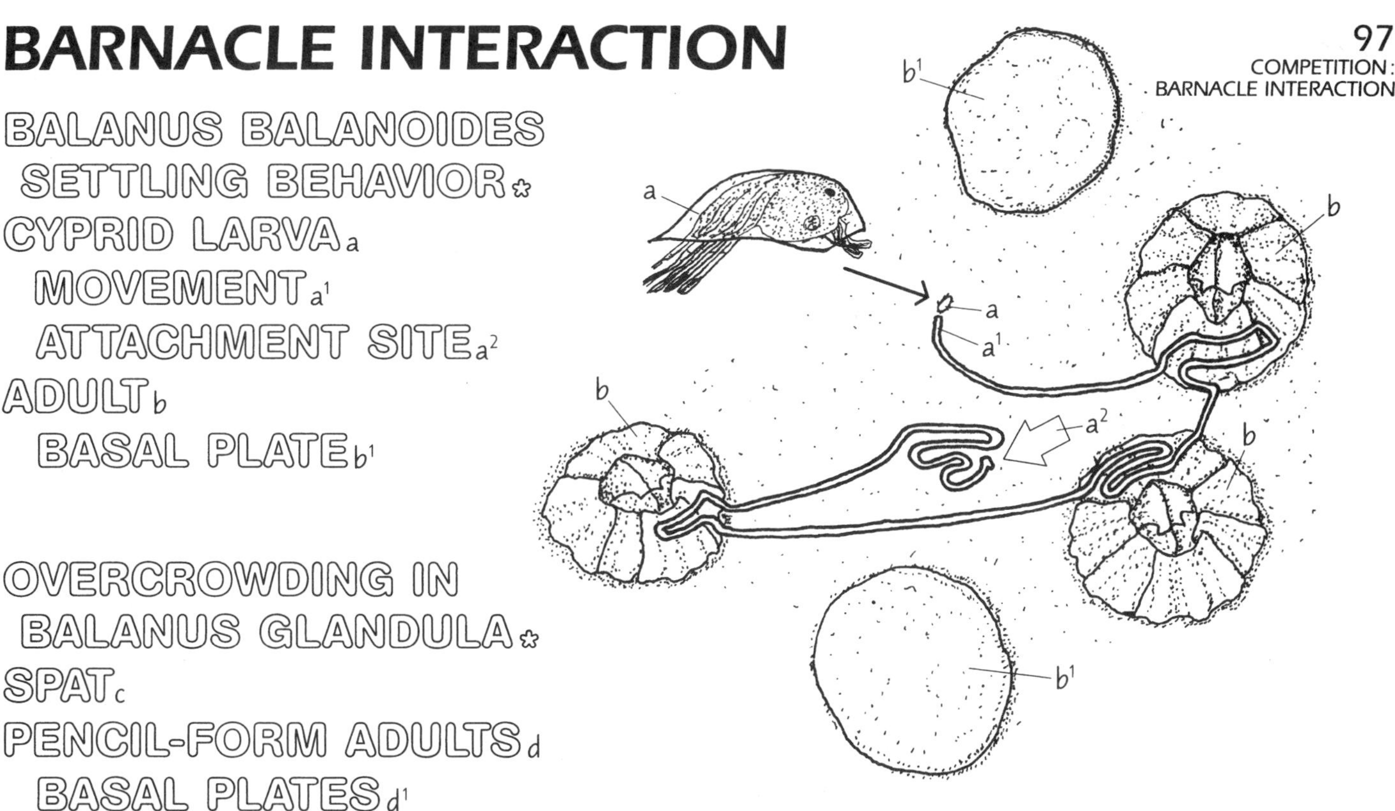

OVERCROWDING IN BALANUS GLANDULA*
SPAT$_c$
PENCIL-FORM ADULTS$_d$
BASAL PLATES$_{d^1}$

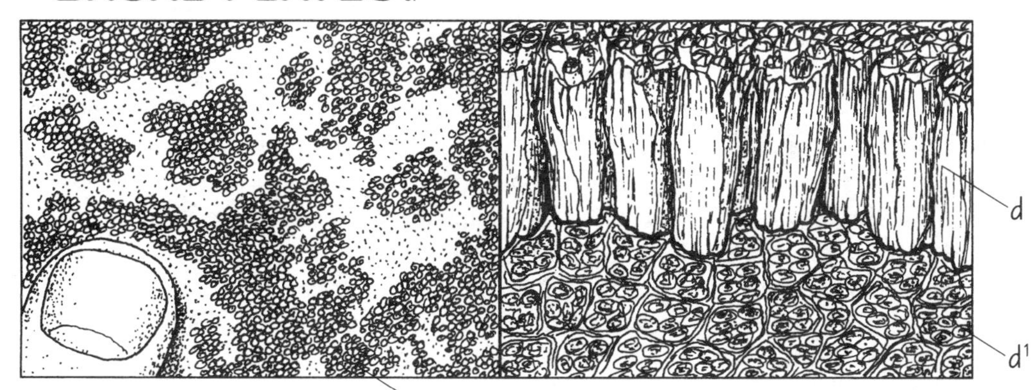

INTERSPECIFIC SETTLEMENT*
TETRACLITA$_e$
BALANUS GLANDULA$_{c^1}$

BALANUS BALANOIDES*
CHTHAMALUS$_f$

SMOTHERING*

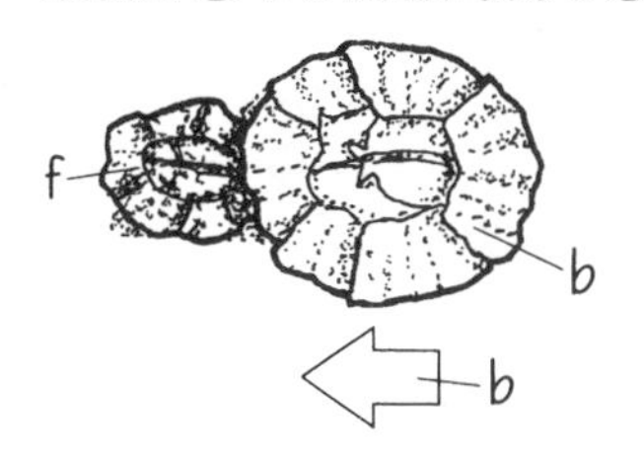

UPROOTING*

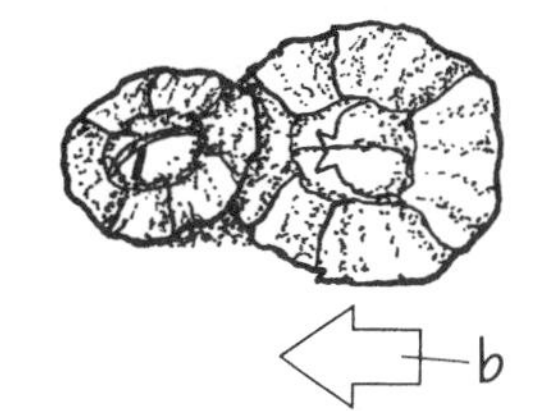

SQUEEZING*

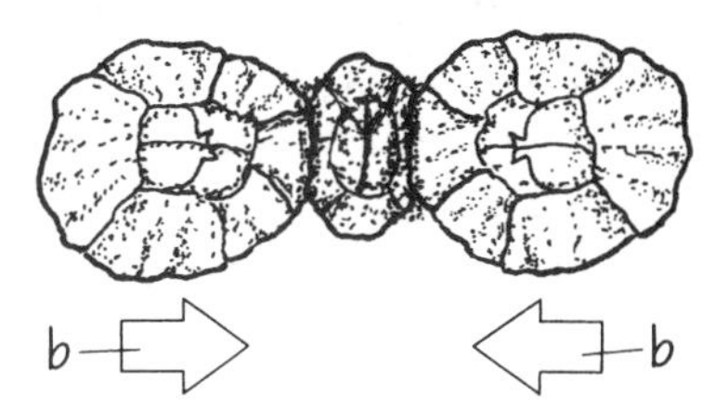

98
COMPETITION: THE FARMER LIMPET

In marine environments where a single resource is highly contested, complex relationships will often develop among competing organisms. One of the most fascinating examples of such a relationship involves the interaction of the large Pacific coast owl limpet with its intertidal competitors for a primary resource: space. The limpet was studied by Dr. John Stimson at Santa Barbara, California. It can grow up to 8 cm (3 in) long and is most commonly found in the midlittoral zone of the rocky intertidal area, where strong wave action occurs. This subhabitat is occupied by mussels and stalked barnacles, which usually cover all open space and crowd other organisms. In such a situation, it would at first seem surprising to find bare patches in the middle of a dense mussel and barnacle clump (Plate 112). Closer inspection usually reveals the presence of a single, large limpet within each clean patch. Stimson determined that the resident limpets "farm" these patches and defend them from encroaching competitors.

Begin by coloring the ventral view of the owl limpet in the upper drawing. This is its actual size. The shell is dark gray or light brown. Now color the limpet on its algal turf, which should be colored a light green or yellow-green. In the lower right magnified view, color the whole circle with this light color, then use a darker color to fill in the radular scrapings. Color the home scar as well.

Limpet patches average about 900 cm^2 (1 ft^2) in area, and their yellow-green color sharply contrasts with the bare rock areas adjacent to the patch. The yellow-green color is due to a low-growing *algal turf* (about 1 mm in height) composed mainly of filamentous blue-green algae on which the limpet's distinctive *radular scrapings* can be seen (Plate 106). These radular scrapings are rectangular and are about one by three millimeters in size. As these large limpets graze on their patches, they leave behind a coarsely cropped algal turf. The radulas of smaller limpets (less than 20 mm, 0.8 in) rasp off the algal turf much closer to the substratum and leave behind what appears to be bare rock. With each high tide the grazing limpet leaves its roosting place, called a *home scar*, and moves in rotation to a different portion of its patch. By this time the algae have sufficiently regrown in the earlier grazed portion, so that the limpet can begin the grazing cycle all over again in much the same way that a farmer rotates his cattle through a series of separate pastures. It was found during four years of observation that many limpets stayed on the same patch throughout that period, which is a relatively long time in such a wave-tossed habitat.

Now color the mussels, stalked barnacles, and the predatory snail both on the algal turf and in the peripheral drawings. If you wish to use natural colors, the mussels would be blue, the barnacles light gray or tan, and the predatory snail brown or orange. Color the owl limpet interactions.

If the patch-dwelling owl limpet is removed, and smaller limpets allowed to graze within the patch, they will graze the algal turf down to bare rock in two weeks, leaving nothing for the large limpet to farm. Also, the *mussels* and *stalked barnacles* that often form the borders of limpet patches are constantly encroaching on the patch. In response to these competitors, the limpet shows a dazzling repertoire of behaviors. When confronted with another limpet (not shown) on its patch, the resident owl limpet backs up and rams its shell into the intruder, repeatedly, until it either loses its grip on the substratum and falls off, or retreats from the patch. To discourage encroaching mussels and barnacles, the limpet uses its large *shell* in bulldozer fashion to push the invaders back, probably dislodging the mussel's *byssal threads* and undercutting the barnacle's *basal attachment*. The farmer limpet saves its most spectacular behavior for predatory gastropods that may venture onto its patch. When a *snail,* which is an occasional predator of small limpets, is encountered, the limpet (putting aside its battering method) raises its shell high off the substratum, in a sort of wind-up, and brings the leading edge quickly down on the front of the predator's foot. The predator responds by releasing its grip on the substratum in an effort to retreat into its shell, and the limpet then eases up, allowing the predator to be washed away by the surf.

More recent work with the owl limpet by Dr. William Wright while at Moss Landing Marine Laboratories and Scripps Institute of Oceanography reveals that almost all of the large, territory-holding limpets are females. The larvae settle amongst the barnacles and mussels and are initially male. The young male owl limpets grow larger and move from the mussel clump to seek territories, changing to females when they reach a certain size.

THE FARMER LIMPET

OWL LIMPET a
SHELL a1
FOOT b
HEAD c
MOUTH d
RADULA e
MANTLE f

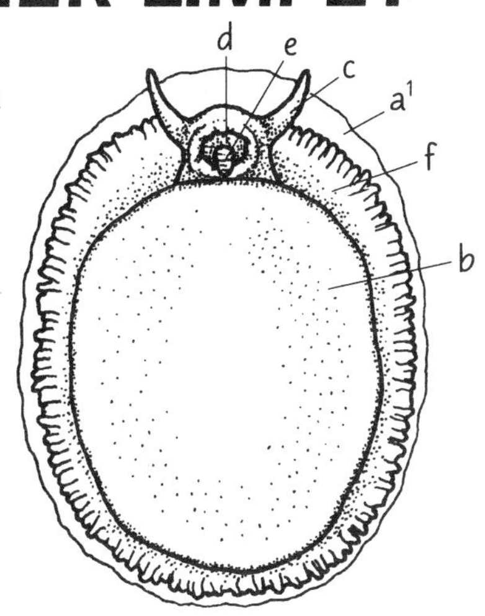

ALGAL TURF g
RADULAR SCRAPING e1
HOME SCAR h
MUSSEL i
BYSSAL THREAD j
STALKED BARNACLE k
BASAL ATTACHMENT l
PREDATORY SNAIL m
BODY n

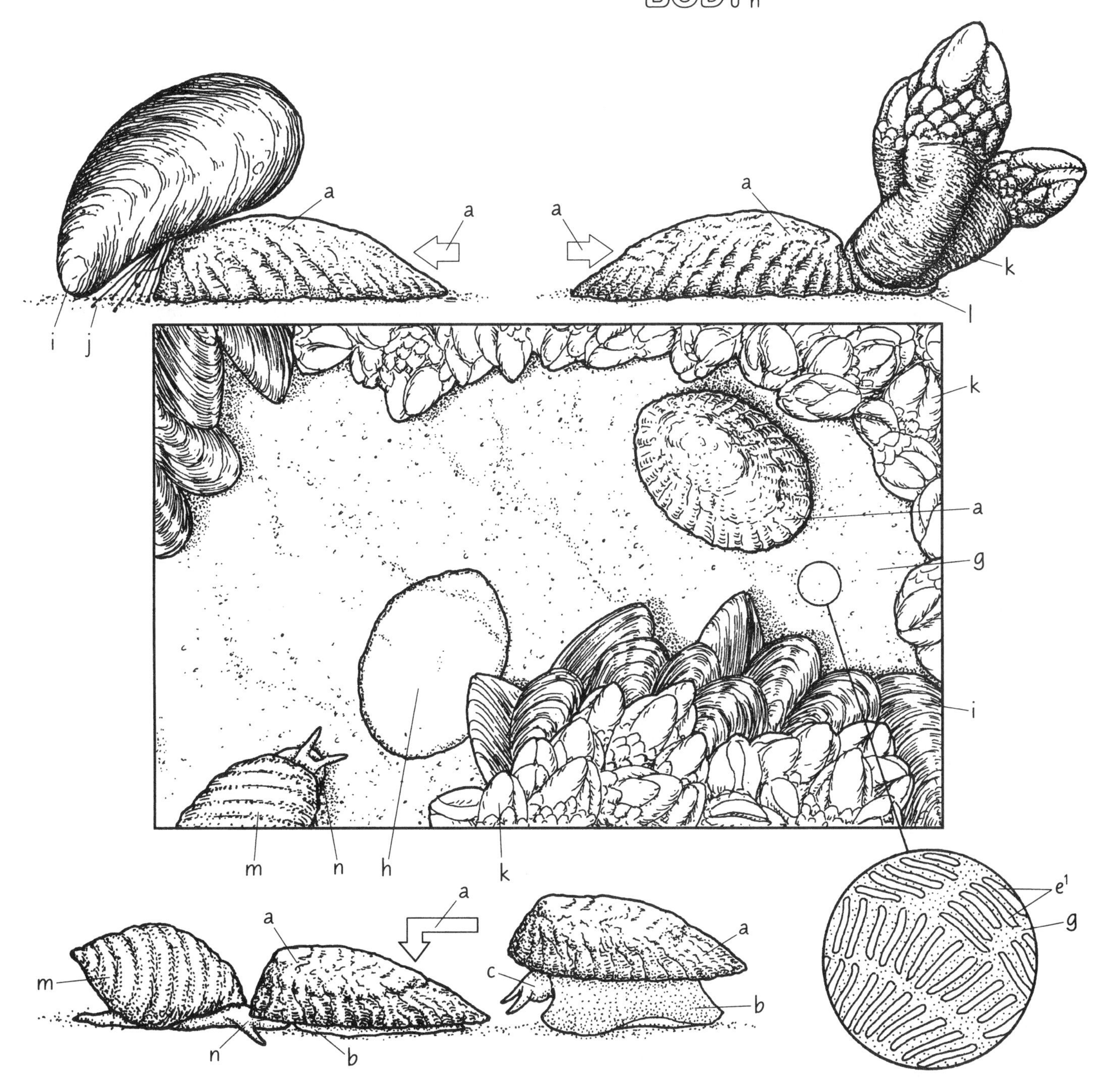

99
SEA PALM STRATEGY

In the rocky intertidal zone, there is constant competition for space, a competition where plants are set against sessile animals. Both the California sea mussel and the sea palm, a brown alga, occupy Pacific Northwest mid-intertidal areas of heavy surf. The mussel is a conspicuous and dominant organism that forms extensive beds which can remain for many years. Mussels are both secured to the rocky substratum and linked to each other by their tough byssal threads (Plates 30, 112). How an annual plant like the sea palm is able to persist in this environment against such competition involves an interesting interplay of biological and physical factors.

Color the sea palm at the upper left of the plate. Next, color the scene at the bottom of the plate, and note that the holdfast of the beached plant contains the remains of barnacles and a piece of an algal colonizer. Finally, color the rectangular diagrams; note that here only one color is used for the entire sea palm.

Sea palms appear in February or March and grow rapidly from April to June. They persist through the summer, but die back in autumn; they are gone by the end of November. The individual sea palm has a large *holdfast* that anchors it in the wave-swept habitat. The *stipe* tapers upward from the holdfast and is crowned with many *blades* that give the alga its palmlike appearance. And like the true palm whose flexible trunk sways in the tropical winds, the sea palm's stipe bends and returns upright after the onslaught of each crashing wave.

The sea palm rarely occurs singly; it is usually present as a tight clump of various-sized individuals. This clumping occurs partly because the plant releases its *spores* during low tides in the spring and summer, causing many spores to land nearby and grow up on the holdfast of the adult plant. The timing of spore release has important consequences.

The expansive *mussel* beds are vulnerable to predation by the *Pacific sea star* (Plate 112) and to the impact of large, wave-borne objects like logs. Logs bash into the mussels, crushing and dislodging individuals and exposing the bed to further erosion by wave action. Large areas of substratum are cleared in this way, and many species of *algae* and *barnacles* quickly *colonize* the newly opened space. If there is a sea palm clump nearby to serve as a source of spores or, if a spore-bearing portion of the sea palm drifts on to this space, young sea palms will grow.

Sea palms may be partly responsible for the clearing of the substratum. They are able to establish themselves and grow on top of some other organisms, including algae and barnacles. The sea palm's holdfast provides the usual secure anchorage on the underlying organism, and, as the holdfast grows, smothers it. While this is occurring, the sea palm's stipe is likewise growing and absorbing more and more impact of the breaking waves. The plant is well able to bear this stress, but the organism to which its holdfast is fixed is now dead, decaying, or increasingly unstable; eventually the whole mass is torn off, leaving the rock bare.

It commonly happens that the sea palm—the conspicuous sporophyte plant—is torn away late in the season by early fall storms. The rocky substratum may appear to remain bare through the winter, but it is not unlikely that the tiny sea palm gametophytes have colonized the area and will provide for the large sporophyte plant to establish itself there the following spring (Plate 73).

In this manner, the clump of sea palms can replace itself annually and persist in an area for several years until the slow, steady growth of the mussel bed overtakes it. The presence of the sea palm in an area is dependent on a nearby source of spores and on fairly frequent interruptions of the mussels' monopoly on space, either by physical or biological agents (wave-borne debris or sea star predation).

SEA PALM STRATEGY

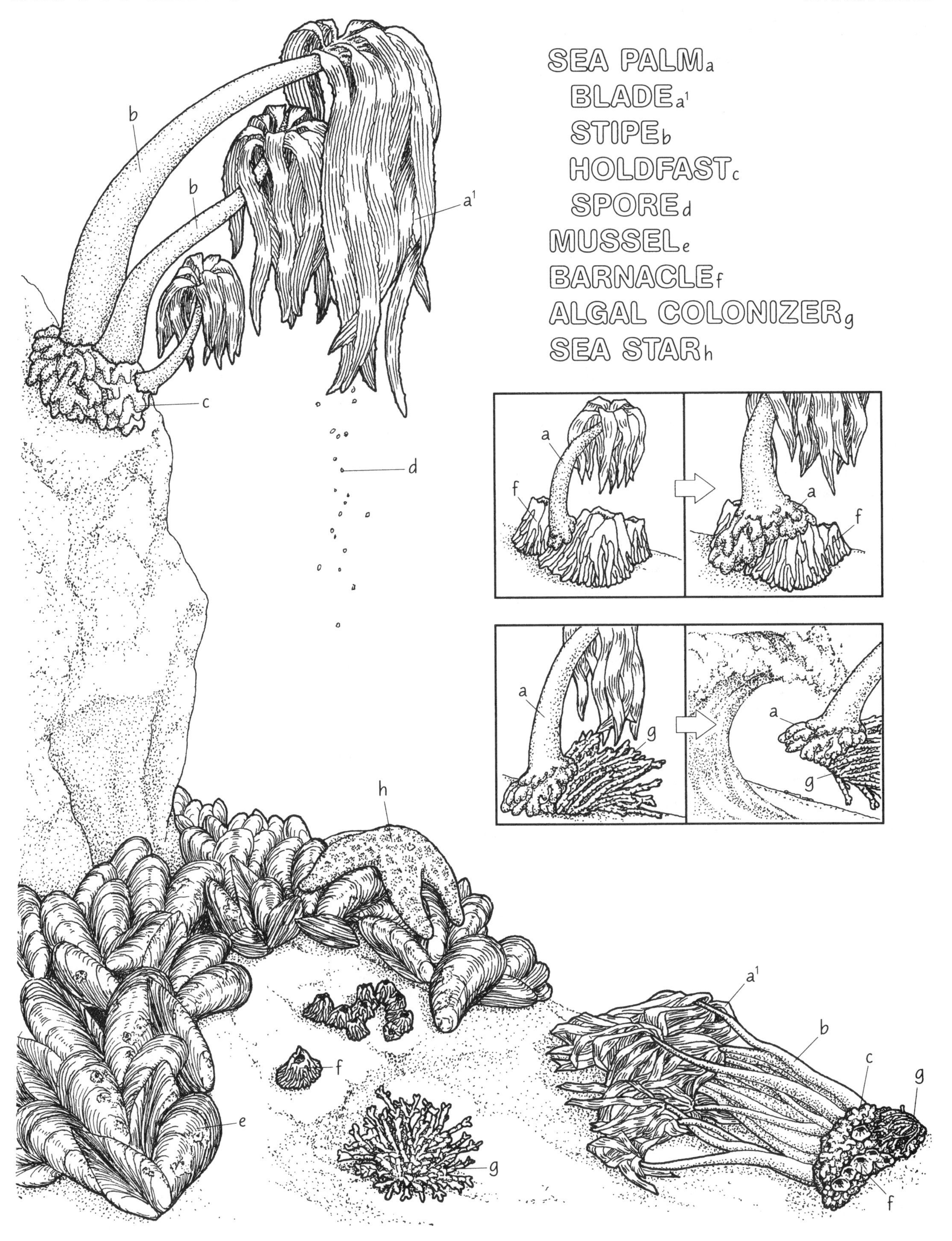

100
DEFENSE MECHANISMS: THE NEMATOCYST POACHER

Most aeolid nudibranchs feed primarily on cnidarians, especially hydroids and sea anemones. Their ability to eat these organisms without succumbing to the cnidarian's nematocysts is quite remarkable. In addition, they not only eat cnidarians and their nematocyst-studded tentacles, but these nematocysts are utilized by the nudibranchs for their own protection. This plate illustrates how the nudibranch accomplishes this task.

Color only the tentacles of the sea anemone at the top left. Color both parts of the arrow (b) suggesting the swallowing of nematocysts from the tentacles to the gut of the nudibranch. Color the cerata of the nudibranch (in life they are often a pale pink). Then color the diagram of the nudibranch gut on the left. After coloring the gut, color the enlargement of the cnidosac, in the lower left corner, and the coiled and discharged nematocysts to its right. You may wish to use gray for the fish predator, shown being warded off by the discharge of nematocysts from the nudibranch's cerata.

Aeolid nudibranchs are relatively common in the lower intertidal and subtidal zones in cold temperate waters along both the Atlantic and Pacific coasts of North America. On the Pacific coast, some species feed on sea anemones, including the aggregating anemone (Plate 96). The nudibranch approaches the anemone and, after initial contact, pulls back, erecting the *cerata* forward over the head. It then moves in to feed. By pulling its body in under the cerata, it resembles another anemone. The nudibranch prefers to feed on the anemone's *tentacles,* and bites off pieces with stout bladelike jaws. In many instances, the anemone is not completely consumed and will regenerate the lost tissue. When the nudibranch ingests the nematocyst-bearing portions of the anemone, the *nematocysts* are swallowed intact and somehow remain undischarged. How the nudibranch prevents nematocyst discharge is a mystery; it may secrete a special mucous substance that immobilizes the discharge mechanisms. Once swallowed, the nematocysts are moved along special ciliary tracks from the *gut* (stomach) into *gut diverticula* (sacs), located in each of the nudibranch's cerata. Aeolid nudibranchs use these gut diverticula to digest their food. However, the nematocysts are not digested, instead they are moved to the tips of the cerata. Here the undischarged nematocysts are maintained in special *cnidosac cells*, located within larger pouches called *cnidosacs* that open to the outside.

When the cerata are roughly touched or pulled off the nudibranch, as would occur when a predatory fish moved in for a meal, special circular muscles around the cnidosac contract and expel the nematocysts. These promptly *discharge*, often into the mouth of the predator, embedding themselves in its tongue and nearby soft tissues. One or two of these encounters will cause a fish to eliminate these nudibranchs from its diet!

The process involved in this defense mechanism is not fully understood. When the cnidosacs empty their contents, the nematocysts are replaced within three to twelve days. Most aeolid nudibranchs employ only certain types of nematocysts from among the several kinds present in their cnidarian prey. There is also some evidence suggesting that only immature nematocysts are utilized for routing to the cnidosacs, perhaps explaining how they are maneuvered through the gut without being discharged.

THE NEMATOCYST POACHER

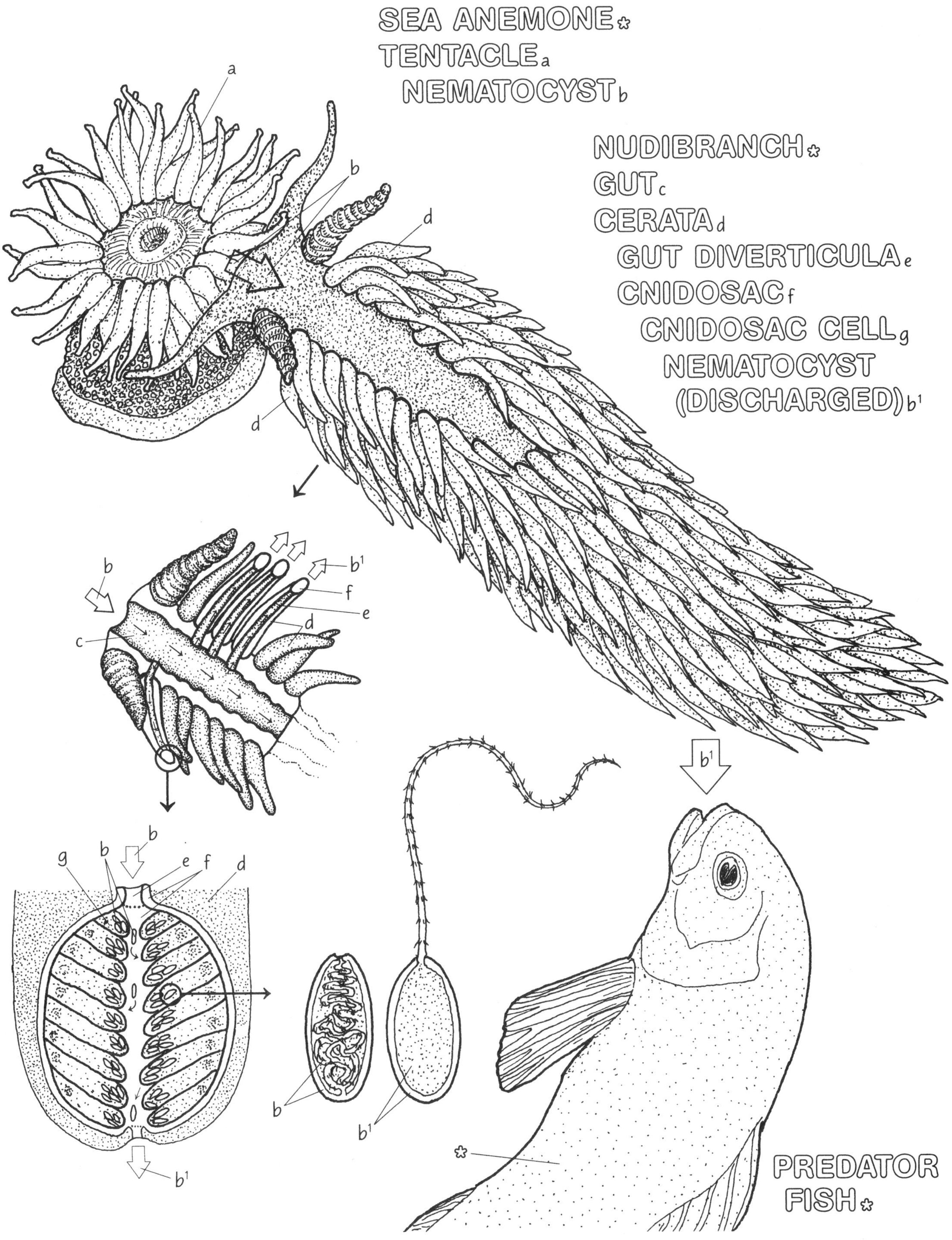
SEA ANEMONE*
TENTACLE a
NEMATOCYST b
NUDIBRANCH*
GUT c
CERATA d
GUT DIVERTICULA e
CNIDOSAC f
CNIDOSAC CELL g
NEMATOCYST (DISCHARGED) b1
a
b
d
d
b
b1
f
e
d
c
b
g
b
e
f
d
b
b1
b
b1
b1
*
PREDATOR FISH*

101
MARINE INVERTEBRATE DEFENSE RESPONSES

The study of marine invertebrate defensive responses has revealed some interesting patterns. Some species will react to local predators but not to previously unencountered ones, though they may be closely related. Within a single group, some invertebrate species will show very similar defensive behaviors, while others lack them entirely. Still more intriguing are the parallel behaviors that have evolved in some species belonging to widely separated groups. In this and the next three plates, we will investigate defense behavior in marine animals.

Color the keyhole limpet and its predator, the ochre sea star. The drawing on the far upper left depicts the normal appearance of the limpet, while the upper middle drawing shows the full response to the sea star. The limpet's shell is light brown, and its body is cream colored. The sea star is purple or yellowish orange.

Sea stars occur on most ocean bottoms and feed on a wide variety of invertebrate prey. The keyhole limpet, when touched by the *tube feet* of the common, intertidal *ochre star*, shows a quick and effective response. The limpet pushes its *shell* upward and throws one fold of its *mantle* up around its shell and another fold down around its *foot*. It also erects its *siphon*, which overlaps the top of the shell. This behavior effectively covers the shell with soft tissue that is difficult for the ochre star to grip. It appears that the ochre star may seek to avoid the limpet's mantle, as contact between the mantle and the sea star's tube feet, when they are on the limpet's shell, causes the star to release its suction hold. The keyhole limpet's mantle may secrete some substance repulsive to the ochre star.

Color the interaction of the sea urchin and the leather star in the middle of the plate. Note that the star has been repulsed by the urchin and still has pedicellariae clinging to its tube feet. The drawing on the far left is a close-up of the urchin's surface. The urchin is purple and the leather star is gray with red blotches.

Looking at the pin-cushion appearance a *sea urchin* puts forward, one wonders how anything could eat it. However, a number of sea stars will readily devour sea urchins if given a chance. One such sea star is the *leather star* (of the west coast of North America) which feeds on the purple sea urchin, especially in shallow water off southern California. If the leather star approaches the sea urchin on a level surface, two responses can occur. First, the urchin may simply move away quite rapidly; urchins have been observed moving three times their normal speed after encountering the tube feet of predatory sea stars. The second response involves the special stalked appendages known as *pedicellariae*, of which there may be two or more types. Pedicellariae are found all over the urchin's body and possess three stout, articulating jaws (Plate 41). When approached by a predatory leather star, the urchin pulls in its *tube feet*, flattens its *spines* against its body, and erects batteries of pedicellariae that grab onto and pinch the skin of the star. In some cases, poison glands may secrete fluid into the wound made by the jaws of the pedicellariae. This counterattack is often sufficient to discourage the leather star, which retreats with its tube feet festooned with the urchin's uprooted, but tenacious, pedicellariae.

Color the bottom illustrations, which demonstrate the defense responses of the sea anemone and feather-duster worm. The drawing at left shows their undisturbed appearance; their defensive postures are shown at right. Feather-duster tentacles are orange or maroon, its tube is tan. The sea anemone has a gray body. The tentacles and oral disc are green.

A more direct, but less colorful mechanism is seen in sea anemones and feather-duster worms. Both of these animals simply hide when danger comes near. A fully erect sea anemone, when disturbed, will retract its *tentacles* into its central cavity; if provoked further, it will expel all the sea water inside and pull down against the substratum (not shown). The feather-duster worms quickly retract their tentacles into their calcium carbonate *tubes* whenever touched. Light-sensitive receptors on their tentacles cause these worms to retract when a shadow passes over or when light is shined directly at them. Feather-duster worms seldom hide inside their tubes for very long, and soon their crown of filter-feeding tentacles is spread open again.

MARINE INVERTEBRATE DEFENSE RESPONSES

KEYHOLE LIMPET*
SHELL$_a$
HEAD$_b$
MANTLE$_c$
SIPHON$_d$
FOOT$_e$

OCHRE STAR$_f$
TUBE FOOT$_g$

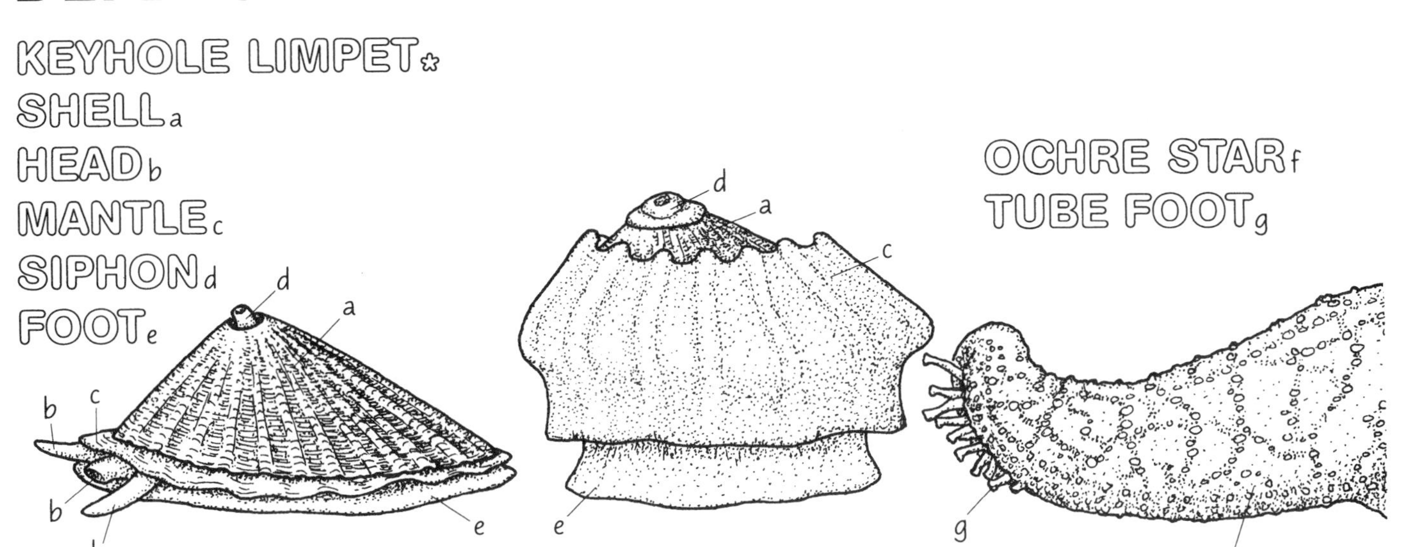

SEA URCHIN$_h$
SPINE$_{h^1}$
TUBE FOOT$_i$
PEDICELLARIA$_j$
POISONOUS PEDICELLARIA$_{j^1}$

LEATHER STAR$_k$

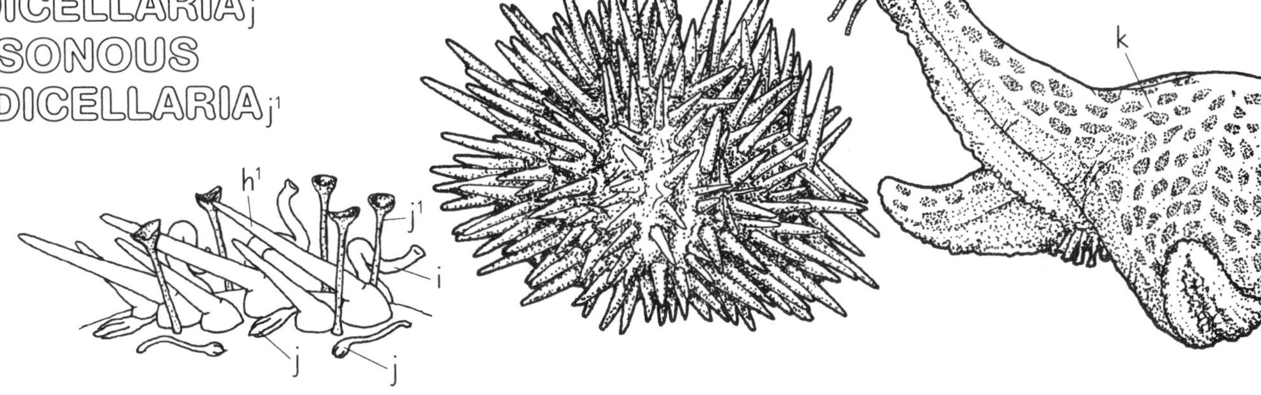

FEATHER-DUSTER WORM*
TUBE$_l$
TENTACLE$_m$

SEA ANEMONE*
BODY$_n$
TENTACLE$_o$
ORAL DISC$_p$

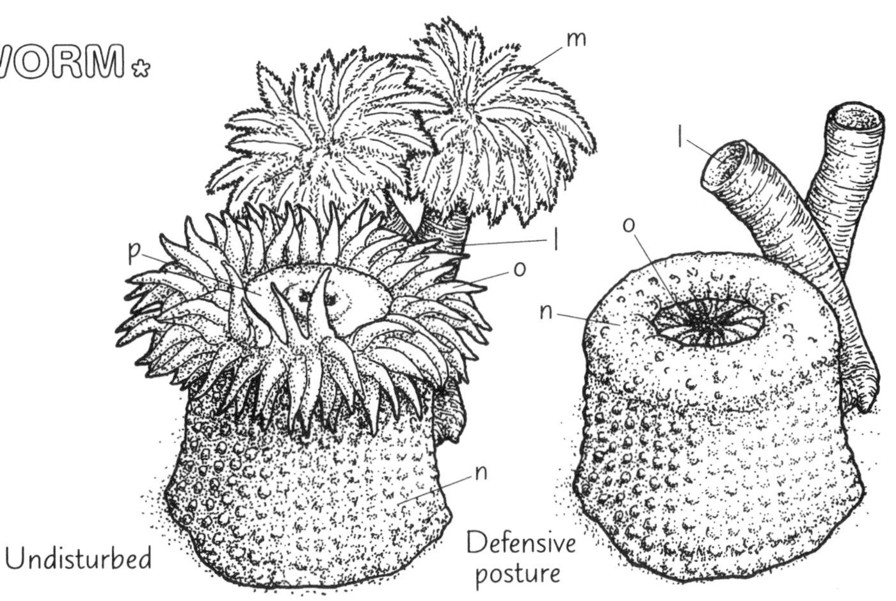

102
DEFENSE MECHANISMS: SPINY TRICKS

The defense mechanisms of fish often involve the use of spines. We have already seen the poisonous spines of the lionfish (Plate 63), stonefish (Plate 64), and stargazer (Plate 46), which provide a deadly and effective deterrent. Spines are utilized in a variety of ingenious mechanisms, as this plate illustrates.

Begin with the normal and inflated (top right) porcupinefish. In the inflated porcupinefish (with erect spines), note that the spines and body both receive the color used for the spines; the fish's body is light grayish brown in life. In the lower illustration of the porcupinefish, the spines lie flat; color the body and spines the color chosen for the body; this should also be a light brown color.

The first, and by far the most spiny fish to be considered, is the porcupinefish. This fish is cosmopolitan in warm waters and can grow quite large (about 1 meter, 3 ft). Under normal circumstances, the porcupinefish goes about its business of hunting among the crevices of coral reefs for molluscs, crustaceans, and echinoderms that it crushes with its stout, beaklike *jaws*. Its protective *spines* are folded nearly flat with their tips facing toward the back. When molested or threatened, the porcupinefish rapidly inflates its *body* by swallowing water and becomes nearly spherical in shape. The once-flattened spines now become fully erect and stick out in a menacing, pin-cushion fashion. Once the danger has passed, the porcupinefish quickly expels the water and resumes its normal activities. The erect spines can cause serious puncture wounds. South Pacific Islanders once used the spiny skins of porcupinefish for helmets.

Color the surgeonfish and the inset magnified view of the hinged spine protruding from its body wall. The fish's body is silver gray and the spine is surrounded by bright orange.

The surgeonfish is another familiar resident of coral reefs. This medium-sized (15–60 cm, 6–24 in) fish nibbles on filamentous algae with its small terminal *mouth*. When bothered by an intruder, this peaceful herbivore erects a pair of lancelike spines located on either side of the base of the tail. The spines are actually pairs of modified scales hinged on the posterior end, so that their sharp inner edge faces forward when erect. When not in use, the spines retract into horizontal grooves. The spines are razor sharp and have inflicted many a nasty wound on unsuspecting fishermen. They may possibly be used to slash other fish, or perhaps only as a threat display. Many surgeonfish species have the spines boldly outlined in a color that contrasts with the surrounding body color. A warning sweep of the tail, flashing these colors and the erect spines often deters a would-be transgressor.

Color the triggerfish and note that the spines under discussion are actually part of the dorsal fin. The inset above the fish shows the spines folded. Be careful to leave the outlined spots and bands on the triggerfish white to maintain the high contrast against its black body. The triggerfish's mouth is outlined in orange and a black and yellow reticulate saddle is found on its back.

When trouble courts the clown triggerfish (25–50 cm, 10–20 in) of the tropical Pacific it quickly finds shelter in a small cave or crevice in the coral reef. Once there, the triggerfish erects the long first spine of its *dorsal fin* and locks it into place from behind with the smaller second dorsal spine, thus wedging itself tightly in place. The triggerfish cannot be removed from its cave without breaking the stout, dorsal spine. To fold the first spine backward, the smaller second spine (the "trigger") must be depressed. The triggerfish will also raise its dorsal fin in any situation of attack or defense. This presents a more formidable posture and makes the fish look bigger.

Now color the shrimpfish's black stripe and the black spines of the sea urchin. The fish's body is transparent.

The last spiny trick considered here is most unusual. The shrimpfish is a small (15 cm, 6 in) inhabitant of coral reefs and sea grass beds of the tropical Pacific. Shrimpfish have a long snout and are very highly compressed laterally. They are encased in a transparent armor consisting of separate plates that taper rearward into a long point which is the first spine of the dorsal fin. Shrimpfish swim by undulations of their fins and always in a vertical position, head down. When feeding on small invertebrates—and especially when it is not moving—this long, thin shape is hard to distinguish from the blades of sea grass. On the coral reef the shrimpfish borrows the spiny protection of long-spined sea urchins. Its small size and elongated form allow the shrimpfish to fit comfortably among the spines—a relationship that deters most predators. This spiny trick is not unique to the shrimpfish, but its shape and the long, black stripe along its body make urchin spines an especially effective hiding place for this species.

SPINY TRICKS

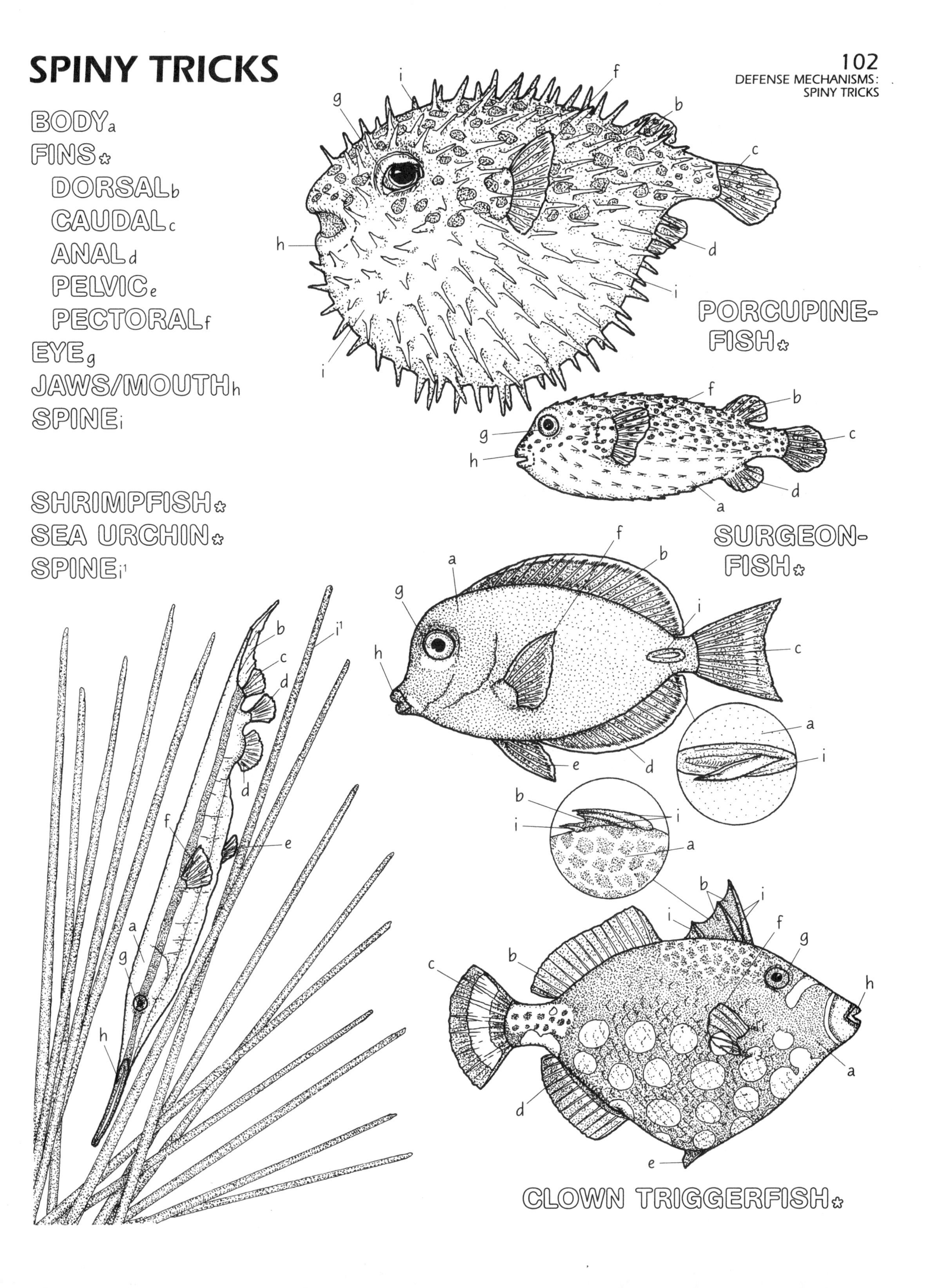

103 DEFENSE MECHANISMS: SAYING NO TO A SEA STAR

Sea stars are unlikely-looking predators. Seen at the beach when the tide is out, they often seem not to move at all. Even when viewed underwater, their movements seem slow and without menace. However, when these movements are sped up by time-lapse photography, the predatory skills of the sea star become quite apparent. The foraging sea star is a relentless, multi-armed predator that slowly but surely overpowers its prey and immobilizes them with its many tube feet. A number of the more mobile marine invertebrates have developed modes of behavior that often allow them to escape sea star predation. In this plate, we will see three ways invertebrates say "no" to a sea star.

Color each of the three escaping invertebrates and the related sea star predator, commencing with the scallop and ending with the anemone. Note that the pink short-spined star is a predator of both the scallop and the cockle, and will trigger an escape response in each animal.

The bivalves known as scallops are not, as one might suspect, sedentary invertebrates. The posterior adductor muscle (the edible part of scallops), which is used in other bivalves to close the shell and keep it closed (Plate 29), is centrally located in the scallop *shell*. It is able to rapidly contract, expelling water out of the scallop. The *mantle* is muscular along the edge and forms openings through which the expelled water is directed out in propulsive *jets*, sending the scallop through the water in the opposite direction. When swimming normally, the scallop moves with the hinge (straight edge of shell) facing backward and appears to be "chewing" its way through the water with rapid clacking of its orange shells. When a predatory sea star, like the *short-spined star*, touches a resting scallop, the startled mollusc leaps from the substratum propelled by jets of water forced out through the front of its shell, as illustrated. The scallop may leap a meter or more, putting it well out of the sea star's grasp.

This escape response isn't foolproof; the scallop may jump straight up and come down right back on top of the sea star.

Another molluscan high-jumper is the Pacific coast cockle. The cockle burrows just below the substratum surface and has a large digging *foot* to reburrow itself if dislodged. If touched on the mantle, foot, or siphons by a predatory sea star, the cockle responds swiftly. The foot extends out from the buff-colored shell and curls back underneath it; then with a powerful thrust, the muscular foot extends its full length away from the shell, giving it a tremendous push against the substratum. The cockle is vaulted several centimeters into the water and may continue for several leaps until it is well out of harm's way. If the cockle is touched by one's finger or other non-sea star material, it will only clamp its valves shut and remain in place. It appears that the cockle's escape response is triggered solely by its sea star predators.

Although sea anemones are thought of as fairly sedentary types, almost all can move about to some extent by a slow creeping movement of the *pedal disc*. Such movement is measured in only millimeters per hour. However, one sea anemone of the Pacific Northwest shows a very rapid escape response when confronted by the *leather star*, a known predator of the sea anemone (Plate 101). When touched by the leather star, the anemone withdraws its cream-colored *tentacles* while at the same time detaching its pedal disc from the rock substratum. In a few seconds, the anemone is completely detached and a cone-shaped projection is visible in the center of the pedal disc. The anemone then begins an awkward but effective swimming movement away from the leather star by the rapid back-and-forth bending of its column-shaped gray *body*. The whole escape takes little more than ten seconds! After moving some distance, the anemone will sit quietly for awhile and, if unmolested, will reattach itself in about 15 minutes.

SAYING NO TO A SEA STAR

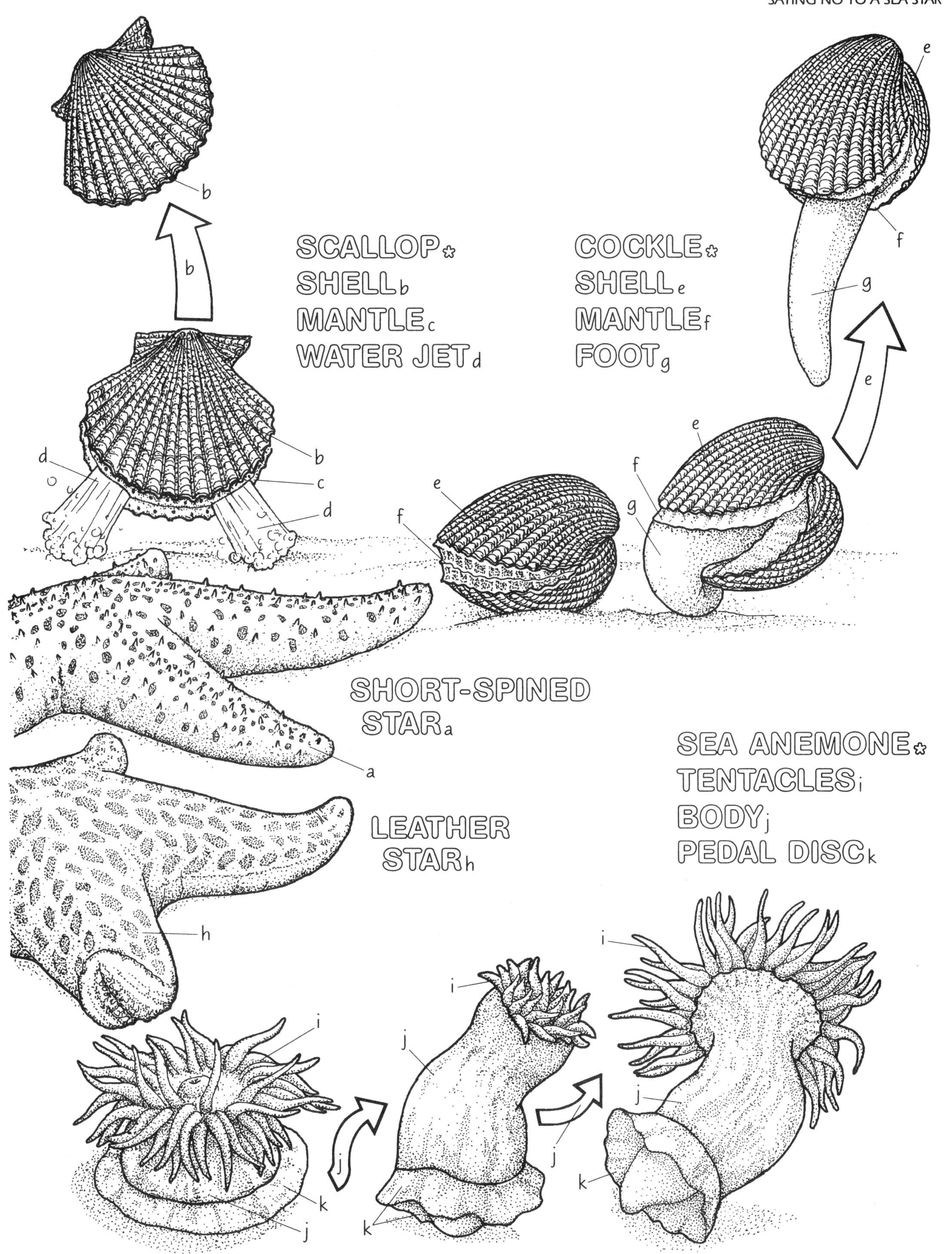

104
ESCAPE RESPONSES: CEPHALOPOD MAGIC

Any consideration of defensive behavior in marine animals must include the cephalopod molluscs with their unparalleled ability to change color almost instantaneously. Such behavior is employed offensively in capturing prey and in mating squabbles, and defensively, as we will see in this plate where three instances of escape behavior, or "cephalopod magic" are introduced.

Color the upper row of illustrations showing defense responses of the squid to the preying dolphin. Save red for the ocelli in the bottom drawing. In the squid, the funnel and ink discharge receive the same dark color (gray or black). Color the ink discharge with a ragged edge to suggest its ghostlike shape. Note that the fleeing squid has blanched (leave it uncolored) and has moved into the background.

The pelagic *squid* does not have a substratum into which it can disappear like the bottom-dwelling octopus. Despite the squid's swiftness, there are pelagic predators, such as *dolphin* and tuna, that can overtake and devour them. The squid, like the octopus and cuttlefish, has an ink sac located near the end of its digestive tract. Ink is emptied from the sac into the rectum, out the anus, and into the mantle cavity. The cephalopod, when pursued, can *discharge* the contents of this ink sac at will and direct a blob of mucus-bound ink into the water through the *funnel*. Simultaneously, the squid will darken its entire body by expansion of chromatophores (Plate 68). Immediately after releasing the ink, the squid pales markedly by contracting the chromotophores. This sequence of events presents the pursuer with a sudden change in the color of its prey followed by the appearance of a dark, somewhat squidlike mass of unknown origin. The resulting confusion often causes the predator to hesitate and gives the squid an opportunity to swim away.

Color the middle illustration of the octopus defense response to a grouper. The octopus in dymantic display is left uncolored except for dark rings around its eyes and a dark margin along its arms.

Octopus species are not only able to change color to match their surroundings, but with the aid of special muscles, can alter their skin texture to more closely match the texture of an irregular background. This cryptic (hiding) ability is extremely effective but is not always utilized.

The common *octopus* is a secretive animal which skulks about the rocky bottom, blending into one background after another. However, if confronted by a moving object larger than itself, it will often go into a *dymantic display* (dymantic=threatening). The animal flattens itself against the bottom and changes from a dark color to uniform paleness over its body with the exception of the area around the eyes and along the margin of the interbrachial web (which joins the upper portion of the octopus' arms). This response presents the potential predator with a pair of greatly accentuated eyes against a pale background of flesh that exaggerates the octopus' size considerably. This image is enough to cause a predator, like the bottom-feeding *grouper*, to pause in confusion, during which time the octopus may give a blast of its ink sac and disappear into a crevice!

Color the bottom illustration of the *Tremoctopus* defense response to a tuna. *Tremoctopus* receives a different color from its two autotomizable arms. The narrow band encircling each ocellus is left uncolored. The ocelli should be colored red.

The small (20 cm, 8 in) *Tremoctopus* is prepared to go to great lengths to avoid capture! This pelagic-dwelling octopus can shed, or autotomize (self-cut), its dorsal pair of arms to avoid capture. These elongated (15 cm, 6 in) *autotomizable arms* are flat and are kept rolled up when *Tremoctopus* is not disturbed. When threatened, the brightly colored arms are unfurled to reveal bright red *ocelli*, or eyespots, outlined in white, lined up along the arm's length. Midway between each pair of ocelli is a zone of weakness or break point (*autotomy plane*). This area contains a special muscular arrangement that will allow *Tremoctopus* to release these flattened, ocellus-bearing segments sequentially into the water. Once detached, the segments expand in size, probably by the relaxation of partially contracted muscles. In the lower drawing, a menacing *tuna* is shown momentarily stunned by the bizarre display of staring, expanding arm segments, during which time the octopus escapes. Later *Tremoctopus* will form new segments at the base of these specialized arms to replace the ones cast off.

CEPHALOPOD MAGIC

DOLPHIN$_a$

SQUID$_b$

FUNNEL$_c$

INK DISCHARGE$_{c^1}$

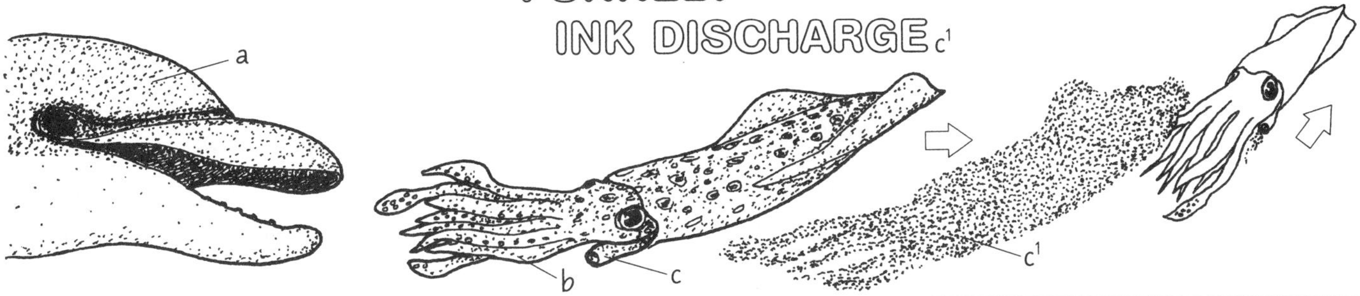

GROUPER$_d$

OCTOPUS$_e$

DYMANTIC DISPLAY$_f$

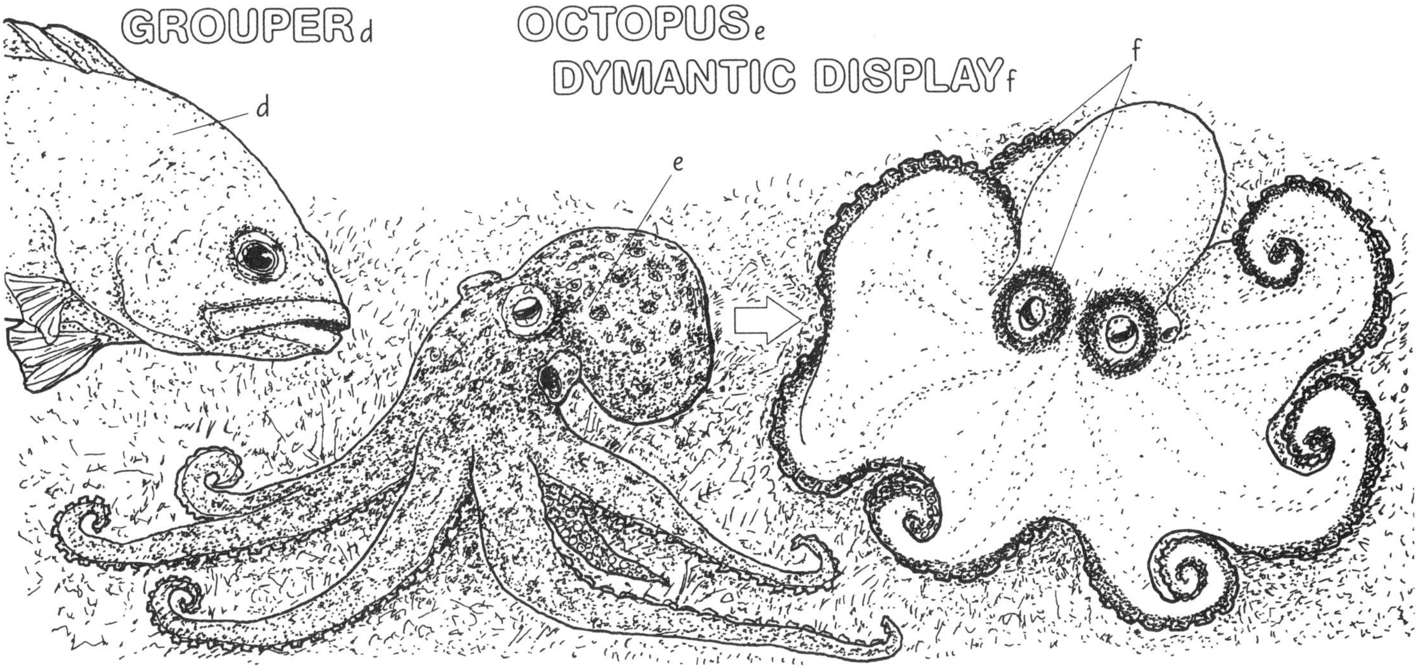

TUNA$_g$

TREMOCTOPUS$_h$

AUTOTOMIZABLE ARMS$_i$

OCELLUS$_j$

AUTOTOMY PLANE$_k$

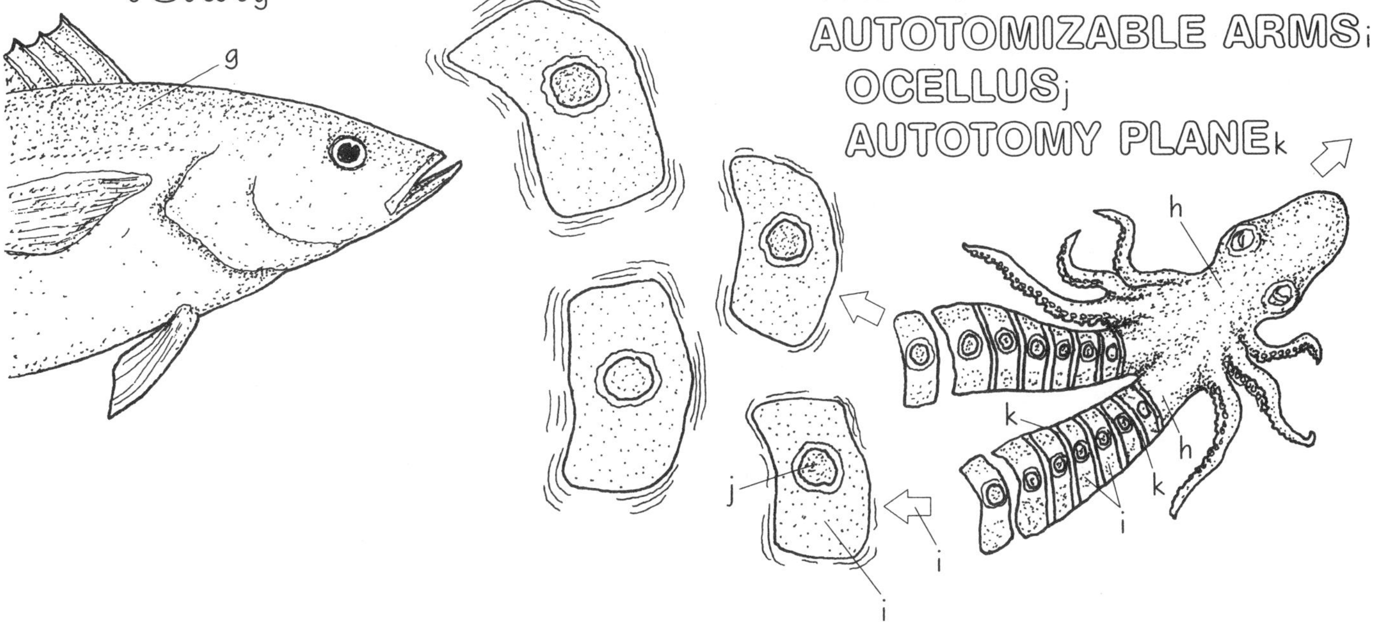

105 FEEDING: MODES OF FILTER FEEDING

The water in many near-shore marine environments is a huge, nutrient-rich "soup." Zooplankton and phytoplankton are suspended throughout the water column, as well as small particles of plant and animal remains (detritus) that have been worn down or broken up by wave action. Many shallow-water marine animals have evolved ways to reap this floating harvest. Collectively, these animals are called filter feeders because they filter food from large volumes of sea water. In this plate, two members of this interesting group are introduced.

In order to remove suspended food material, a filter feeder must either generate a water current over its collecting device or live where the water moves over the filter. Many animals have cilia that beat rapidly to create a feeding current. These animals also have a filter, covered with a sticky mucus, to trap the suspended food. This type of filter feeding is called ciliary-mucus feeding. The ciliary-mucus combination has been worked to perfection by the bivalve molluscs, sea squirts, and many of the filter-feeding polychaete worms (Plate 28).

One polychaete that departs from this pattern is the odd-looking but efficient feeder known as *Chaetopterus variopedatus*.

Color the *Chaetopterus* tube and note that it is cut in a longitudinal section and only its edge is colored. The interior of the tube receives the color for sea water, which also covers the openings of the tubes. Use six light colors for the worm itself. Note that the mucous net is located on the front third of the body and overlaps the body, parapodia, cup, and ciliated groove.

Chaetopterus lives in a tan, U-shaped parchment *tube*. Instead of employing cilia, it generates a feeding current with three paired *fanlike parapodia* located on its back in the middle of its light gray *body*. These three fans beat rhythmically and draw food-bearing *water* through the tube.

The specialized *parapodia* on the twelfth body segment are *winglike* and have many mucous-producing glands, which are employed in the production of a *mucous net*. The "wings" of these parapodia arch upward against the tube and secrete a film of mucus that is carried rearward by the current and forms a net. The end of the net is caught and held by a ciliated *cup* located in front of the fanlike parapodia. This mucous net strains suspended plankton and detritus from the water as it is pumped through the tube. The end of the food-laden net is rolled into a ball by the ciliated cup until the ball is about three millimeters, (0.12 in) in diameter. Then, the winglike parapodia release the front of the net, and the cup rolls it all up and deposits the entire *food ball* in a special *ciliated groove*. The cilia in the groove move the ball forward to the *mouth* where it is swallowed. Once completed, the whole process begins anew with the secretion of another mucous net.

Color the sand dollar. The oral surface, mouth, and food groove are to be colored on both the large sand dollar and on the diagram to the left. Then color the spines and tube feet (these are not shown on the large sand dollar). Finally, color the sand dollars in the sand dollar bed.

The Pacific sand dollar goes one step further than *Chaetopterus*. It uses local water movement, like tidal currents, as its feeding current. In sandy areas with a steady *water current*, the sand dollar burrows part-way into the sediment and then hoists itself up on edge so that its body projects above the surface. Directing its flattened *oral surface* into the current, the sand dollar catches small particles in weak feeding currents created by cilia at the base of the *spines*. These particles are enveloped in mucus and transported to the food grooves that lead to the mouth. Thus, it was once thought this organism was strictly a ciliary-mucus-feeder. However, when Dr. Patricia Timko took a closer look at feeding sand dollars, she found food particles in the gut much too large to have been transported by the weak ciliary currents. By close observation, she determined that the sand dollar could and did capture large food particles like diatom chains (Plate 19), and even more active prey like barnacle larvae. In fact, the Pacific sand dollar appears to feed non-selectively on whatever is suspended in the water floating above it, including phytoplankton, zooplankton, and detritus. When an active, suspended animal like a crustacean larva is carried onto the oral surface by the moving water current, the small (4 mm, 0.15 in) violet spines fold around it to form a cone-shaped trap. The prey is then passed along by the spines and *tube feet* to the nearest food groove. Once in the food groove, the prey is passed along in a coordinated, rowing motion by the stubby tube feet that line its side. The prey is delivered to the mouth where it is thoroughly chewed by the jaws and then swallowed.

MODES OF FILTER FEEDING

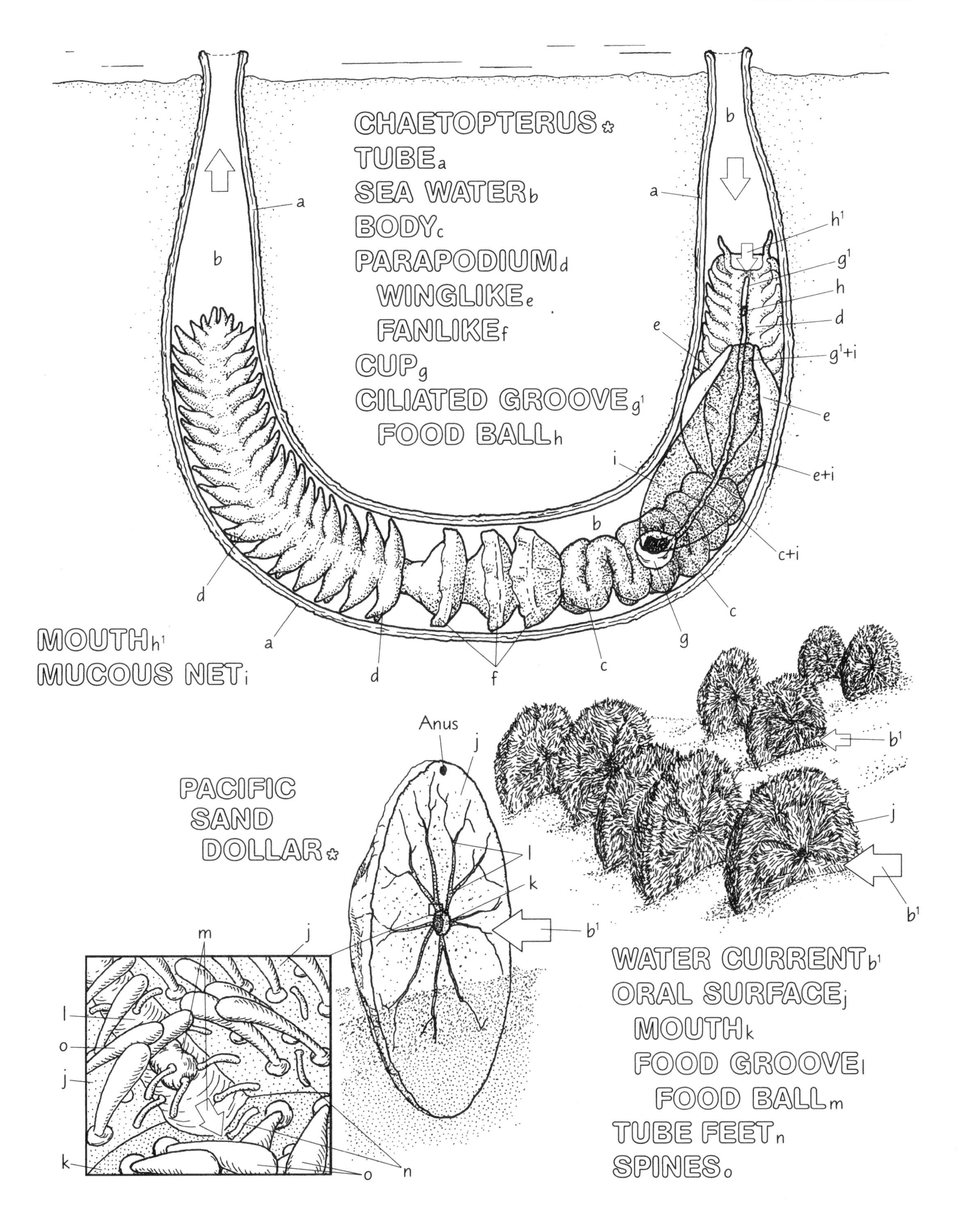

106
FEEDING: MOLLUSCAN RADULA

The radula is one of the most remarkable feeding structures in the animal kingdom. A radula is basically a flexible file used as a feeding implement in all molluscan classes except the primarily filter-feeding bivalves (Plate 30). The radula demonstrates its greatest functional diversity in the gastropod molluscs (Plates 31, 32). Members of this class use the radula to brush over the substratum, to grasp and bite, to grasp and suck, to tear up flesh, to bore through shells, and even to harpoon prey (Plate 108).

Color the generalized area of the radula teeth in the ring-top shell snail (upper left), a gastropod common in California kelp beds. Next, color the parts of the enlarged radula area in the central illustration. Note that separate sets of antagonistic (working in opposition) muscles are given the color of the structure that they operate.

In its most general form, the radula consists of a tough *membranous belt* on which are mounted successive transverse rows of sharp *radula teeth*. The teeth are formed of a complex bio-polymer of carbohydrate and protein called chitin which can be very hard. This toothed belt is part of a complex accessory feeding apparatus located just inside the snail's *mouth*. Because the radula teeth are so sharp, the radula is confined to a *radula sac* when not in use, so as not to injure the snail's mouth. When the snail wishes to feed, the tooth-bearing membranous belt is pulled from the protective radula sac, and the radula is placed against the substratum. The whole feeding apparatus is manipulated by special muscles and cartilage.

As demonstrated in the drawing, the toothed membranous belt rides over the *odontophore cartilage*. A *protractor muscle* attached to the lower end of the belt pulls it down and around the odontophore cartilage; the radula teeth are facing forward and fold down and slide back over the substratum easily. Once the belt is fully protracted (pulled around as far as it will go), then the membranous belt is retracted, that is, pulled back around the odontophore cartilage by the *retractor muscle* attached to its upper end. This is the effective stroke of the radula (as shown by the directional arrows). As the radula teeth come in contact with the substratum, they are pulled erect by friction and their sharp cutting edges rasp the surface, tearing pieces free. The pieces are carried into the mouth and swallowed. When the snail is finished feeding, the retractor muscle attached to the odontophore cartilage is contracted, returning the radula to its radula sac.

The continued antagonistic action of the protractor and retractor muscles attached to the membranous belt produces a to-and-fro movement that allows the radula to be used like a flexible file on the surface of the substratum. Possession of this marvelous tool allows gastropods to be most effective feeders, as anyone who has snails in their garden can readily testify.

With feeding, the radula teeth become eroded. New rows of teeth are formed in the radula sac throughout the animal's life, and the membranous belt grows forward at the rate of several transverse rows of radula teeth per day. The worn, anterior teeth continuously fall off and the anterior membranous belt is resorbed (assimilated) by the snail.

Color only one transverse row of radula teeth in the illustrations of the herbivore and carnivore radula. The herbivorous abalone is chewing on an alga. The oyster drill is a carnivore, rasping its way to an oyster meal.

The size of the radula and the shape and number of radula teeth vary considerably among different gastropods; and this is, in part, related to their feeding habits.

The radula of the *abalone* is an example of the feeding tool of a plant eater. The abalone's radula has many radula teeth per transverse row. The *central* and *lateral teeth* are hooked and sharp, and are used to cut through the abalone's algal (kelp) food. The many small *marginal teeth* collectively act as a broom to sweep the dislodged algal pieces into the snail's mouth.

In contrast, the carnivorous *oyster drill* uses its radula to drill through (rasp) the shells of oysters and other molluscs. There are only three radula teeth per transverse row. The central tooth is very large and has three long, sharp cusps (cutting edges). These are used in drilling the shell, which is first softened by the acid secretion of a special gland located on the snail's foot. The two lateral teeth are hooked and protrude at right angles to the central tooth. These operate as "meat hooks" to grasp and pull the pieces of tissue from the body of the prey into the mouth once the *hole* is gouged through the shell. The oyster drill takes approximately eight hours to drill through a shell two millimeters (0.08 in) thick.

MOLLUSCAN RADULA

MOUTH/GUT$_a$
RADULA SAC$_b$
RADULA TEETH$_c$
CENTRAL$_d$ LATERAL$_e$ MARGINAL$_f$
MEMBRANOUS BELT$_g$
BELT PROTRACTOR$_{g^1}$
BELT RETRACTOR$_{g^2}$
ODONTOPHORE CARTILAGE$_h$
CARTILAGE PROTRACTOR$_{h^1}$
CARTILAGE RETRACTOR$_{h^2}$
GOUGED HOLE$_i$

RING-TOP SHELL SNAIL$_*$

c

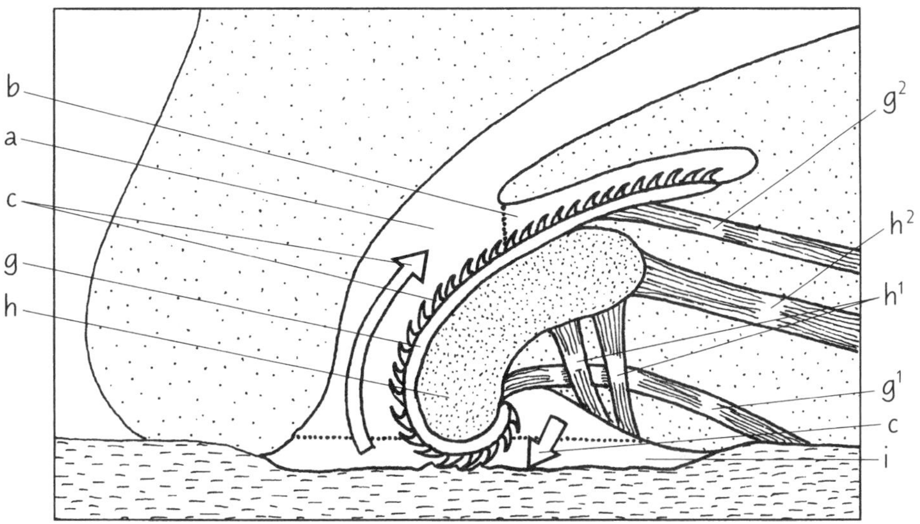

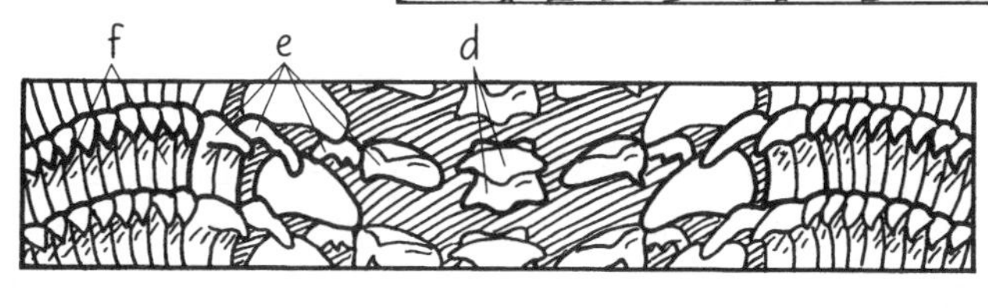

HERBIVORE$_j$

CARNIVORE$_k$

OYSTER DRILL$_k$

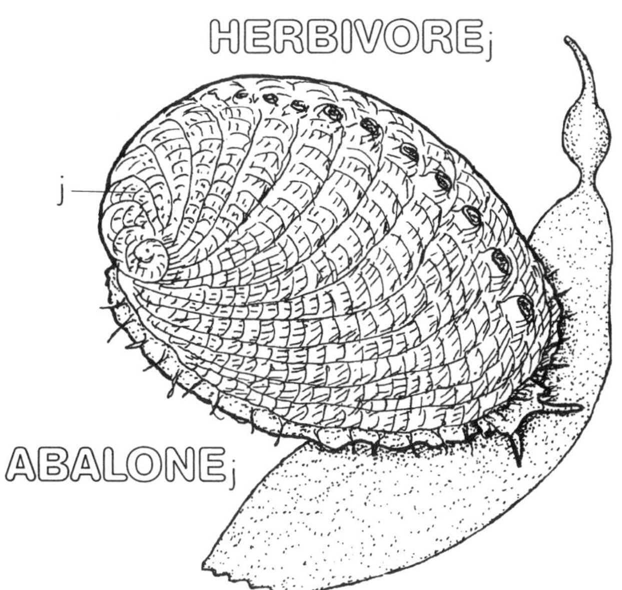

ABALONE$_j$

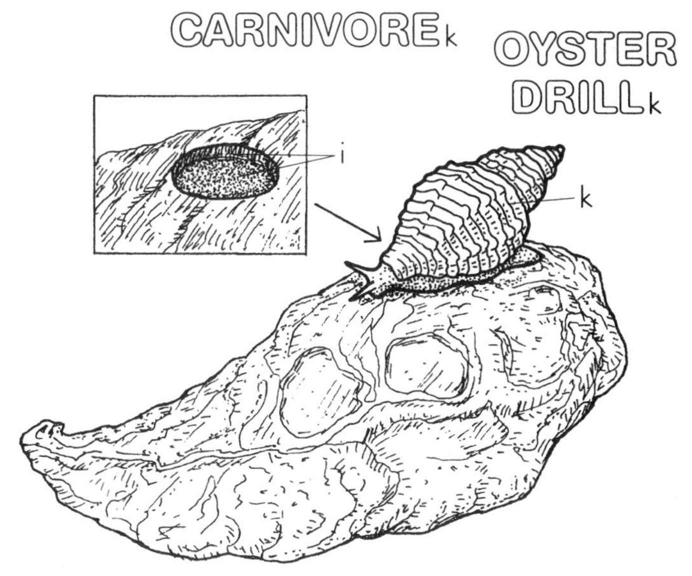

107 FEEDING: INVERTEBRATE HERBIVORES

The large marine plants present a bountiful source of food for marine invertebrate plant eaters, or herbivores. In this plate, four invertebrate herbivores from the west coast of North America are introduced.

Start on the left, and color the sea urchins and the giant kelp plant. Next move to the upper right and color the enlarged piece of drift kelp and the beach hoppers. Note that the drift kelp was detached from the giant kelp plant below. Proceed down the right side of the page as each herbivore is discussed. The chiton's shell plates and the coralline algae receive the same color. If you wish, leave the thinner lines on the shell plates blank, as they are often white in life.

Sea urchins are a common sight in southern California kelp forests. They may be seen on the surface of rocks or in crevices where they wait for drifting algae to pass. They grab the algae with their suckered tube feet and devour it with their five-jawed Aristotle's lantern (the jaw structure). However, if drift algae is swept away by storms or is in insufficient supply, the hungry sea urchins will forage actively for food.

The most vulnerable point on the *giant kelp* plant is the stipe just above the holdfast (Plates 11, 21). Hungry sea urchins will climb up the holdfast and eat through the stipes, sometimes releasing a 20 meter (65 ft) plant from its contact with the bottom. The huge kelp plant will wash either out to sea or up on the beach. Normally the sea urchins are in balance with their plant food, but population explosions, often mediated by human pollution, have occurred. In these instances, sea urchin hordes or "fronts" have moved through and devastated acres of kelp forest habitat.

The *drift algae* that wash up on the beach feed huge populations of *beach hoppers*. These herbivores are large (to 4 cm, 1.5 in) amphipod crustaceans (Plate 9, 35), which burrow in the sand high on the beach by day and emerge to feed during nocturnal low tides. These animals are equipped with strong, biting mouth parts and devour copious amounts of drift algae during a single tide. They appear to prefer the kelp species, perhaps because they are more tender and easier to chew. They also prefer the freshest algae on the beach. Beach hoppers play a significant role in breaking up large seaweeds into much smaller pieces that can then be used by other marine animals.

Another kelp lover is the small (15 mm, 0.6 in) *limpet*, that both dines and makes its home on the *feather-boa kelp*. This kelp grows in the low rocky intertidal and subtidal zones. It has many straplike stipes that may reach a length of six meters (20 ft). The limpet feeds along the mid-rib of the kelp, leaving a depression or *grazed area* where it has eaten. These grazed areas are excavated by the limpet with its specialized radula. It lives in one of these depressions, called a home scar.

One of the most strikingly colored intertidal animals of the west coast is the lined chiton. The *shell plates* of this small (5 cm, 2 in) mollusc are usually light red, marked with zigzag lines of alternating colors in combinations of dark and light red, dark or light blue and red, and frequently white and red. The lined chiton is one of the few animals that feeds on tough *coralline algae* (Plate 20). These red algae sequester calcium carbonate (lime) from sea water and use it to impregnate their cell walls, giving the algae a crusty, plasterlike texture that discourages most herbivores. The lined chiton feeds on the coralline algae and perhaps even utilizes the algae's pigments in its shell. The colors in the chiton's shell plates often match the coralline algae with which it is associated, effectively camouflaging the chiton from visual predators.

INVERTEBRATE HERBIVORES

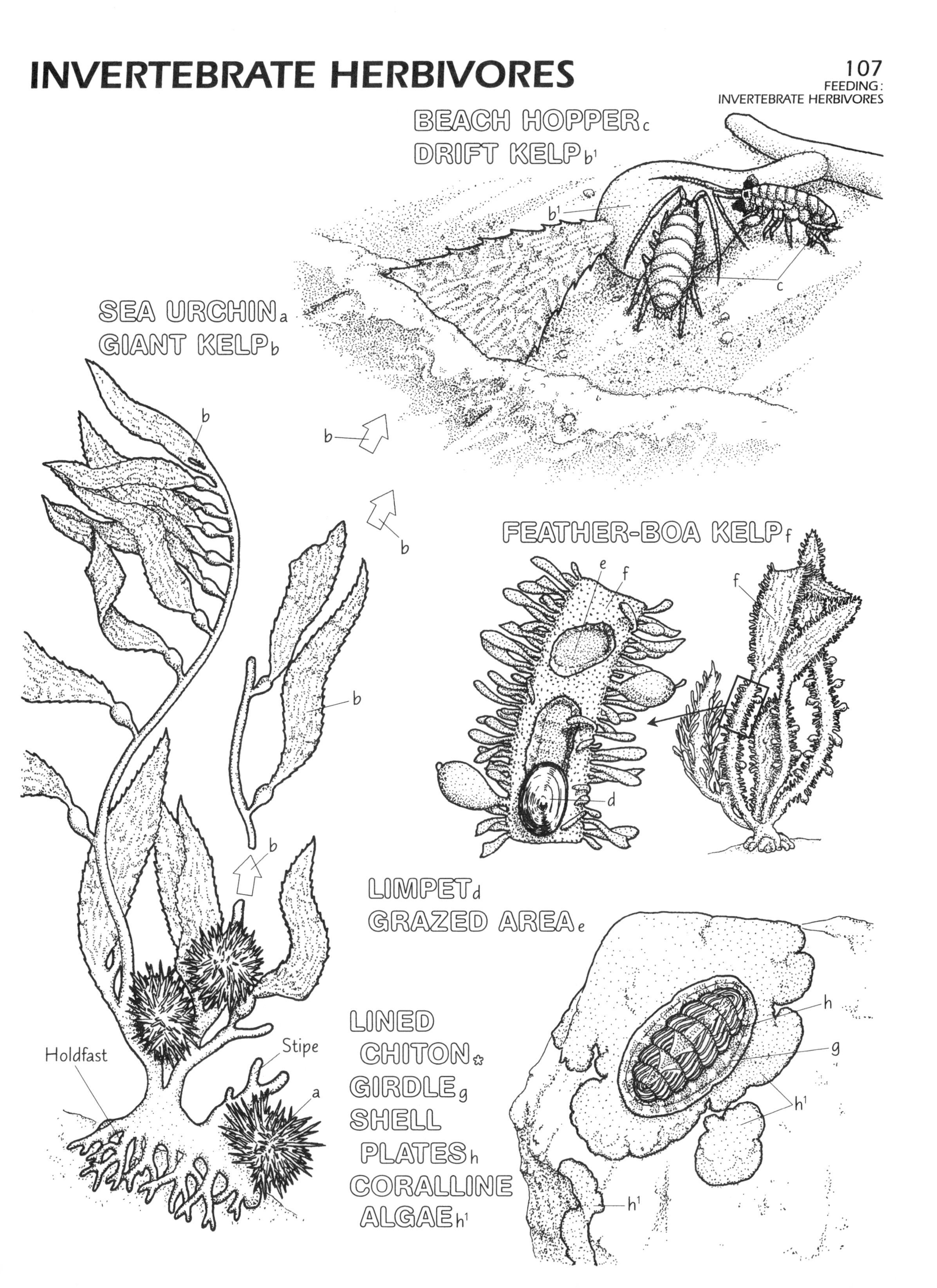

108 FEEDING: INVERTEBRATE PREDATORS

There is a tendency to classify all invertebrates as sluggish, plodding, unexciting animals. This fallacy can be quickly dispelled with many examples, such as the predatory invertebrates. The techniques of these animals compare equally with or exceed many of the predaceous methods of their vertebrate counterparts.

Begin by coloring the cone snail, and then color each predator separately as it is discussed in the text. The victims can be left uncolored or colored gray in the five examples shown.

The cone snails employ "harpoons" in dispatching their prey. These snails (various species, 2–25 cm, 0.8–10 in) are most abundant in tropical and subtropical Pacific and Atlantic shallow-water habitats. They feed on a variety of prey, ranging from polychaete worms to small fish. The "harpoon" is actually a single, specialized *radula tooth* that is equipped with a spearlike barb at the tip. The tooth has a groove along its length through which the snail injects a quick-acting poison from a gland located adjacent to the radula sac. When a prey animal is located, the *proboscis* is moved in close and the radula tooth "harpoon" is fired into the prey by the contraction of a large muscular bulb. The prey animal is quickly subdued by the poison. The radula tooth remains attached in fish-eating species and is "reeled" in by retraction of the proboscis. The fish is swallowed whole by the mouth, which is capable of incredible distension. Those species that specialize in capturing fish have a poison that is potentially fatal to humans. Thus, cone snails and their beautiful *shells* are best appreciated from a distance.

Another aggressive invertebrate is the crustacean known as the *mantis shrimp*. They are so named because their stalked eyes and large, elevated *claws* give them a striking resemblance to the insect known as the praying mantis. There are approximately 300 species of mantis shrimp, ranging in size from 5 to 30 cm (2–12 in). They are found in burrows on soft bottoms or in rock or coral crevices. Most species feed on soft-bodied invertebrates like snails or shrimp, but some catch fish. Mantis shrimp usually lie in wait at the front of their burrows until prey comes near; then they quickly swim out and slash the prey with their large claws. The last, bladelike segment of the claw is fitted either with sharp spines or a knife edge. The "blade" folds backward into a depression in the claw segment behind it. Mantis shrimp have nasty temperaments and have earned the name "split thumb" from fishermen trying to remove them from their nets.

A more subtle, but no less lethal predator is the large (15 cm, 6 in diameter) beaded *sea anemone* shown here with a bat star in the embrace of its *tentacles*. This predator waits for prey to venture near, and then its tentacles discharge hundreds or thousands of nematocysts into the victim. It can quickly subdue even large prey such as sea stars, although they are a rare catch for anemones.

Stretched across the page with a lobster in its grasp, is the *arm* of a large *octopus*. Crabs and lobsters are common prey for the eight-armed octopus. When captured, the victim is pinned down with the *sucker*-bearing arms and deftly bitten by the beak; then poison is secreted into the site of the bite by modified salivary glands. The octopus floods the victim with digestive enzymes, sucks in the partially digested tissues, and discards the shells as an empty husk. For humans seeking to catch octopuses, the bite of this animal is of much greater concern than encirclement by tentacles.

The predatory habits of the *crown-of-thorns sea star* continue to be a concern in the tropical Pacific. This sea star eats live coral and prefers the species which are the principal reef builders (Plates 12, 13). Like many sea stars, the crown-of-thorns can evert its stomach onto its prey, such as the brain coral shown here, and secrete digestive enzymes onto the exposed coral tissue. The prey is digested outside the star's body and absorbed by the everted stomach. The crown-of-thorns is also quite agile and can consume the fast-growing staghorn corals. Normally only a few of these stars occur per hectare of reef. However, population explosions were noted in the 1950s and have continued to occur and re-occur throughout the western Pacific. Hundreds of sea stars are found where one or two would be normal. The result is the wholesale destruction of large portions of reef habitat. Time of reef recovery varies from seven to forty years and can be much longer if new outbreaks occur before complete recovery is accomplished. The cause of these sea star outbreaks is hotly debated and as yet undetermined. Some believe these are natural population cycles, while others point to human degradation of tropical habitat, both marine and terrestrial. Whatever the cause, the degradation of coral reefs worldwide is a very real phenomenon, and their conservation and restoration has become a global priority.

INVERTEBRATE PREDATORS

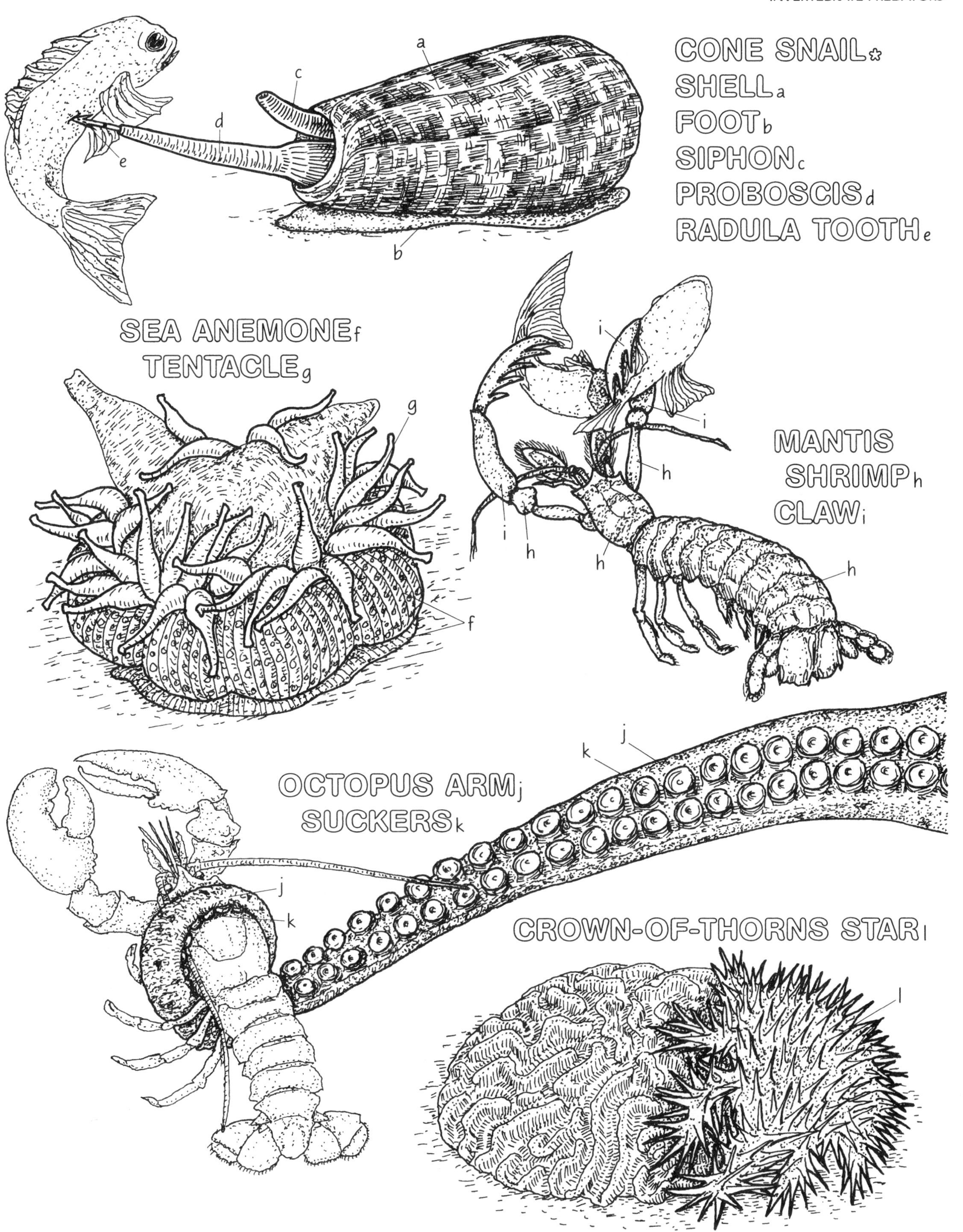

109
FEEDING IN BONY FISHES: ATTACKERS AND AMBUSHERS

Seeing a marine habitat such as a coral reef for the first time, one is invariably entranced by the beauty and variety of fish present (Plate 47). Upon seeing so many different kinds of fish in such numbers, one wonders how the habitat can support such a wide diversity of fish without severe competition. Part of the answer lies in the variety of ways marine fishes feed. In the next three plates, different kinds of feeding strategies of marine fish will be explored, along with the corresponding range of behavioral and morphological adaptations.

Color each predatory fish separately as it is discussed in the text. The prey fish can be left uncolored or colored gray. The shark on the right side of the page, although an attack-type predator in its own right, is left uncolored. Bluefish are blue-green above and paler on the ventral surface. Note that the frogfish has a lure, which is a modified dorsal fin spine, and receives the dorsal fin color.

The most well-known feeding strategy is that of the pursuing carnivore that "runs down" its prey (usually other fish or active invertebrates), attacks, and devours them. Such fish are usually highly visible. One example is the great barracuda which is found worldwide in warm and temperate waters except the eastern Pacific and Mediterranean. This silver-white fish is very long (up to 3 meters, 10 ft), streamlined, muscular, and possesses a large falcate (scythe-shaped) *caudal fin,* which is used for quick acceleration and rapid swimming over short distances. The barracuda is a crafty, curious fish and often approaches divers, but always remains just out of reach. Any quick movement toward it results in a swift departure. The barracuda has long *jaws* with long sharp teeth (many of which curve backward) for seizing and holding prey fishes. When a fish is too large, the barracuda may bite it in two and return to gulp the pieces.

This slashing attack is also characteristic of the bluefish schools of the western Atlantic. So swift and menacing is this medium-sized (75 cm, 30 in) predator that it is accused of tearing into schools of terrified bait fish and killing far more than it can eat.

Other pursuing predators are the jacks, which are capable of amazing acceleration and speed. They have been observed stealing meals out of the mouths of sharks feeding on fish tossed from boats. Pictured here is the greater amberjack, known in all tropical and subtropical seas. The amberjack usually has a brass-colored stripe along the body at the level of the eye; it is olive to brownish above the stripe and silvery white below. This fish can get quite large; there are hook-and-line records of 1.4 meters (4.6 ft) and over 63.5 kilograms (140 lb). The amberjack exemplifies the body form of a fast-swimming pelagic fish with its torpedo-shaped *body* and large lunate caudal fin. Jacks tend to swim in groups and range over large areas. Though not residents of coral reefs, they will swim over reefs and feed on local fishes.

A more leisurely predation strategy is to "sit and wait." The odd-looking trumpetfish (Plate 47) employs this tactic, as does the lizardfish. Appropriately named, this slender, lizard-shaped fish has a big *mouth* full of reptilelike teeth. The fish illustrated here (the rockspear lizardfish) occurs on both sides of the Atlantic and reaches a length of 33 cm (13 in). It is basically silver-white in color with mottled reds and browns, and it blends in with the mud and sand bottoms that it frequents. Lizardfish sit quietly on the ocean floor, sometimes partially buried, and wait for small fish to swim overhead, at which time they dart upward and seize their prey in their well-equipped mouths.

A slight variation on this ambush technique involves the trick of aggressive mimicry (Plate 64) and is seen in a strange group called the frogfish. These fishes, represented here by the splitlure frogfish, have a modified first dorsal fin spine on their snouts called an illicium. The illicium functions as a moveable fishing rod with an attached *lure*. The lure, or esca, is wiggled about enticingly to attract unsuspecting fish. The frogfish remains immobile on its peculiar stumpy *pelvic fins*, looking like a black-streaked, light brown sponge on a coralline-algae-encrusted rock. When a fish comes in to grab the lure, the frogfish sucks the entire prey into its cavernous mouth by rapidly expanding its mouth cavity. Studies of related frogfish species have shown that they can expand their mouth cavity to 12 times its normal resting volume within 6 to 10 milliseconds. This is one of the fastest of all capture mechanisms in the animal kingdom.

ATTACKERS AND AMBUSHERS

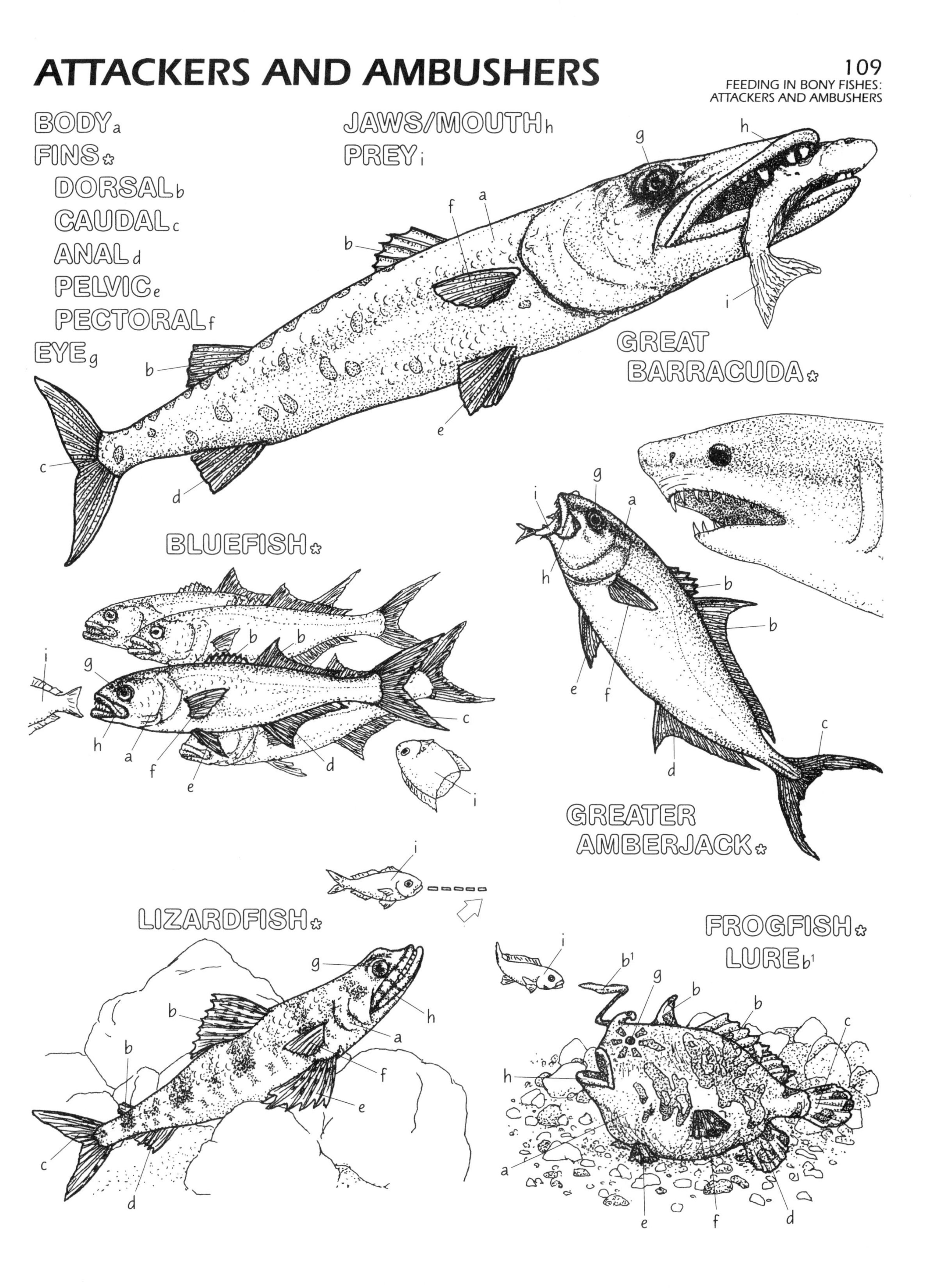

110
FEEDING IN BONY FISHES: PICKERS, PROBERS, AND SUCKERS

In this and the next plate, fishes that feed on a variety of prey besides other fish will be considered. Many fishes ("generalists") feed somewhat opportunistically, on whatever is available, and many others will eat a small variety of prey species. Placing fishes into feeding categories is sometimes arbitrary, and there may be considerable overlap between categories.

Begin by coloring the California sheephead and its prey. Color the drawing of the pipefish and the inset to the left showing its mouth and prey. Follow the same procedure for the filefish and the butterflyfish. Finally, color all views of the triggerfish and its sea urchin prey.

The California sheephead is a familiar sight in the kelp forests of California (Plate 11). This large (up to 1 meter, 3 ft) fish feeds on invertebrates and is considered a generalist. The male California sheephead is dark gray in color, with a pink-orange midsection and white lower jaw. The female is solid pink-orange. Sheepheads have stout *jaws* and large protruding *teeth* for crushing a variety of *prey*. They are known to feed on sea urchins, sand dollars, mussels, scallops, abalone, lobsters, crabs, hermit crabs, octopus, tube-dwelling polychaetes, or any small- to moderate-sized marine invertebrate they find in the kelp habitat.

Prey selection is much more limited for the pipefish. Pipefishes and the closely related seahorses are delicate suctorial feeders. They use their tubelike snouts and small *mouths* to ingest food by a rapid intake of water, as if sucking on a straw. They have prehensile grasping tails that they use to anchor themselves on algae and other substrata. Their *eyes* can move independently, like those of a chameleon, allowing them to scan the water for their small prey from their anchored perches. Pipefishes are poor swimmers; their *caudal fins* are reduced and the *dorsal* and *pectoral fins* provide the main swimming thrust. Pelvic fins are absent. There are approximately 150 species of pipefishes, found mostly in the shallow waters of tropical and subtropical seas; a few are freshwater inhabitants. They are generally small (15 cm, 6 in), with the largest reaching 50 cm (20 in) and are brown or green above and cream-colored ventrally.

The reef filefish of the tropical Pacific feeds on small reef invertebrates by probing into crevices with its elongated snout and using its small, sharp, incisorlike teeth to pick out prey. This small (10–30 cm, 4–12 in), brightly colored (green with orange spots) fish is often seen among the branches of various species of coral and is known to feed primarily on coral polyps. Filefish are closely related to triggerfish; they have a similarly large first dorsal spine, but lack the triggerfish's locking mechanism (Plate 102). Their skin is rough to the touch and was once utilized as a kind of sandpaper.

The bright yellow forceps butterflyfish of Pacific coral reefs also picks and probes for food. Sporting the familiar butterflyfish eyespot (Plate 63), this small fish (10–15 cm, 4–6 in) has an elongated gray-brown snout and pointed jaws with sharp teeth that are used like a pair of forceps to extract small morsels from tight quarters. Its principal food items are the tentacles of tube-dwelling polychaetes, coral polyps, and tube feet and pedicellariae plucked from between the spines of sea urchins. Like many butterflyfishes, this species raises its stout dorsal fin spines as a defense mechanism, presenting an unappetizing mouthful to swallow.

The queen triggerfish of the Caribbean Sea is a relatively large fish (up to 26 cm, 10 in) and a generalist feeding on invertebrates. The bands around the mouth and eyes are light blue and the background is yellowish brown. Instead of the large jaws of the sheephead, it has a smaller mouth with short, stout jaws, each with eight protruding incisorlike teeth. It uses this equipment to render the hard portions of its molluscan and crustacean prey into small pieces. The queen triggerfish prefers the long-spined sea urchin, which presents a formidable defensive posture. Triggerfishes have a tough leathery hide made of bony scales that provide a flexible armor against the sea urchin's spines. The queen triggerfish lifts the urchin by a single spine, carries it up off the substratum, and drops it. The urchin usually lands oral side down, but after repeated upheavals the urchin will eventually fall with its vulnerable oral area exposed. The queen triggerfish then darts in between the shorter oral spines, quickly bites through the soft mouth area, and proceeds to eat the urchin from the inside out.

PICKERS, PROBERS, AND SUCKERS

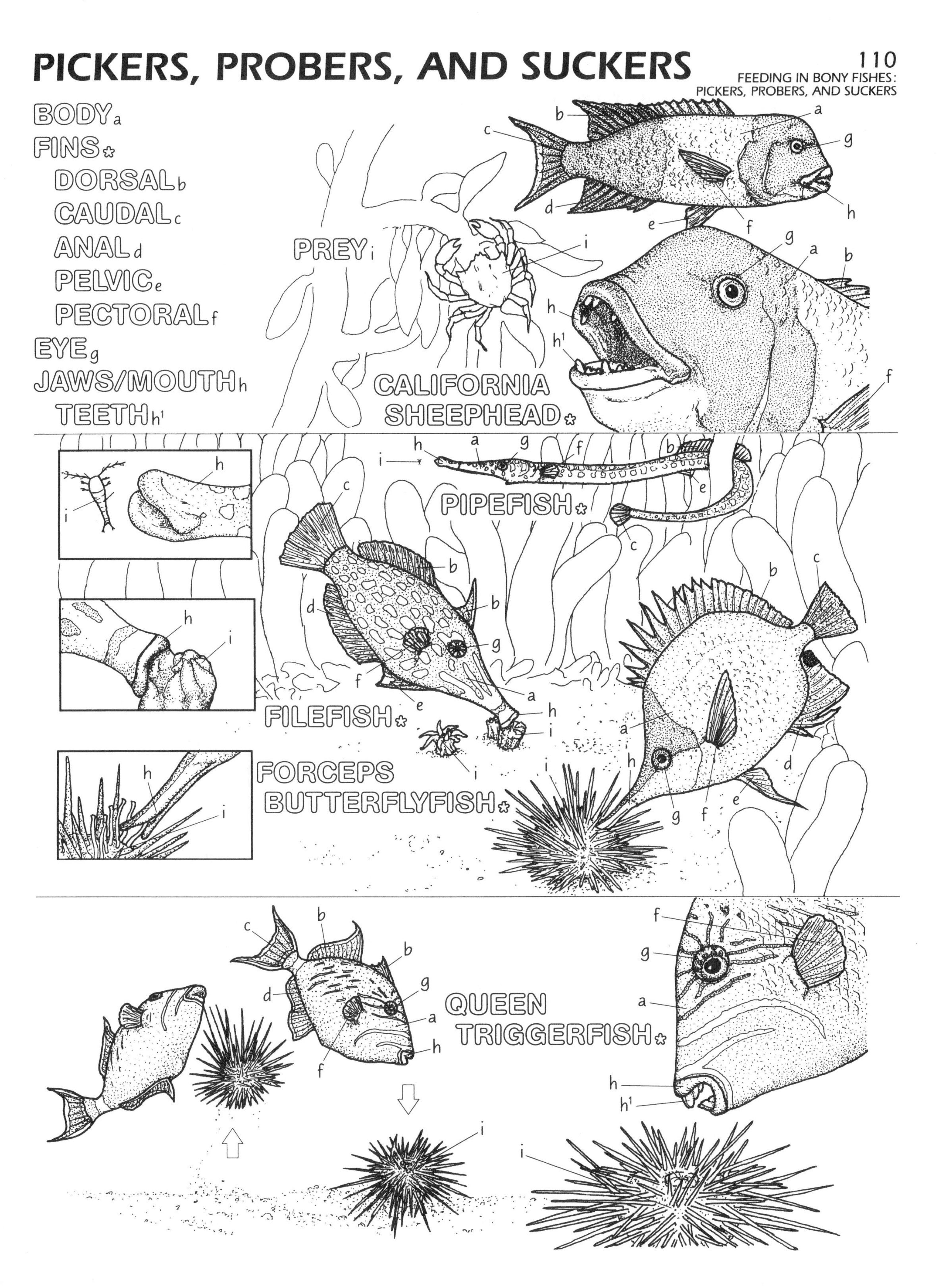

111
FEEDING IN BONY FISHES: GRAZERS AND GRUBBERS

Some fishes, like the barracuda or the sheephead, attack, chase, and consume prey whole. Other fishes, like the reef filefish and forceps butterflyfish, probe and pick away at parts of their prey. In this plate, fish that graze (like the continual browsing of cows or sheep), and those that grub through soft substrata seeking their prey, are considered.

Color each fish and the diagram of its mouth. Note that the trunkfish lacks pelvic fins. The bat ray's mouth is not visible in the larger view; in the smaller view, it is shown from the ventral surface. The bat ray lacks caudal and anal fins.

The northern anchovy represents a group of grazers that utilize the vast pasture of the plankton. These small fish (to 22.5 cm, 9 in) swim through the plankton-rich water with their large *mouths* open and trap the plankton on their gill rakers. They will occasionally go after a larger plankter and snap it up. Anchovy species are found in all the world's oceans, and are important as bait and food fish.

The parrotfishes of Caribbean and Pacific coral reefs are also grazers. These medium to large electric blue fish (25–100 cm, 10–40 in) are primarily herbivores, feeding on the low growth of algae found on coral rock. The parrotfish's *teeth* are fused into a stout beak with sharp cutting edges. These are used to scrape the algal growth from coral rock, and sometimes to ingest living coral as well, supposedly for the plant cells contained in the coral polyps (Plate 91). The ingested algae and coral are ground up by a "pharyngeal mill" consisting of molarlike teeth on the floor and roof of the throat. This feeding activity turns coral rock into coral sand and, in local areas, can account for substantial sediment increase and coral reef destruction. A single parrotfish may turn a ton of coral reef into sand in a year's time.

Many marine invertebrates live within the soft substrata that form the dominant types of ocean bottom. This infauna includes clams, crustaceans, polychaetes, and many others. Frequently, the predator's problem is simply finding them. Fishes that feed on these buried, bottom-living forms are often called grubbers, suggesting the chore of sorting through sand or mud to find food.

The brownish-colored trunkfishes are sophisticated grubbers. Instead of mucking about in the substratum, they swim almost vertically above it and direct jets of water from their mouths downward onto the surface of the substratum. This removes the upper sediment layer and exposes the buried prey, which are then consumed. Trunkfish feed primarily on small crustaceans and polychaete worms.

The gray-brown bat ray is a most determined grubber. This large, 1 meter (3 ft) wide, 95 kg (210 lb) elasmobranch fish uses its winglike *pectoral fins* to fan away the substratum and expose burrowed prey. It feeds primarily on bivalves and crustaceans, which are readily crushed by its stout *jaws* and flat pavementlike teeth. The bat ray has been reported to use the undersurface of its wing like a giant suction cup to suck buried prey out of their burrows.

More grubbing finesse is demonstrated by the familiar goatfish of tropical waters. These medium-sized fish (25–50 cm, 10–25 in) possess a pair of chemosensory appendages called *barbels* that hang from their lower jaws like chin whiskers. The flexible barbels are moved rapidly over the substratum or through it, detecting buried *prey* (small crustaceans and worms), which are quickly excavated and devoured by the small, downward-facing mouth. There are approximately 50 species of goatfish, and most of them are found inshore in shallow water. Individual species show specific feeding preferences for sand or mud substrata and also for feeding times (some are nocturnal feeders). Many species are characterized by group feeding in moderate-sized schools of 25 to 50 individuals. Goatfish are yellow on the dorsal surface and silver-white below.

GRAZERS AND GRUBBERS

BODY$_a$
FINS*
DORSAL$_b$
CAUDAL$_c$
ANAL$_d$
PELVIC$_e$
PECTORAL$_f$
EYE$_g$
JAWS/MOUTH$_h$
TEETH$_{h^1}$
BARBELS$_i$
PREY$_j$

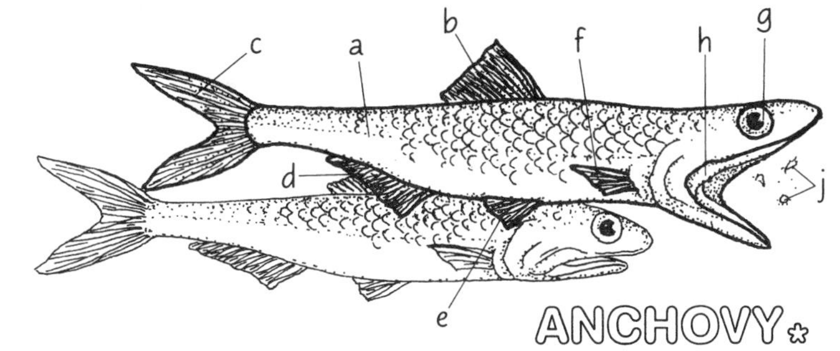

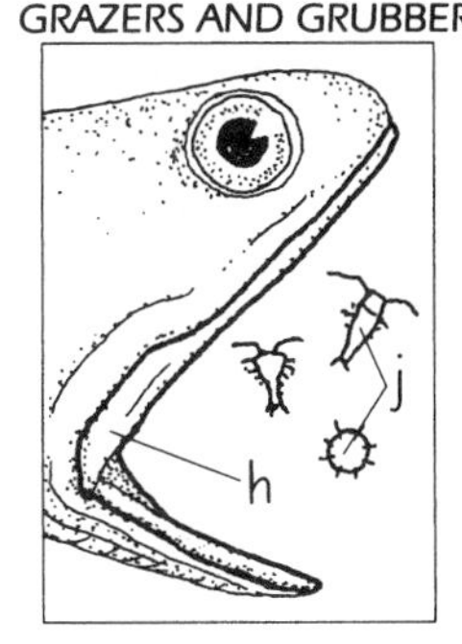

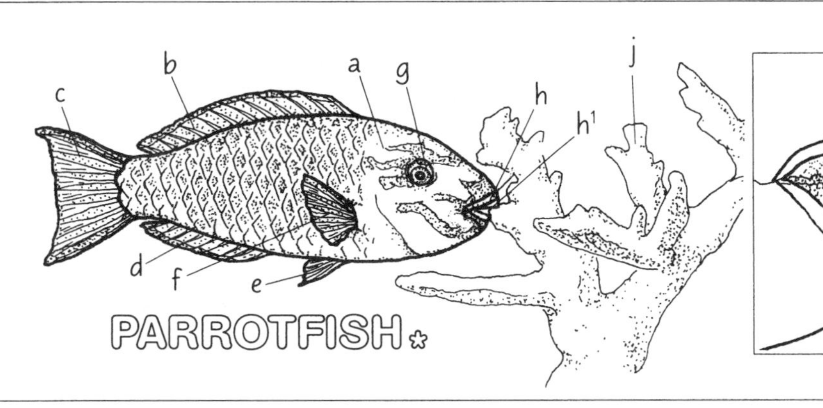

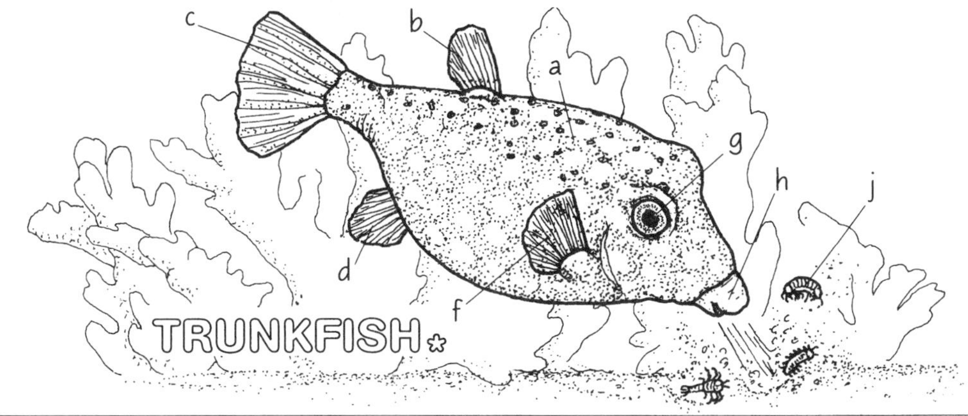

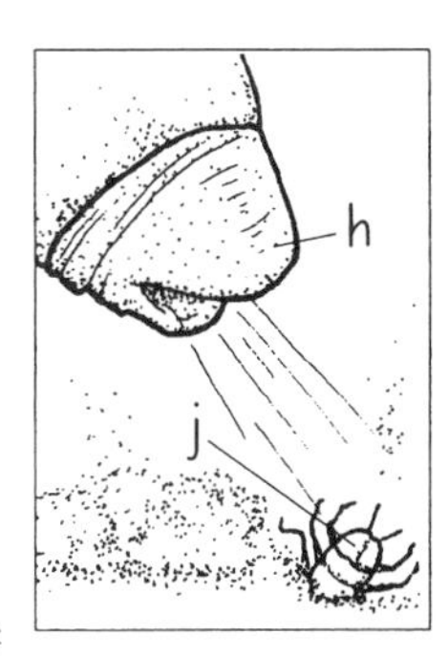

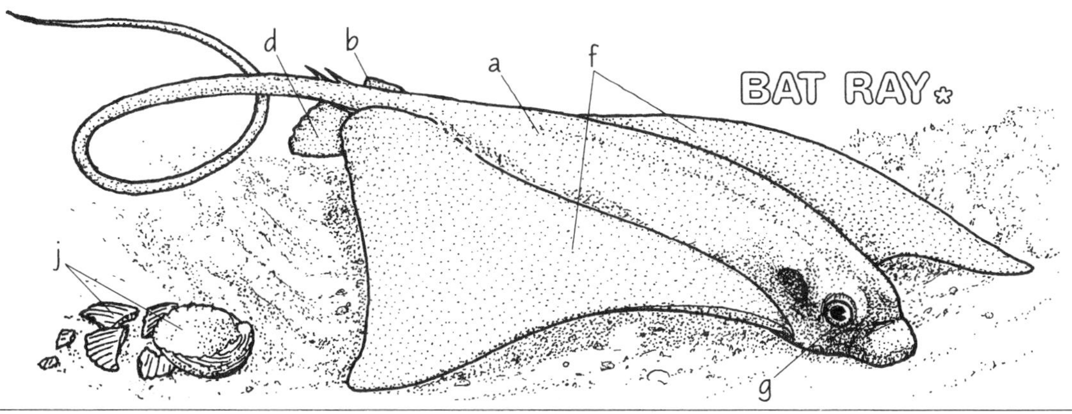

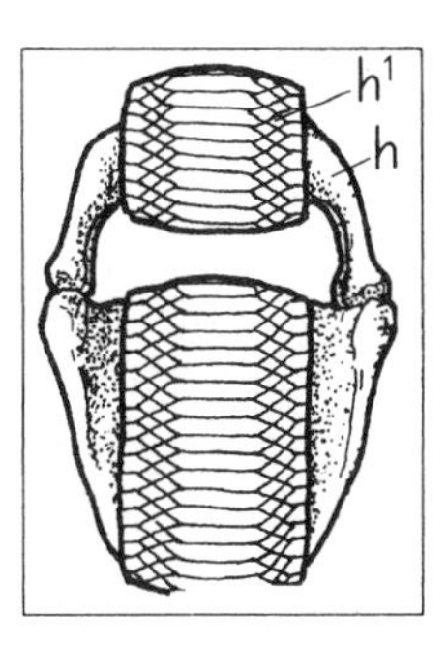

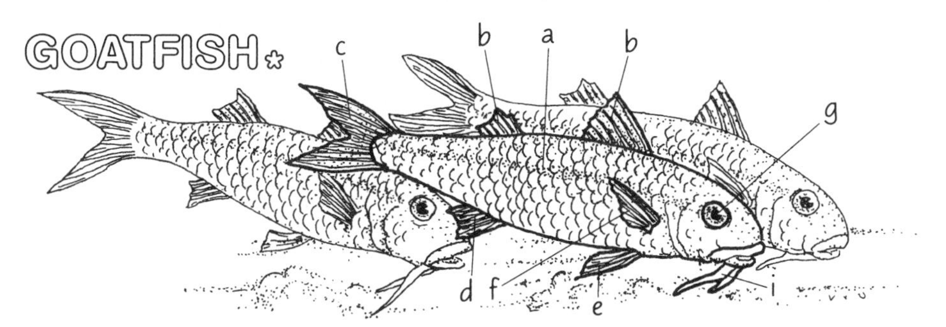

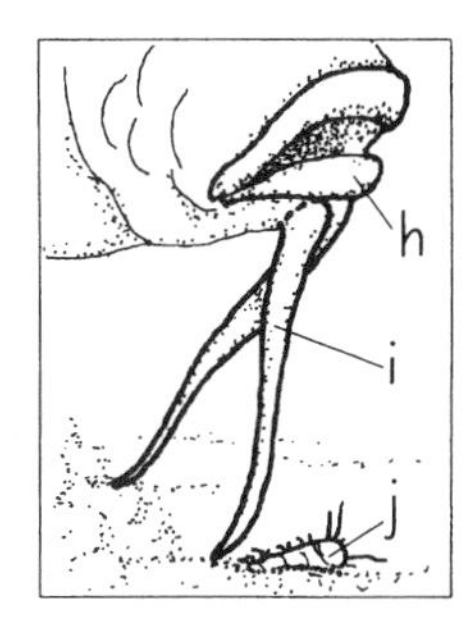

112 THE SEA STAR/ MUSSEL INTERACTION

In this plate we will see how the interaction between a predator, the sea star, and its preferred prey, the mussel, plays a key role in developing and maintaining intertidal zonation. Rocky intertidal zonation patterns (Plate 4) are formed by successful individual species that can withstand the physical stress and interspecies competition of this habitat.

Read through the entire text. The illustration represents the same habitat with the sea star (top) and without (bottom). Begin by coloring the sea stars (purple or orange) first and then the mussels (dark blue) in both illustrations. Color the remaining organisms as they are mentioned. The difference in diversity of organisms (and color) will indicate the importance of the sea star's predatory role.

Along the Pacific coast of North America, the California sea *mussel* occurs only in areas exposed to wave action, where it forms a conspicuous band in the middle intertidal zone. The upper limit of this solid band of mussels seems to be determined by exposure to air. The mussel band is absent from areas uncovered by tides for periods beyond its exposure tolerance. The lower limit of the band of mussels is determined by the predatory behavior of the Pacific *sea star, Pisaster ochraceus*. The sea star can capture and eat almost all the invertebrates of the middle intertidal, but it prefers the mussel over all other potential prey. When the tide is in, the sea star moves upward and preys on the lower mussels. However, the sea star must successfully find and eat a mussel and return to a lower position before the tide goes out; it cannot tolerate the length of exposure to air encountered at the level of the mussel band. Mussels that occur above the sea star's reach are safe from this dominant predator and form the mussel band.

In the middle intertidal zone, below the mussel band, a variety of other organisms that are tolerant of wave action occur, including large algae like the *feather-boa kelp* and *sea palms*. Large patches of the *stalked barnacle* and the solitary giant *green sea anemones*, as well as herbivorous molluscs like the *limpets* and *chitons* are found here.

The predatory interaction between the sea star and the mussel has been studied for over thirty years by Dr. Robert Paine of the University of Washington. Dr. Paine selected wave-exposed, rocky intertidal habitats with distinct mussel bands along the Washington coast. In certain areas, he removed all the sea stars, leaving adjacent areas unaltered in order to compare the growth of the mussel band with and without sea star predation.

In the areas where the sea star had been removed, the mussel band began to extend itself downward. This increase in its width was caused by the growth of the mussels already present, and by the settling of new mussel larvae from the plankton (Plate 75). The mussel larvae settled on the shells of the adults and among the filaments of *red algae* growing near the mussel band, and metamorphosed into *juveniles*. In the absence of sea star predation, the population of mussels grew and extended its range downward. As the mussels colonized the lower intertidal zone, attached organisms such as stalked barnacles and sea anemones were encircled and gradually squeezed to death. Large algae were slowly overgrown and smothered, and herbivorous invertebrates were deprived of food and attachment space.

In the adjacent areas where no sea stars had been removed, the width of the mussel band remained unchanged. When sea stars were allowed to re-invade the sea star removal areas, their predation began to push the lower edge of the mussel band upward toward its original position.

Thus, the sea star's predation on mussels restricts their extension downward into lower zones. This prevents the mussel from monopolizing the available space in the middle and low intertidal zones, and assures space for the attachment and growth of other intertidal organisms.

Dr. Paine's work represents one of the few long term data sets in marine ecology. His extended observations reveal that the conspicuous band of mussels in the middle rocky intertidal zones represents a complex and dynamic environment impacted by local, regional, and global factors. The sea star effectively limits mussels to the middle intertidal zone over the long term, provided the sea star successfully recruits from the plankton and remains present in sufficient abundance. The mussels likewise are dependent on larval recruitment to maintain their presence. If they recruit too heavily, the mussel band becomes several mussels deep and is ultimately unable to maintain firm attachment to solid substrate against the onslaught of storm waves, especially the repeated high intensity storms of an El Niño (Plate 2). The storm waves also bring logs and other large floating objects that smash into the mussels, causing extensive damage. Large patches of the mussels are removed, leaving bare substrate for all to colonize. Dr. Paine determined that the mussels would eventually reclaim their former habitat by virtue of their superior competitive ability, and such recovery took seven years on average.

THE SEA STAR/ MUSSEL INTERACTION

SEA STAR$_a$
MUSSEL$_b$
JUVENILE MUSSEL$_{b^1}$
FEATHER-BOA KELP$_c$
SEA PALM$_d$
STALKED BARNACLE$_e$
GREEN SEA ANEMONE$_f$
LIMPET$_g$
CHITON$_h$
RED ALGA$_i$

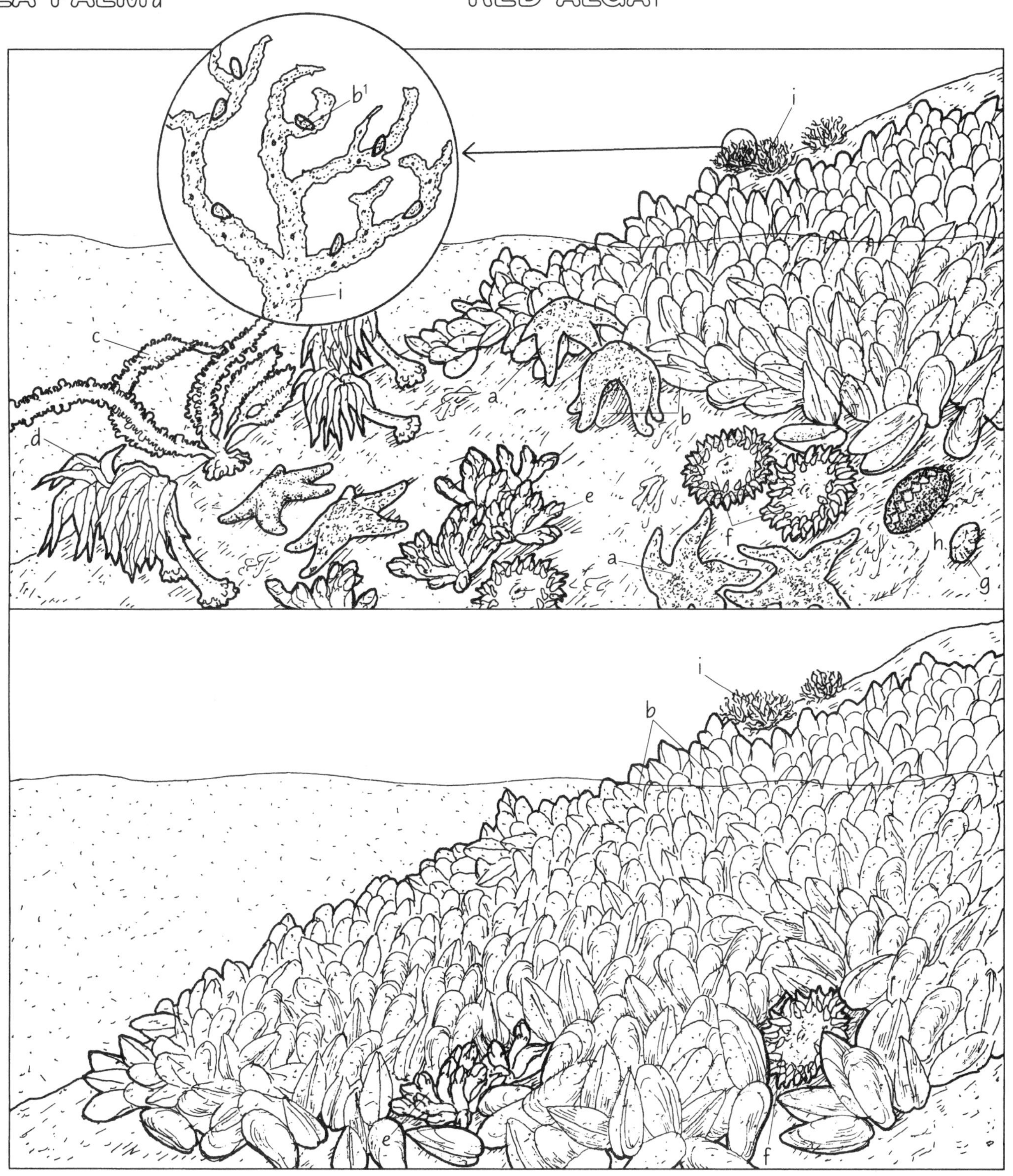

113
SEA OTTERS

Read the entire text first before coloring this plate. The left frame shows the sea otter at work in a kelp bed. The right frame represents an area of former kelp bed habitat.

The *sea otter* is commonly seen in the *kelp* forests along the central California coast. This whiskered, playful marine mammal plays a leading role in a drama that has both biological and historical implications. The sea otter once occurred in a continuous arc along the west coast of North America from Baja California to Alaska and the Aleutian Islands. It was hunted relentlessly in the eighteenth and nineteenth centuries by Russian fur traders for its fine, thick fur. The fur hunters eliminated the otters from much of their historical geographical range. Along the west coast of the United States, only a small colony survived, unnoticed until the 1930s, near Big Sur, California. This colony of sea otters increased in numbers and spread north and south along the central California coast until the 1970s, when it slowed to a standstill. The reasons for the slowed expansion are still undetermined, although a number of physical and biological explanations have been proffered. The population apparently stabilized between 2000 and 2500 animals in the late 1990s. Now, at the turn of the century, the otter population appears to be experiencing an as yet unexplained decline.

Sea otters lack the insulating blubber layer found in most other marine mammals, and therefore, they must consume up to 25 percent of their own body weight in food each day to maintain their body heat in the cold Pacific water. Because of this requirement, sea otters spend a good deal of their day hunting and eating large invertebrates and occasionally fish. When they enter a new area, they first devour the largest, most abundant invertebrate prey. In a kelp forest, this diet consists mostly of *abalone* and *sea urchins*. In kelp forests where otters have been feeding regularly, the only remaining sea urchins and abalone occur under rocks and in deep crevices well out of the reach of the otters. Broken abalone shells and empty urchin *tests* (skeletons) litter the bottom and attest to the sea otter's hunting skill.

The absence of sea otters in kelp forests has been shown to have significant biological ramifications. In the otter's northern range in the Gulf of Alaska, kelp forests once inhabited by the sea otters have disappeared and not returned. Divers discovered these barren areas were inhabited by sea urchins and patches of encrusting *coralline algae* that are unpalatable to the urchins (Plate 20). Apparently when the otters were removed by hunters in the nineteenth century, the urchins increased in number and devastated the kelp forest, creating the "urchin barren." After eliminating the kelp forest, the urchins prevented kelp forest re-establishment by their continuous foraging. Evidence that this scenario is accurate comes from islands in Alaska where the sea otters have been re-introduced. In a period of 10 to 15 years, the feeding otters have reduced the sea urchin population to a few small individuals hiding in crevices, and the kelp forest has re-established and is flourishing. In a startling recent development, these Alaskan otters have become prey for killer whales. The orcas have turned to the less-preferred otters due to drastic decline in their normal prey, the sea lions.

Many conservationists hoped that the return of the sea otter to the California coastline would significantly alter much of the shallow near-shore environment. The hope was that areas previously dominated by sea urchins would be opened up for the establishment of kelp forests or other rich algal habitats populated by many organisms (Plate 11). Obviously, now that the spread of the sea otter has slowed, this possibility seems remote. In central California where the otter has re-established and remained, the predicted alteration of the sea urchin population has occurred and kelp forests flourish.

The occurrence of the very strong 1982–83 El Niño (Plate 2) demonstrated that the fate of the kelp forest environment is ultimately linked to large scale processes that occur well beyond its boundaries. The great winter storms generated by El Niño destroyed large tracts of kelp forest habitat. The subsequent failure of upwelling the following spring starved recruiting kelp plants of nutrients, further impacting recovery. These extreme physical events occur on roughly a 10 year cycle. Recovery of the kelp forest and re-establishment of the community associated with it, including sea urchins and otters, is likewise subject to this cycle.

SEA OTTERS

SEA OTTER$_a$
KELP$_b$
ABALONE$_c$

SEA URCHIN$_d$
TEST$_{d^1}$
CORALLINE ALGAE$_e$

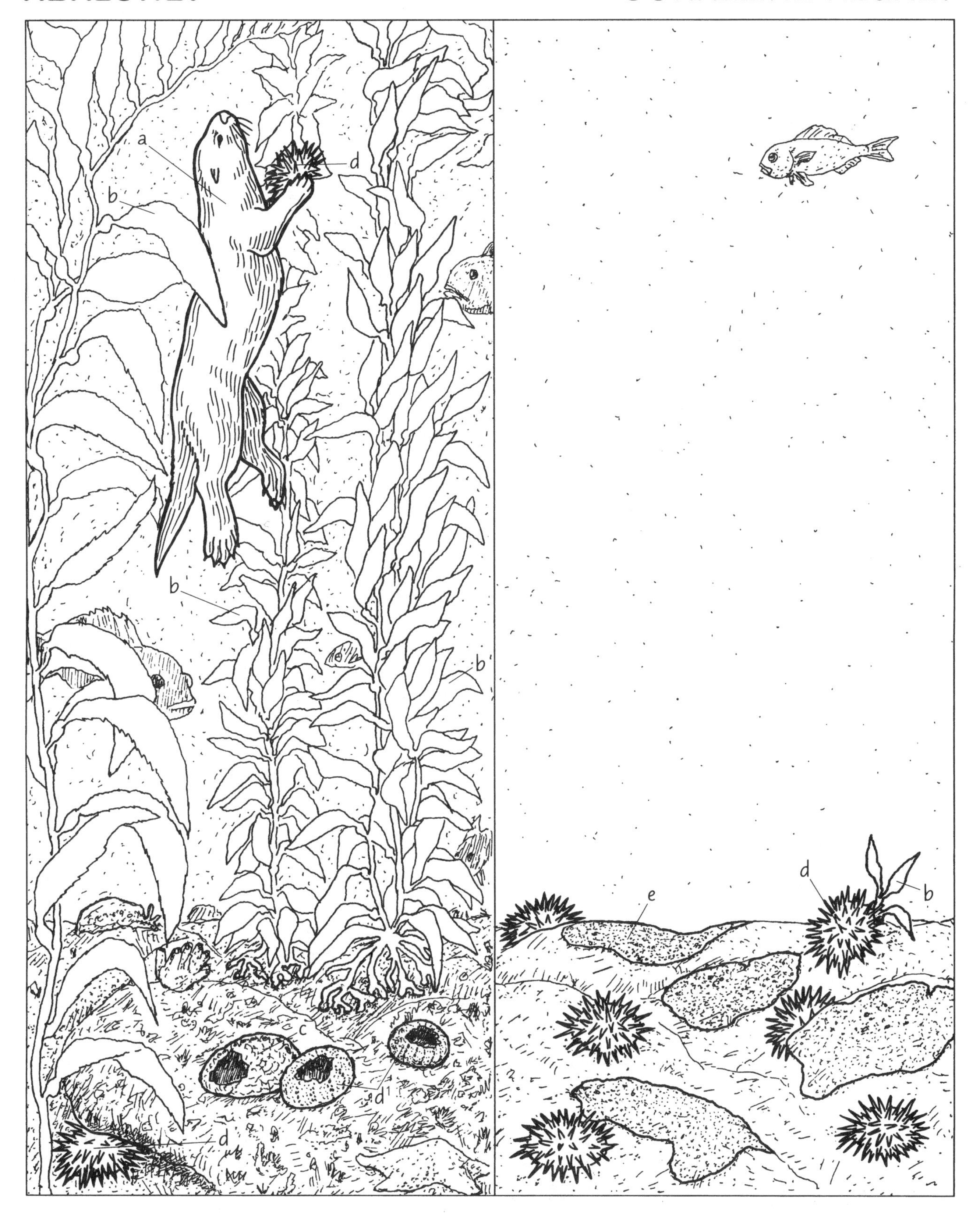

114
NEW TOOLS TO STUDY THE OCEAN

Historically, marine scientists have explored the depths of the ocean second hand, using various kinds of nets and dredges. To sample the deepest ocean bottom, these heavy, clumsy pieces of gear would have to be lowered and recovered great distances through the water column, sometimes requiring 24 hours to collect a single sample. The organisms sampled would sustain considerable damage and disfigurement from the changes in pressure and from being compacted in the net. The softer, more delicate organisms would often disintegrate and pass through the net, never to be seen.

Color the tending vessel on the surface. Then, color each oceanographic vehicle as it is mentioned in the text. The vehicles are not drawn to scale.

In the latter quarter of the twentieth century, new tools have been developed to study the ocean. Mini-submersibles with substantial depth capabilities allow scientists to observe deep-sea animals first hand. *Alvin*, Woods Hole Oceanographic Institution's (WHOI) pioneering (1964) workhorse submersible, has been joined by a number of deep-diving vessels. *Alvin* has undergone substantial modification and retrofitting over its long career and even survived a sinking mishap that occurred when sea water poured into its open hatch during a botched launching. *Alvin* was retrieved from the deep-sea bottom by a nuclear submarine belonging to the United States Navy, and went on to be instrumental in the discovery of the Galapagos hydrothermal vent communities (Plate 17) and in locating the final resting place of the ill-fated *Titanic*.

For shallower work, smaller, lighter submersibles, like Delta Oceanographics' *Delta*, have been perfected that do not require a tending vessel for monitoring, launching, and retrieval. Only a standard ship's winch is required to launch and retrieve *Delta*. This small sub operates on battery power and can dive all day on an overnight charge to depths up to 400 meters (1300 ft). *Delta* has been involved in many oceanographic studies and helped locate HMS *Lusitania*, which was sunk in shallow water by German U-boats off the coast of Scotland.

Remotely operated vehicles (ROVs) are not hampered by on-board human limitations, and have become an invaluable tool in the scientist's quest for an understanding of the sea. It is estimated that 98 percent of the ocean can be reached by this new generation of vehicles. ROVs are equipped with several small electric motors that allow speed and maneuverability, which can be precisely controlled by a pilot on the *tending vessel* above via an oceanographic cable. High resolution video cameras provide crystal clear images of what the ROV is "seeing" for the pilot and scientists, and at the same time, record the images for later reference. ROVs, like the *Ventana* of Monterey Bay Aquarium Research Institute, can also capture midwater and bottom animals, some in special pressurized traps that allow the animals to be studied intact on the surface. Navigating through poorly studied depths, the ROVs have made truly pioneering observations of the behaviors of midwater animals, and have discovered scores of previously unknown species.

ROVs are sometimes used in tandem with manned submersibles. WHOI's small, maneuverable ROV, *Jason* is often sent into tight quarters to get a "closer look" with its video camera for *Alvin*.

The latest oceanographic vessels brought into play are the autonomous underwater vehicles (*AUVs*). These small, relatively inexpensive vehicles ($16,000 to $50,000 plus, depending on installed equipment, compared to millions for an ROV) are computer programmed and sent down untethered to perform various tasks and observations. Successful operations to date have utilized sidescan sonar to map bottom topography, water column profilers such as fluorometers to measure chlorophyll, and thermal sensors to measure temperature. Chores such as mapping specific sections of the ocean's bottom or measuring temperature fields around hydrothermal vents can be accomplished inexpensively. AUVs can be programmed to resurface for collection or to communicate via two-way satellite links with shore-based researchers to transmit data and receive directions. In 1996, a pair of AUVs was successfully operated in a coordinated study of the mixing of two water currents. At the end of each day of the study, data from the AUVs were downloaded and integrated into an elaborate model of the current system, and new sampling profiles were generated and programmed into the AUVs for the next day's operation. One future plan calls for the long term stationing of AUVs in place on the bottom of the ocean, programmed to return to a bottom docking site to recharge, and then continue their programmed data gathering. Initial experimentation with such a system has gone well, and innovative technology that will allow long distance underwater communication and control of AUVs is now being tested. AUVs stand poised to play a major role in the future study of the oceans.

NEW TOOLS TO STUDY THE OCEAN

TENDING VESSEL a
VEHICLES *
ALVIN b
DELTA c
VENTANA d
JASON e
AUV f

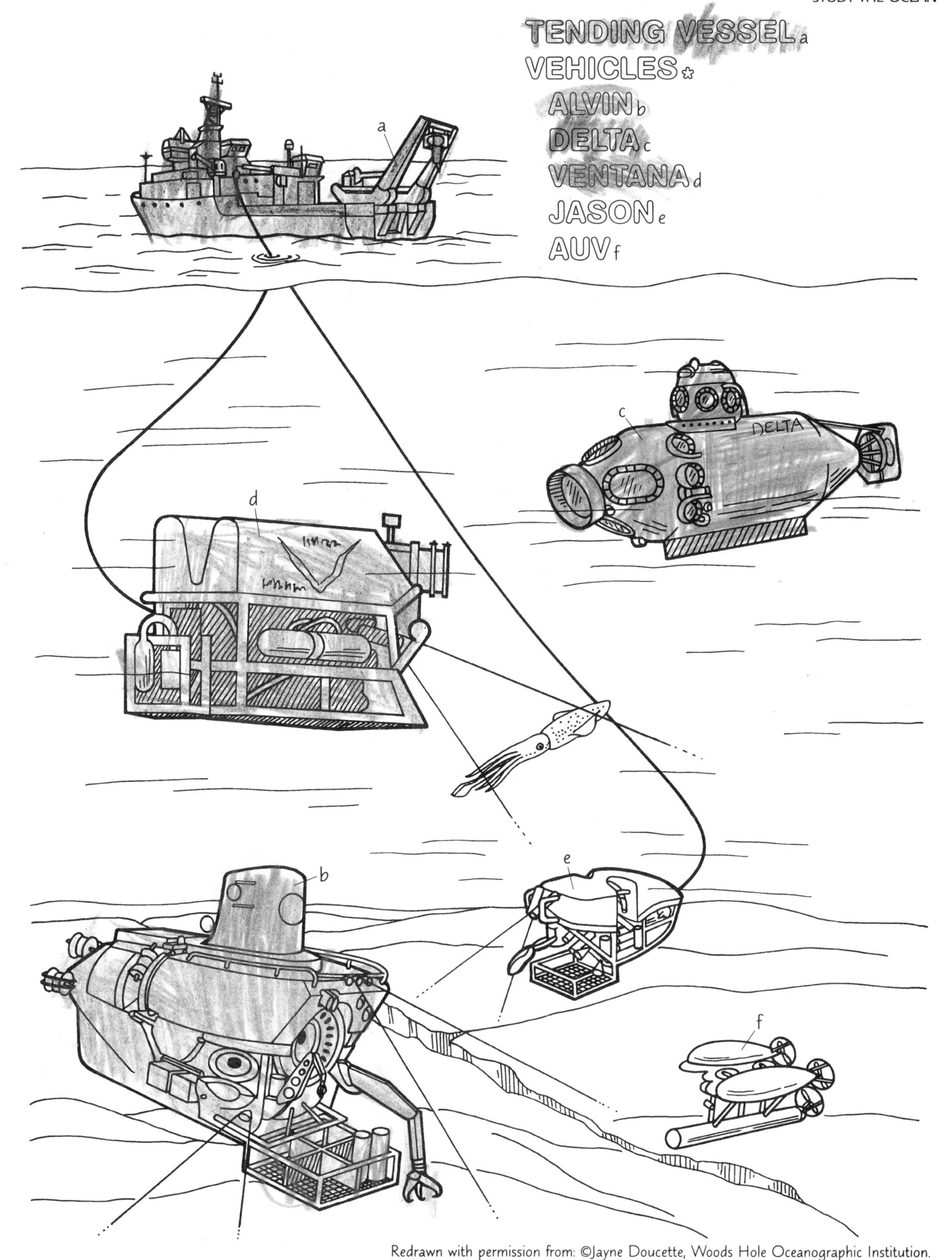

Redrawn with permission from: ©Jayne Doucette, Woods Hole Oceanographic Institution.

115
NEW TOOLS TO STUDY MARINE ANIMALS

The previous plate introduced manned and unmanned vehicles to explore the ocean. Equally impressive is the new generation of instruments used to study marine animals. Traditionally, scientists studying salmon and marine mammals took advantage of the animals' reproductive behaviors that brought them back to specific spawning streams and rookeries. Here, animals could be tagged during one reproductive season and then monitored for their return the following season. These methods provided information on mortality and growth, but told the scientists nothing about where the animals had been and what they were doing during the intervening period. Gathering accurate information for free-ranging animals like swordfish and tunas was virtually impossible. With modern tracking technology, scientists can monitor more marine animal populations and can also track the behavior and physiology of individuals.

Color the animals and the modern data gathering devices as they are discussed in the text.

Scientists studying *great white sharks* entice the shark to swallow a small transmitting instrument package, which houses a temperature (*thermistor*) and depth recorder, by wrapping it in a hunk of elephant seal or horse meat. Once ingested, the cigar sized (22 cm long by 3.5 cm diameter, 8.5 in by 1.5 in) transmitter sends out a signal that can be monitored using hydrophones, allowing the scientists to follow the shark in a small boat. The information sent includes the shark's depth and internal temperature. Data gathered by these instruments have provided scientists with their first insights into great white shark foraging and territorial behavior. The internal body temperate data allowed the researchers to verify the hypothesis that the great white shark maintains an internal core temperature considerably above that of the ambient sea water.

Scientists from the Tuna Research and Conservation Center in Monterey, California (a joint program of Stanford University's Hopkins Marine Station and Monterey Bay Aquarium) have worked with engineers to develop a pair of ingenious tags to study large pelagic fishes. The "*pop-up*" tag is an external device, weighing only 70 grams (2.5 ounces), anchored by a dart in the muscle below the fish's second dorsal fin. It contains a tiny microprocessor and satellite transmitter that can store and transmit data. The buoyant device is designed to pop off the fish by producing an electric current at a preprogrammed time that accelerates corrosion of the metal clasp fastening the tag to the fish. The tag floats (pops up) to the surface where it then transmits its location via satellite to the shore-based researchers. These data provide the scientists with their first look at large scale movement patterns of these economically important and severely threatened species. One of the pop-up tags recorded a Pacific *blue marlin* that traveled from Hawaii to the Galapagos Islands, a distance of more than 4500 km (2800 miles), in 90 days. Pop-up tags can be programmed to release at any time interval and have been deployed for up to a year. In an initial tagging experiment mounted with the cooperation of the fishing community of Cape Hatteras, North Carolina, 35 of 37 pop-up tags attached to bluefin tuna successfully transmitted data.

The second instrument used by the tuna researchers is the *archival tag*. Initially developed for marine mammals, the tag has been made smaller and is now used to study fish ranging in size from large pelagic fish like *bluefin tuna* to smaller tunas and plaice. The archival tag has a 10 cm (4 in) long piece that is implanted into the body cavity of the fish with a 15–20 cm (6–8 in) external stalk. It gathers and records data on the fish's swimming depth, its body temperature, the temperature of the surrounding water, and light intensity. The light data provide information on sunrise and sunset from which the scientists can calculate the fish's latitude and longitude (geoposition). The tags can be set to sample over a range of intervals for up to one year. Unlike the data-transmitting pop-up tag, the archival tag data can only be retrieved if the fish is caught and the tag recovered. Despite this liability, data obtained from archival tags have already provided initial crucial evidence concerning the movement and stock structure of Atlantic bluefin tuna. Ultimately, with continued tagging and data recovery, more enlightened and effective management of these magnificent fish can be achieved.

Marine mammal biologists have known for years that many large pinnipeds such as elephant seals and fur seals forage far at sea for extended periods, but could only speculate on the hunting regimen employed by these animals. Long-term *time* and *depth recorders,* slightly smaller than a soft drink can, are now employed with wondrous results. The recorders are attached with epoxy to the heads of *elephant seals* and recovered when they return to shore. Data provided by these instruments reveal that an individual elephant seal will dive to depths up to 900 meters (3000 ft), remain submerged for periods up to 45 minutes, surface for an interval of only a few minutes, and immediately repeat the dive profile. This exhausting feeding behavior has been recorded for female elephant seals to last continuously for several weeks, interrupted only by sporadic, half-hour naps at the surface.

NEW TOOLS TO STUDY MARINE ANIMALS

DATA RECORDERS*
THERMISTOR a
POP-UP TAG b
ARCHIVAL TAG c
TIME-DEPTH RECORDER d

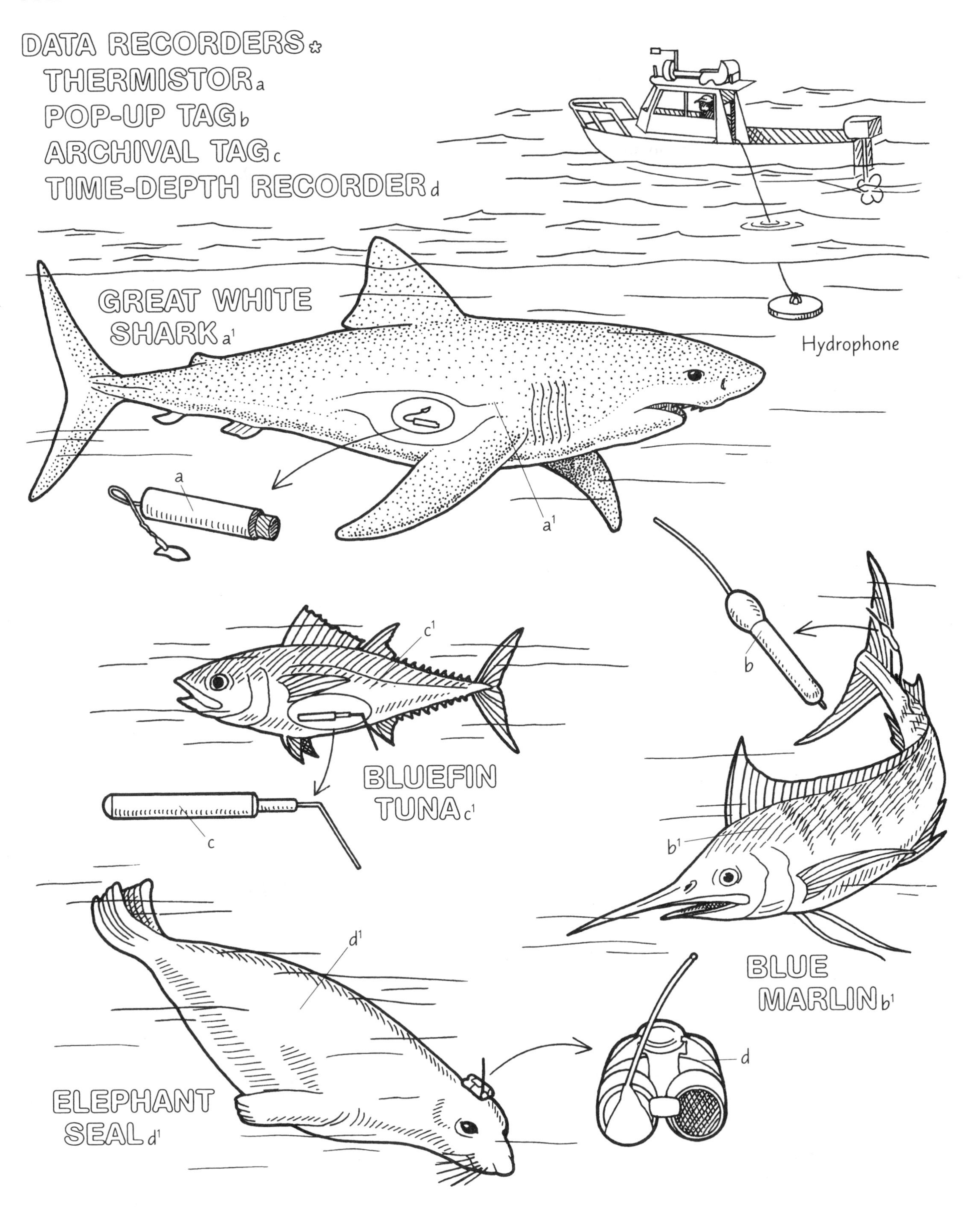

GETTING THE MOST OUT OF COLOR

This book involves coloring. Lots of it. You will be using color to identify a structure and link it to its name (title). Color will be used to differentiate one structure from another, and to show relationships among structures. You will give an aesthetic quality to the plates you have colored. What you have colored you will remember for years based partly on the colors you selected. This brief introduction on the use and character of color will give real support to your coloring goals by providing you with a basic understanding of colors and color matching. It will also provide you with the ability to extend a basic collection of twelve hues to thirty-six or more colors.

What color will you choose? On what basis will you choose it? How many values of a color do you need and how many do you *have*? How can you extend your coloring pen/pencil set to make far more colors than you have? Finally, how can you plan the coloring of each plate to get a really pleasing result? Read on.

PRINCIPLES OF COLOR

Sunlight is white light. White light contains all of the colors in the visible spectrum. Visible light represents a very small band in an immensely large band of radiant energy, most of which is not visible to the human eye. If one places a prism in sunlight, an array or spectrum of colors emerge. Light is the essence of color, yet in itself, it is not a color. Without light, there is no color. Night is the absence of light and therefore the absence of color.

Color vision is based on reflectance. White light, as we have mentioned, is composed of all colors. When light strikes an object such as a lemon, most of the spectrum colors in the light are absorbed by the lemon. A small amount is reflected off of the surface of the lemon—the reflected light. This is the color we perceive. It is the color of the object. In the case of the lemon, the reflected color is yellow.

A good example of a spectrum or sequence of color bands can be seen in a rainbow. Rainbows appear when the sun is shining and it is raining. When the white light of the sun passes through raindrops, the light is bent or refracted. When white light is refracted (as by a prism or by raindrops), the colors of the spectrum separate and become visible. Each color of the spectrum has a different wavelength or characteristic. Simply stated, the rainbow spectrum begins with violet and moves to red, then orange, yellow, green, blue, and back to violet. If we bend the rainbow into a circle and join the violets, we have a color wheel.

To appreciate these color changes, color the rainbow below using the colors indicated. Then color the wheel below the rainbow in the same sequence as the rainbow, starting at the notch with violet.

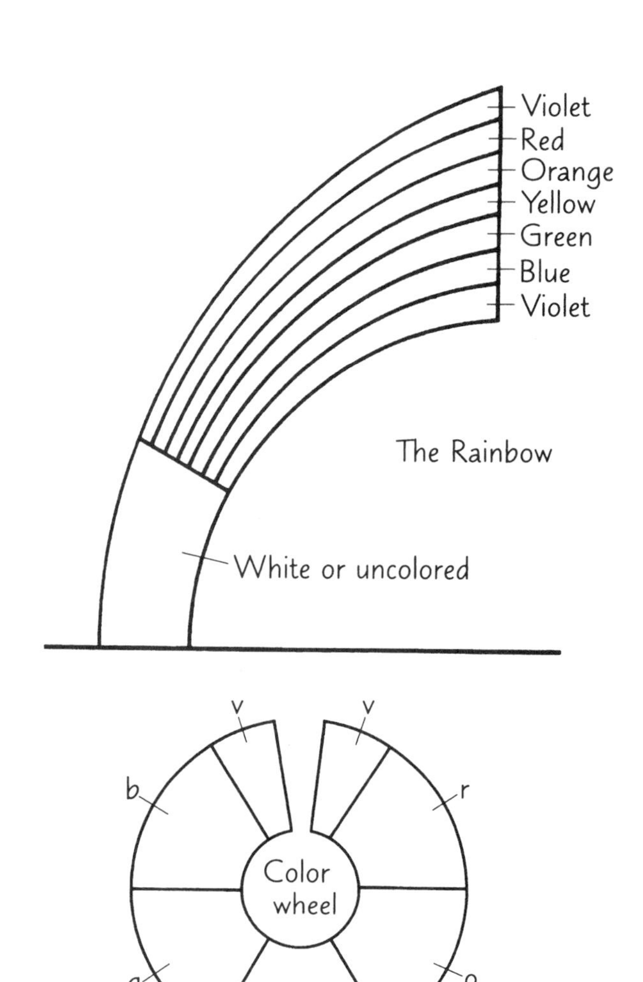

There are three **primary** colors in the spectrum:

Primary colors cannot be created by mixing other colors. They can be combined (mixed) to make other colors.

By mixing two primary colors you create what is called a **secondary** color:

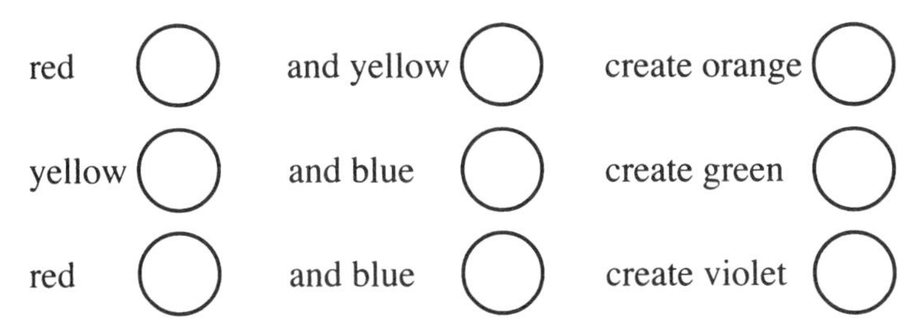

This processing can be continued by mixing a primary and a secondary color, creating what is known as a **tertiary** color. Tertiary colors have simple names based on the colors combined. Thus mixing red and orange creates the tertiary color red orange. There are six tertiary colors.

Below we have another color wheel made up of three concentric circles. The color wheel is divided into six wedges, each marked with a primary or secondary color.

Color each wedge completely with the color indicated. Begin with the primary colors. For the secondary colors, try mixing the primaries instead of using the secondary colors that you may have. This may not turn out well with coloring pens, in which case you may have to use the secondary colors.

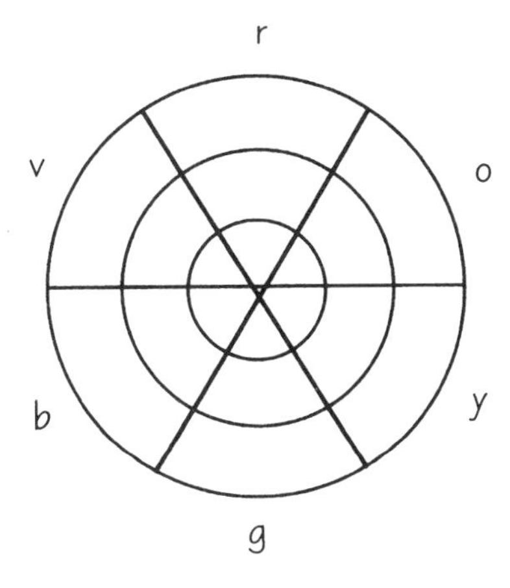

Color is known as **hue**. A pure color means maximum **intensity** or **saturation** of color.

Every pure hue (color) has another set of characteristics known as **value**. Value is the lightness or darkness of a color. Each color has a range of value that extends from very light (near white) to very dark (near black). When we lighten a color, we **tint** it. When we darken a color, we **shade** it. For example, red is a saturated color of maximum intensity. Pink is a tint of red; burgundy is a shade of red.

On your color wheel, color over all of the colors in the outer band or circle with white (or the closest to white that you have). Again, pencils work better than pens in this exercise. Now color all of the colors in the inner band or circle with black, but not enough to obscure the color.

You have now tinted and shaded the primary and secondary colors. There are now three colors for each hue on the wheel. *You can use a color many times by changing its value*. This fact has importance to you as you select various tints and shades of a single color for relating similar structures or related processes in the plates you are working.

Pure or intense colors have different value. Look at the pure colors on your color wheel; notice that blue has a darker value than yellow. Each color has its own value.

Below is a black/white value scale consisting of 11 boxes arranged in a horizontal line (identified as number 1). It is called a gray scale. Starting with white (w) at far left, we have added 10% of black to each square progressively until we have pure, 100% black (b) at far right.

Below this scale, numbered 2, there is another 11-box scale that is blank. Set aside your six primary and secondary colors. One at a time, place the point of one of the pencils/pens over the gray scale and move it across until you find a gray that has the same value (darkness). Fill the box in under that gray with the matching color. In the event that more than one color has the same value, color the space under that square.

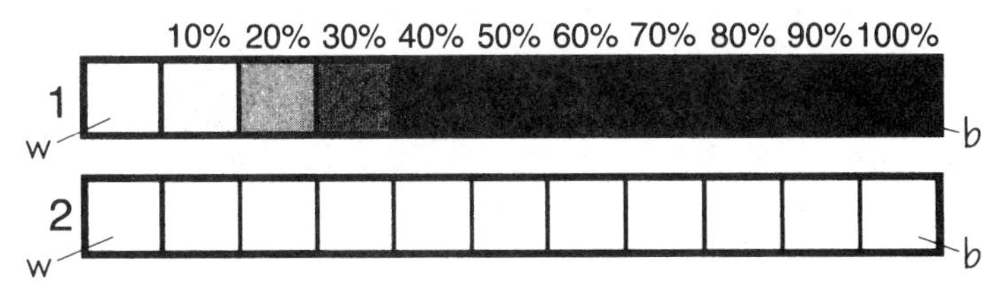

In the series of boxes identified as 3, 4, and 5, you can make your own value scale from three colors. Leave the boxes at far LEFT uncolored (white, w), and fill in the box at far RIGHT with black (b). Locate one pure (intense) hue from scale 2 and color the same box on scale 3. To the left of the hue, progressively tint the boxes until you reach the white box. To the right, progressively shade the color until you reach black. Repeat the process with two different colors in boxes 4 and 5.

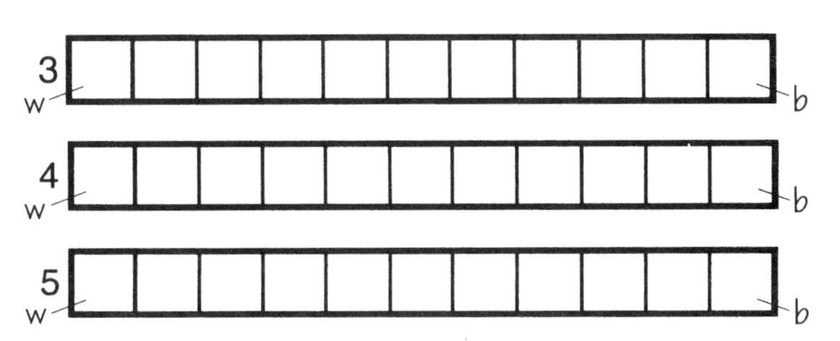

HOW TO USE COLORS

Our next step is to understand how to use color. Colors do many things visually and psychologically. One can create a sense of quiet relaxation, emotional stress, or intellectual excitement. Through color combinations the artist can make one color look like another color, or make a color look brighter than it actually is.

We associate colors with physical phenomena. Colors that are associated with the sun and fire are called "warm colors." Warm colors, such as red, yellow, and orange, visually **advance** or come forward in a scene or painting. "Cool colors" are associated with ice and water; they are blue and green, and they visually **recede**. We cool the far distance in a painting to create **atmospheric** perspective.

Combinations of colors can have many effects. When we use colors that are next to each other on the color wheel, they create a sense of harmony and are called harmonious or **analogous colors**. Analogous colors have a restful nature. An example of harmonious colors on your color wheel would be red, violet, and blue. Place a patch of these colors on the page next to this paragraph to see the harmony. Pick two more harmonious color schemes and place them on the margin.

Color combinations that use colors located far from each other are contrasting in their nature. Contrasting colors create a greater sense of emotion than harmonious colors do. If we use contrasting colors that are an equal distance apart we have a **triad**. Primary colors are triadic. Red, yellow, and blue will create strong contrast. Secondary colors are also triadic.

Colors directly across the color wheel from each other are **complementary colors**. On the color wheel, red and green are complementary colors. Yellow-violet, and blue-orange are also complementary. When complementary colors are placed next to each other, they intensify the color of each; red is a brighter red and green is a brighter green. This is known as simultaneous contrast.

In the boxes below, color the primary colors on the top bar and their complementary secondary color on the lower bar.

If the colors you used are pure hues, you should be able to observe the effects of simultaneous contrast. Artists like Vincent Van Gogh, Paul Gauguin, Toulouse-Lautrec were masters of color contrast. Black and white also create simultaneous contrast.

It is interesting to note that while placing complementary colors next to each other, they brighten the color of each. When we mix complementary colors, they dull or neutralize each other.

With the above insights, you can extend your enjoyment and your skill in coloring the plates in this book. Happy Coloring!

-Jay and Christine Golik
Napa Valley College, California

APPENDIX OF SCIENTIFIC NAMES

2
Anchovy, *Engraulis,* Family Engraulidae
Sardine, *Sardinops,* Family Clupeidae

4
Isopod, *Ligia*
Green alga, *Enteromorpha*
Periwinkle, *Littorina*
Barnacle, *Balanus*
Brown alga, *Alaria*
Brown alga, *Laminaria*

5
Limpet, *Lottia digitalis*
Green-lined shore crab, *Pachygrapsus crassipes*
Mussel, *Mytilus*
Aggregating anemone, *Anthopleura elegantissima*
Mossy chiton, *Mopalia*
Hydroid, *Aglaophenia latirostris*
Purple sea urchin, *Strongylocentrotus purpuratus*

6
Surf grass, *Phyllospadix*
Solitary coral, *Balanophyllia elegans*
Giant green anemone, *Anthopleura xanthogrammica*
Hermit crab, *Pagurus*
Tidepool sculpin, *Oligocottus,* Family Cottidae
Brittle star, Class Ophiuroidea
Broken-back shrimp, *Heptacarpus*
Dunce cap limpet, *Acmaea mitra*
Six-rayed sea star, *Leptasterias*
Polychaete worm, *Amphitrite*
Rock crab, *Cancer*

7
Cord grass, *Spartina alterniflora*
Mussel, *Ischadium demissum*
Menhaden, *Brevoortia tyrannus,* Family Culpeidae
Oyster, *Crassostrea virginica*
Clam, *Mya arenaria*
Grass shrimp, *Palaemonetes pugio*
Striped bass, *Roccus saxatilis,* Family Serranidae
Blue crab, *Callinectes sapidus*
Fiddler crab, *Uca*

8
Lug worm, *Arenicola*
Gaper clam, *Tresus nuttalli*
Bent-nose clam, *Macoma nasuta*
Mud snail, *Nassarius obsoletus*
Ghost shrimp, *Callianassa californiensis*
Moon snail, *Polinices lewisii*
Basket cockle, *Clinocardium nuttallii*
Short-spined sea star, *Pisaster brevispinus*

9
Sand crab, *Emerita*
Bean clam, *Donax variabilis*
Razor clam, *Siliqua patula*
Bristle worm, *Nereis*
Rove beetle, Family Staphalinidae
Ghost crab, *Ocypode*
Swimming crab, *Portunus*
Surfperch, *Amphistichus argenteus,* Family Embiotocidae
Beach hopper, *Megalorchestia*
Sanderling, *Calidris alba*

10
Pacific sand dollar, *Dendraster excentricus*
Sand star, *Astropecten*
Moon snail, *Polinices*
Elbow crab, *Heterocrypta occidentalis*
Sanddab, *Citharichthys sordidus,* Family Bothidae
Angel shark, *Squatina californica,* Family Squatinidae
Brittle star, Class Ophiuroidea
Hermit crab, *Pagurus armatus*
Pismo clam, *Tivela stultorum*
Sea cockle, *Amiantis callosa*
Heart urchin, *Lovenia*
Polychaete worm, *Nepthys*

11
Giant kelp, *Macrocystis pyrifera*
Palm kelp, *Eisenia arborea*
Red algae, Rhodophyta
Sea urchin, *Strongylocentrotus*
Sea hare, *Aplysia*
Abalone, *Haliotis*
Sea bat, *Asterina miniata*
Sunflower star, *Pycnopodia helianthoides*
Sheephead, *Semicossyphus pulcher,* Family Labridae
Sea lion, *Zalophus californianus*
Rockfish, *Sebastes,* Family Scorpaenidae
Sea otter, *Enhydra lutris*

13
Elkhorn coral, *Acropora palmata*
Star coral, *Monastrea annularis*
Brain coral, *Colophyllia natans*
Plate coral, *Agaricia*
Sea fan, *Gorgonia*
Grouper, Family Serranidae
Butterflyfish, Family Chaetodontidae
Damselfish, Family Pomacentridae
Parrotfish, Family Scaridae
Cleaner shrimp, *Stenopus*
Squirrelfish, Family Holocentridae
Grunt, Family Pomadasyidae
Moray eel, Family Muraenidae
Feather star, Class Crinoidea
Spiny lobster, *Panulirus argus*
Sea urchin, *Diadema antillarum*

14
By-the-wind sailor, *Velella velella*
Portuguese man-of-war, *Physalia physalis*
Violet snail, *Janthina*

Copepod, *Calanus*
Euphausiid, *Euphausia*
Arrow worm, *Sagitta*
Herring, *Clupea harengus,* Family Clupeidae
Albacore, *Thunnus alalunga,* Family Scombridae
Blue shark, *Prionace glauca,* Family Carcharhinidae
Blue whale, *Balaenoptera musculus*
Squid, *Loligo*

15
Calycophoran siphonophore, *Muggiaea*
Lobate ctenophore, *Deiopea*
Polychaete, *Tomopteris*
Cock-eyed squid, *Histoteuthis heteropsis*

16
Brittle star, *Ophiomusium*
Sea cucumber, *Scotoplanes*
Amphipod, *Hirondella gigas*
Vampire squid, *Vampyroteuthis infernalis*

17
Vent clam, *Calyptogena magnifica*
Mussel, *Bathymodiolus thermophilus*
Shrimp, *Alvinocaris*
Crab, *Bythograea thermydron*
Pompeii worm, *Alvinella pompejana*
Vent worm, *Riftia pachyptila*

18
Red mangrove, *Rhizophora mangle*
Turtle grass, *Thalassia testudinum*
Surf grass, *Phyllospadix*

19
Pennate diatom, *Pleurosigma*
Centric diatom, *Coscinodiscus*, *Chaetocerus* spp.
Armored dinoflagellate, *Peridinium*
Naked dinoflagellate, *Gymnodinium*

20
Green algae, Chlorophyta
Sea lettuce, *Ulva*
Red algae, Rhodophyta
Salt sac, *Halosaccion glandiforme*
Red alga, *Smithora naiadum*
Coralline red algae genera: *Lithothanium, Corallina, Calliarthron*
Pepper dulce, *Laurencia spectabilis*

21
Brown algae, Phaeophyta
Rockweed, *Fucus*
Oarweed, *Laminaria*
Bull kelp, *Nereocystis luetkeana*
Feather-boa kelp, *Egregia menziesii*
Kelp, *Lessoniopsis*

22
Encrusting sponge, *Haliclona permollis*
Pecten encrusting sponge, *Mycale adhaerens*
Tubular sponge, *Callyspongia*
Boring sponge, *Cliona celata*

23
Green anemone, *Anthopleura xanthogrammica*
Sea anemone, *Metridium*
Hydranth, *Tubularia*
Coral polyp, generalized

24
Hydromedusa, *Polyorchis,* Class Hydrozoa
Jellyfish, *Aurelia,* Class Scyphozoa
Jellyfish, *Pelagia,* Class Scyphozoa
Jellyfish, *Haliclystus,* Class Scyphozoa

25
Flatworm, *Notoplana*, Phylum Platyhelminthes
Ribbon worm, *Tubulanus sexlineatus,* Phylum Nemertea
Peanut worm, Phylum Sipuncula
Nematode/round worm, Phylum Nematoda

26
Innkeeper worm, *Urechis caupo,* Phylum Echiura
Pea crab, *Scleroplax granulata*
Polychaete worm, *Hesperonoe adventor*
Goby fish, *Clevelandia ios,* Family Gobidae
Clam, *Cryptomya californica*

27
Polychaete worm, Class Polychaeta, Phylum Annelida
Clam worm, *Nereis*
Carnivorous worm, *Glycera*
Lug worm, *Arenicola*

28
Fan worm, *Sabella*
Tentacle-feeding worm, *Amphitrite*

29
Cockle clam, *Clinocardium nuttallii*

30
Scallop, *Pecten*
Mussel, *Mytilus edulis*
Cockle, *Clinocardium nuttallii*
Bent-nosed clam, *Macoma nasuta*
Softshell clam, *Mya arenaria*

31
Tulip snail, *Fasciolaria tulipa*
Abalone, *Haliotis*
Moon snail, *Polinices lewisii*
Cowry, *Cypraea*

32
Dorid nudibranch, *Diaulula sandiegenesis*
Aeolid nudibranch, *Hermissenda crassicornis*
Sea hare, *Aplysia*

33
Chambered nautilus, *Nautilus*

34
Squid, *Loligo opalescens*
Octopus, *Octopus*

35
Acorn barnacle, *Balanus*
Copepod, *Calanus*

Stalked barnacle, *Lepas*
Amphipod, generalized gammaridian
Isopod, *Ligia pallasii*

36
Shrimp, *Heptacarpus*
Hermit crab, *Pagurus*
Sand crab, *Emerita*
Lobster, *Homarus americanus*

37
True crab, Tribe Brachyura
Cancer crab, *Cancer antennarius*
Shore crab, *Pachygrapsus crassipes*
Blue crab, *Callinectes sapidus*
Box crab, *Calappa*

38
Phoronid worm, Phylum Phoronida
Lamp shell/brachiopod, Phylum Brachiopoda
Bryozoan, Phylum Bryozoa

39
Sea star, *Pisaster giganteus*

40
Brittle star, Class Ophiuroidea
Feather star, Class Crinoidea

41
Sea urchin, *Strongylocentrotus*
Sand dollar, *Dendraster excentricus*
Sea cucumber (deposit feeder), *Parastichopus*
Sea cucumber (filter feeder), *Cucumaria miniata*

42
Sea squirt (solitary), *Ciona intestinalis*
Sea grape, *Molgula manhattensis*
Compound tunicate, *Botryllus*
Larvacean tunicate, *Oikopleura*

43
Grouper/sea bass, Family Serranidae

44
Sea bass, Family Serranidae
Moray eel, Family Muraenidae
Tuna, Family Scombridae
Barracuda, Family Sphyraenidae
Butterflyfish, Family Chaetodontidae
Triggerfish, Family Balistidae
Electric fish, Family Gymnotidae
Seahorse, Family Syngnathidae

45
Flying fish, *Cypselurus,* Family Exocoetidae
Northern herring, *Clupea harengus,* Family Clupeidae
Swordfish, *Xiphias gladius,* Family Xiphiidae
Sunfish, *Mola mola,* Family Molidae
Albacore, *Thunnus alalunga,* Family Scombridae

46
Tidepool sculpin, *Oligocottus maculosus,* Family Cottidae
Sea robin, *Prionotus,* Family Triglidae
Stargazer, *Astroscopus,* Family Uranoscopidae
Starry flounder, *Platichthys stellatus,* Family Pleuronectidae

47
Trumpetfish, *Aulostomus chinensis,* Family Aulostomidae
Coral grouper, *Cephalopholis minatus,* Family Serranidae
Golden boxfish, *Ostracion tuberculatus,* Family Ostraciontidae
Moray eel, *Gymnothorax,* Family Muraenidae
Butterflyfish, *Chaetodon auriga,* Family Chaetodontidae

48
Hatchetfish, *Argyropelecus*, Family Sternoptychidae
Lanternfish, *Myctophum,* Family Myctophidae
Pacific viperfish, *Chauliodus macouni,* Family Chauliodontidae
Black devil, *Melanocetus johnsonii,* Family Melanocetidae
Pelican gulper, *Eurypharynx pelecanoides,* Family Eurypharyngidae
Tripodfish, *Bathypterois viridensis,* Family Bathypteroidae

49
Grouper/sea bass, Family Serranidae
Spiny dogfish shark, Family Squalidae

50
Spiny dogfish shark, *Squalus acanthias,* Family Squalidae
Skate, *Raja binoculata,* Family Rajidae
Stingray, *Dasyatis americana,* Family Dasyatidae

51
Basking shark, *Cetorhinus maximus,* Family Cetorhinidae
Hammerhead shark, *Sphyrna mokarran,* Family Sphyrnidae
Great white shark, *Carcharodon carcharias,* Family Lamnidae
Thresher shark, *Alopias vulpinus,* Family Alopidae

52
Manta ray, *Manta birostris,* Family Mobulidae
Spotted eagle ray, *Aetobatus narinari,* Family Myliobatidae
Sawfish, *Pristis,* Family Pristidae
Electric ray, *Torpedo nobiliana,* Family Torpedinidae

53
Green sea turtle, *Chelonia mydas*
Yellow-bellied sea snake, *Pelamis platurus*

54
Iguana, *Amblyrhynchus cristatus*
Saltwater crocodile, *Crocodylas porosus*

55
Sea gull, *Larus*
Heron, *Ardea*
Oystercatcher, *Haematopus*

56
Great blue heron, *Ardea herodias*
Sanderling, *Calidris alba*
Long-billed curlew, *Numerius americanus*
Ruddy turnstone, *Arenaria interpres*
Black oystercatcher, *Haematopus bachmani*

57
Double crested cormorant, *Phalacrocorax auritas*
Black skimmer, *Rhynchops niger*
Great black-backed gull, *Larus marinus*
Little gull, *Larus minutus*
Brown pelican, *Pelacanus occidentalis*

58
Royal albatross, *Diomede epomophora*
Magnificent frigatebird, *Fregata magnificens*
Great skua, *Catharacta skua*
Horned puffin, *Fratercula corniculata*
Emperor penguin, *Aptenodytes forsteri*

59
Sea otter, *Enhydra lutris*, Family Mustelidae
California sea lion, *Zalophus californianus,* Order Pinnipedia
Dugong, *Dugong dugon,* Order Sirenia
Bottlenose dolphin, *Tursiops truncatus*, Order Cetacea

60
Fur seal, *Callorhinus ursinus,* Family Otaridae
Harbor seal, *Phoca vitulina,* Family Phocidae
Walrus, *Odobenus rosmarus,* Family Odobenidae
Elephant seal, *Mirounga angustirostris,* Family Phocidae

61
Toothed whales, Suborder Odontoceti
Dolphin, *Tursiops*
Sperm whale, *Physeter catodon*
Blue whale, *Balaenoptera musculus*
Killer whale, *Orcinus orca*

62
Baleen whales, Suborder Mysticeti
Right whale, *Eubalaena glacialis*
Humpback whale, *Megaptera novaeangliae*
Gray whale, *Eschrictius robustus*

63
Garibaldi, *Hypsypops rubicundus,* Family Pomacentridae
Lionfish, *Pterois lunulata,* Family Scorpaenidae
Koran angelfish, *Pomacanthus semicirculatus*, Family Chaetodontidae
Copperband butterflyfish, *Chelmon rostratus,* Family Chaetodontidae

64
Common mackerel, *Scomber scomber,* Family Scombridae
Grouper, *Epinephelus,* Family Serranidae
Clown anemonefish, *Amphiprion tricinctus,* Family Pomacentridae
Stonefish, *Synanceia horrida,* Family Scorpaenidae

65
Flatfish, Family Pleuronectidae

66
Peppermint shrimp, *Hippolysmata grahami*
Cuttlefish, *Sepia officinalis*
Nudibranch, *Chromodoris*

67
Red sponge nudibranch, *Rostanga pulchra*
Red sponge, *Ophlitaspongia pennata*
Isopod, *Idothea montereyensis*
Red alga, *Plocamium*
Limpet, *Lottia digitalis*
Stalked barnacle, *Pollicipes polymerus*
Decorator crab, *Pugettia richii*

68
Fiddler crab, *Uca*

69
Dinoflagellate, *Noctiluca*
Fireworm, *Odontosyllis enopla*
Comb jelly, *Euplokamis dunlapae*
Whale krill, *Euphausia*
Firefly squid, *Wataseni scintillans*

70
Lanternfish, *Myctophum,* Family Myctophidae
Hatchetfish, *Argyropelecus,* Family Sternopthychidae
Stomiatoid, *Idiacanthus,* Family Stomiatoidae
Flashlight fish, *Photoblepharon,* Family Anomalopidae

71
Humpback whale, *Megaptera novaengliae*
Singing toadfish, *Porichthys notatus*, Family Batrachoididae
Pistol shrimp, *Alpheus*

72
Armored dinoflagellate, *Gonyaulax*

73
Green alga, *Ulva*
Bull kelp, *Nereocystis luetkeana*
Nori, *Porphyra*

74
Turtle grass, *Thalassia testudinum*
Sea star, Class Asteroidea
White-plumed sea anemone, *Metridium*
Coral, Class Anthozoa
Polychaete worm, *Autolytus prolifer*

75
Brittle star, Class Ophiuroidea

Porcelain crab, *Petrolisthes*
Sunfish, *Mola mola,* Family Molidae
Polychaete worm, *Nereis*

76
Hydrozoan polyp colony, *Obelia*
Jellyfish, *Aurelia*
Sea anemone, *Epiactis prolifera*

77
Polychaete worm, *Spirorbis*
Clam worm, *Nereis*
Palolo worm, *Palola viridis*

78
Virginia oyster, *Crassostrea virginica*

79
Abalone, *Haliotis*
Moon snail, *Polinices*
Whelk, *Nucella emarginata*
Dorid nudibranch, *Archidoris*

80
Octopus, *Octopus vulgaris*
Squid, *Loligo opalescens*
Paper nautilus, *Argonauta argo*

81
Barnacle, *Balanus*
Copepod, *Calanus*

82
Generalized gammaridian
Amphipod, *Caprella*
Mantis shrimp (stomatopod), *Squilla*
Spiny lobster, *Panulirus interruptus*
Shrimp, *Heptacarpus*

83
Red rock crab, *Cancer productus*

84
Sea urchin, Class Echinoidea
Six-rayed sea star, *Leptasterias hexactis*

85
Horn shark, *Heterodontus,* Family Heterodontidae
Spotted dogfish shark, *Scyliorhinus caniculus,* Family Scyliorhinidae
Skate, *Raja binoculata,* Family Rajidae
Smoothhound shark, *Mustelus manazo,* Family Carcharhinidae

86
Surfperch, *Amphistichus,* Family Embiotocidae
Gafftopsail catfish, *Barge marinus,* Family Ariidae
Kurtus, *Kurtus,* Family Kurtidae
Seahorse, *Hippocampus,* Family Syngnathidae

87
Sockeye salmon, *Oncorhynchus nerka,* Family Salmonidae
Grunion, *Leuresthes tenuis,* Family Atherinidae
Damselfish, *Pomacentrus*, Family Pomacentridae

88
Northern herring, *Clupea harengus,* Family Clupeidae
European eel, *Anguilla anguilla,* Family Anguillidae

89
California gray whale, *Eschrichtius robustus*

90
Elephant seal, *Mirounga angustirostris*

91
Sea slug, *Elysia viridis*
Green alga, *Codium*
Giant clam, *Tridacna*
Green sea anemone, *Anthopleura xanthogrammica*
Elkhorn coral, *Acropora palmata*
Brain coral, *Colophyllia natans*

92
Cleaner shrimp, *Periclimenes pedersoni*
Sea anemone, *Bartholomea annulata*
Cleaner wrasse, *Labroides dimidiatus,* Family Labridae
False cleaner blenny, *Aspidontus taeniatus,* Family Blennidae

93
Anemonefish, *Amphiprion clarkii,* Family Pomacentridae
Sea anemone, *Heteractis magnifica*
Butterflyfish, *Chaetodon lunula,* Family Chaetodontidae

94
Sea anemone, *Stylobates aeneus*
Hermit crab, *Parapagurus dofleini*
Southern stingray, *Dasyatis americana*, Family Dasyatidae
Bar jack, *Caranx ruber,* Family Carangidae
Sea bat, *Asterina miniata*
Commensal polychaete, *Ophiodromus pugettensis*
Pistol shrimp, *Alpheus lottini*
Crab, *Trapezia*
Coral, *Pocillopora*
Crown-of-thorns sea star, *Acanthaster planci*

95
Flying fish, *Cypselurus*, Family Exocoetidae
Copepod, *Penella exocoeti*
Commensal barnacle, *Conchoderma virgatum*
Pearl fish, *Carapus bermudensis* Family Carapodidae
Sea cucumber, *Actinopyaa agassizi*
Sacculinid barnacle, *Sacculina*
Crab host, *Carcinus*

96
Aggregating anemone, *Anthopleura elegantissima*

97
Atlantic barnacle, *Balanus balanoides*
Pacific barnacle, *Balanus glandula*
Thatched barnacle, *Tetraclita squamosa*

98
Limpet, *Lottia gigantea*
Mussel, *Mytilus californianus*
Stalked barnacle, *Pollicipes polymerus*
Snail, *Nucella emarginata*

99
Sea palm, *Postelsia palmaeformis*
California sea mussel, *Mytilus californianus*
Sea star, *Pisaster ochraceus*
Barnacle, *Balanus cariosus*

100
Aeolid nudibranch, *Aeolidia papillosa*
Aggregating anemone, *Anthopleura elegantissima*

101
Ochre star, *Pisaster ochraceus*
Keyhole limpet, *Diodora aspera*
Leather star, *Dermasterias imbricata*
Sea urchin, *Strongylocentrotus purpuratus*
Feather-duster worm, *Eudistylia*
Sea anemone, *Anthopleura xanthogrammica*

102
Porcupinefish, *Diodon hystrix,* Family Diodontidae
Surgeonfish, *Acanthurus glaucopareius,* Family Acanthuridae
Clown triggerfish, *Balistoides niger,* Family Balistidae
Shrimpfish, *Aeoliscus strigatus,* Family Centriscidae
Long-spined sea urchin, *Echinothrix diadema*

103
Scallop, *Pecten*
Short-spined star, *Pisaster brevispinus*
Cockle, *Clinocardium nuttallii*
Sea anemone, *Stomphia coccinea*
Leather star, *Dermasterias imbricata*

104
Squid, *Loligo opalescens*
Common octopus, *Octopus vulgaris*
Pelagic octopus, *Tremoctopus*

105
Pacific sand dollar, *Dendraster excentricus*
Polychaete worm, *Chaetopterus variopedatus*

106
Ringed-top shell snail, *Calliostoma annulatum*
Abalone, *Haliotis*
Oyster drill, *Urosalpinx cinerea*

107
Sea urchin, *Strongylocentrotus*
Giant kelp, *Macrocystis*
Beach hopper, *Megalorchestia*
Limpet, *Notoacmea insessa*
Feather-boa kelp, *Egregia menziesii*
Lined chiton, *Tonicella lineata*

108
Cone snail, *Conus*
Mantis shrimp, Order Stomatopoda
Sea anemone, *Urticina*
Bat star, *Asterina miniata*
Crown-of-thorns sea star, *Acanthaster planci*

109
Great barracuda, *Sphyraena barracuda,* Family Sphyraenidae
Bluefish, *Pomatomus salatrix,* Family Pomatomidae
Greater amberjack, *Seriola dumerili,* Family Carangidae
Splitlure frogfish, *Antennarius scaber,* Family Antennariidae
Rockspear lizardfish, *Synodus synodus*, Family Synodontidae

110
Sheephead, *Semicossyphus pulcher,* Family Labridae
Pipefish, *Syngnathus,* Family Syngnathidae
Reef filefish, *Oxymonacanthus longirostris,* Family Monacanthidae
Forceps butterflyfish, *Forcipiger flavissiimus,* Family Chaetodontidae
Queen triggerfish, *Balistes vetula,* Family Balistidae
Sea urchin, *Diadema antillarum*

111
Northern anchovy, *Engraulis mordax,* Family Engraulidae
Parrotfish, *Scarus*, Family Scaridae
Trunkfish, *Ostracion,* Family Ostraciontidae
Bat stingray, *Myliobatus californicus,* Family Myliobatidae
Goatfish, *Mulloidichthys,* Family Mullidae

112
Sea star, *Pisaster ochraceus*
Mussel, *Mytilus californianus*
Feather-boa kelp, *Egregia menziesii*
Sea palm, *Postelsia palmaeformis*
Stalked barnacle, *Pollicipes polymerus*
Green sea anemone, *Anthopleura xanthogrammica*
Chiton, *Katharina tunicata*
Limpet, *Lottia* sp.
Red alga, *Endocladia muricata*

113
Sea otter, *Enhydra lutris*
Sea urchin, *Strongylocentrotus*
Abalone, *Haliotis*

115
Great white shark, *Carcharodon carcharias,* Family Lamnidae
Blue marlin, *Makaira nigricans*, Family Istiophoridae
Bluefin tuna, *Thunnus thynnus*, Family Scombridae
Elephant seal, *Mirounga angustirostris*

INDEX